Vol. 11

Frontiers in Nanobiomedical Research

CANCER THERAPEUTICS AND IMAGING

Molecular and Cellular Engineering and Nanobiomedicine

Frontiers in Nanobiomedical Research

ISSN: 2251-3965

This new book series "Frontiers in Nanobiomedical Research" (FNBMR) effectively facilitates interactions and collaborations between the nano and medical research communities, who are focused on solving key problems based on emerging nanoscience and technologies, by publishing scientific monographs, graduate level texts, and reference materials, including handbooks and major reference works. These books will address the applications of a broad spectrum of materials, chemical, biological and medical sciences, and engineering. In addition, the book series publications summarize the wealth of experimental results in both nanomaterials and biomedicine and introduce new aspects of nanoscience relevant to biological and medical applications. The series also demonstrates new trends in the field and cultivates a new research community in both frontier research and higher education.

Published

Vol. 11 *Cancer Therapeutics and Imaging: Molecular and Cellular Engineering and Nanobiomedicine*
edited by Kaushal Rege and Sheba Goklany (Arizona State University, USA)

Vol. 10 *Tissue Engineering and Nanotheranostics*
edited by Donglu Shi (University of Cincinnati, USA) and Qing Liu (Tongji University, China)

Vol. 9 *The World Scientific Encyclopedia of Nanomedicine and Bioengineering II: Bioimplants, Regenerative Medicine, and Nano-Cancer Diagnosis and Phototherapy (A 3-Volume Set)*
edited by Donglu Shi (University of Cincinnati, USA), Maoquan Chu (Tongji University, China) and Jiang Chang (Chinese Academy of Sciences, China)

Vol. 8 *Bioengineering in Wound Healing: A Systems Approach*
edited by Martin L. Yarmush (Rutgers University, USA & Harvard Medical School, USA) and Alexander Goldberg (Tel Aviv University, Israel)

Vol. 7 *Multiscale Technologies for Cryomedicine: Implementation from Nano to Macroscale*
edited by Xiaoming He (The Ohio State University, USA) and John C. Bischof (University of Minnesota, USA)

For further details, please visit: http://www.worldscientific.com/series/fnbmr

(Continued at the end of the book)

CANCER THERAPEUTICS AND IMAGING

Molecular and Cellular Engineering and Nanobiomedicine

editors

Kaushal Rege

Sheba Goklany

Arizona State University, USA

Vol. 11

Frontiers in Nanobiomedical Research

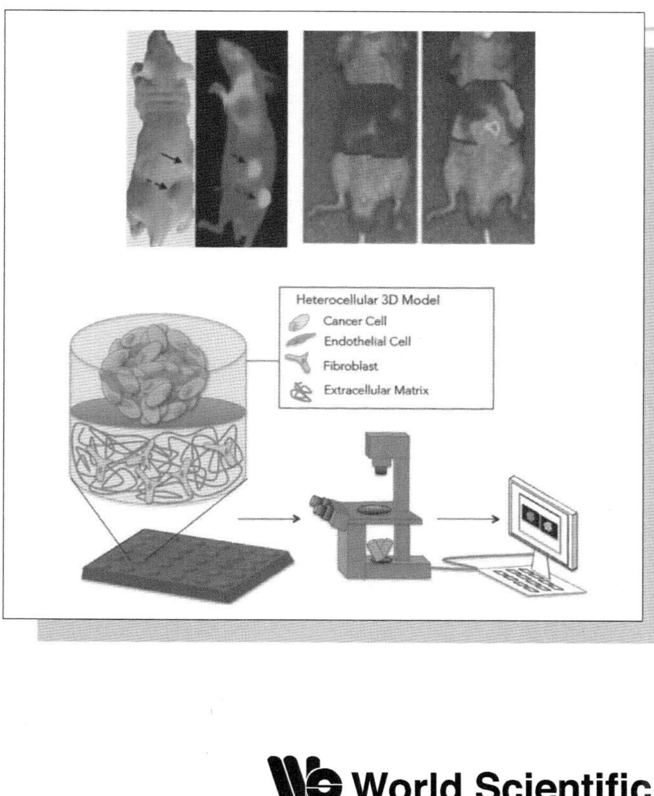

World Scientific

NEW JERSEY • LONDON • SINGAPORE • BEIJING • SHANGHAI • HONG KONG • TAIPEI • CHENNAI • TOKYO

Published by

World Scientific Publishing Co. Pte. Ltd.
5 Toh Tuck Link, Singapore 596224
USA office: 27 Warren Street, Suite 401-402, Hackensack, NJ 07601
UK office: 57 Shelton Street, Covent Garden, London WC2H 9HE

Library of Congress Cataloging-in-Publication Data
Names: Rege, Kaushal, editor. | Goklany, Sheba, editor.
Title: Cancer therapeutics and imaging : molecular and cellular engineering and nanobiomedicine / editors, Kaushal Rege, Sheba Goklany.
Other titles: Frontiers in nanobiomedical research ; v. 11. 2251-3965
Description: New Jersey : World Scientific, 2017. | Series: Frontiers in nanobiomedical research ; volume 11 | Includes bibliographical references.
Identifiers: LCCN 2017039468 | ISBN 9789813222540 (hardcover : alk. paper)
Subjects: | MESH: Neoplasms--therapy | Neoplasms--diagnostic imaging | Nanoparticles | Nanotechnology | Drug Delivery Systems | Genetic Therapy
Classification: LCC RC261 | NLM QZ 266 | DDC 616.99/4--dc23
LC record available at https://lccn.loc.gov/2017039468

British Library Cataloguing-in-Publication Data
A catalogue record for this book is available from the British Library.

For any available supplementary material, please visit
http://www.worldscientific.com/worldscibooks/10.1142/10504#t=suppl

Typeset by Stallion Press
Email: enquiries@stallionpress.com

Printed in Singapore

CONTENTS

Chapter 1

OPTICALLY MODULATED THERANOSTIC NANOPARTICLES

*Niksa Valim and Amit Joshi**

Department of Radiology, Baylor College of Medicine, Houston, TX 77030, USA
**ajoshi@mcw.edu*

Abstract

Nanoparticles are rapidly advancing as drug carriers for directing therapies in a spatio-temporally controlled manner. By virtue of their size ranging from tens to hundreds of nanometers, nanoparticles can preferentially trap in tumors and aberrant vascular interstitium, and can be targeted to specific molecular pathologies. Theranostic nanoparticles is an umbrella term designating the carriers, which combine both the drug/therapy transport with an imaging contrast. With suitable targeting, theranostic nanoparticles can be used for molecular imaging, and potentially for monitoring therapy response.[1] To enhance therapy specificity to the disease site and reduce off-target effects, external and direct spatial-temporal control on therapy delivery is desirable.[2] Multiple approaches for modulating theranostic nanoparticles have been reported, *viz*. magnetic fields, sonic energy, and optical radiation. Optical modulation in the range of visible and near infrared (NIR) wavelengths is a valuable tool for theranostic nanoparticle modulation because of its non-ionizing and safe nature, tissue penetration of multiple centimeters, easy multiplexing by including multiple interrogation wavelengths/ multiple fluorophores, low-cost, and advances in miniaturization of point of care optical illumination/detection devices. Optical stimulation is routinely used for noninvasive visualization of exogenous fluorescent markers, fluorescent proteins, or for triggering the delivery of drugs in humans and in laboratory animals.[3] In this chapter, recently reported optically modulated theranostic nanoparticle platforms are described with a brief discussion on their biomedical applications and potential of translation to clinic. The nanoparticle platforms are organized according to the synthesis materials, with focus restricted to platforms exhibiting external optical control on diagnostic/therapeutic triggering.

1.1 Polymeric nanoparticles

Polymeric nanoparticles composed of large chain organic molecules and synthetic/natural polymers comprise the largest and best-studied class of theranostic nanocarrier systems because of their following unique characteristics; (i) they have controllable physiochemical properties such as particle size, surface charge, and degradation rate, (ii) they can transfer different types of therapeutic or imaging agents, incorporating their unique hydrophilicity or hydrophobility profiles within the controllable nanoparticle architecture, and (iii) their surfaces can be modified for binding to targeting moieties *via* relatively simple and reproducible chemistry techniques.[4–6]

Polymeric nanoparticles encapsulating photosensitizers were demonstrated as one of the first truly optically modulated theranostic nano-platforms.[7,8] Their imaging and therapeutic applications are respectively due to strong fluorescence emission and generation of cytotoxic singlet oxygen from the photosensitizers under irradiation of light with appropriate wavelength. While photosensitizers such as porphyrin derivatives have been used for imaging and therapy for at least a decade,[9,10] some of their intrinsic characteristics limit their widespread application. First, most photosensitizers are barely soluble in physiological conditions. Second, they may cause severe phototoxicity and photosensitivity in patients' skin and eyes when exposed to the sun up to several months after treatment. Therefore, photosensitizers require suitable delivery carriers to modulate their action in space and time, especially for their intravenous administration. Multiple polymeric nanoparticles have emerged for this purpose to date.

Weissleder and co-workers reported the production of poly nanoparticles (lactic-co-glycolic acid) that encapsulate meso-tetraphenylporpholactol photosensitizer as a theranostic nano-platform.[7] The encapsulation of the photosensitizer stabilized them for systematic administration. Upon cellular internalization, the photosensitizer is released and the irradiation by visible light to the target area activated it for fluorescence emission and generation of singlet oxygen. Nanoparticles induced significant cytotoxicity to the cancer cells both *in vitro* and *in vivo*. Further advantage of these nanoparticles was their bio-degradability.

Kwon and co-workers reported polymeric self-assembled nanoparticles based on hydrophobic biocompatible/biodegradable glycol chitosan (HGC).[11,12] These particles were shown to circulate through the blood stream for up to three days and preferentially accumulated in tumors *via* enhanced permeability and retention (EPR) effect. These particles were loaded with paclitaxel for cancer chemotherapy. Additionally, the HGC nanoparticles were labeled with Cy5.5, which is a near infrared fluorescent dye, for imaging. These polymeric nanoparticles were used for simultaneous visualization and treatment in SCC7 tumors *in vivo*. Figure 1.1

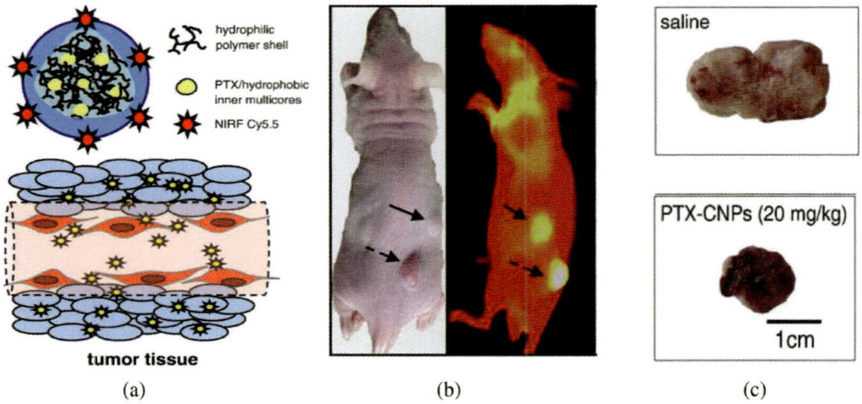

Figure 1.1 (a) Schematic diagram of chitosan-based nanoparticles, (b) *in vivo* images of SCC7 tumor. The black arrow indicates intravenous injection of pactitaxel loaded-HGC nanoparticles labeled with Cy5.5, and (c) representative images of excised tumors treated with Saline and PTX-CNP for 18 days (Reproduced with permission from Ref. 11, copyright © Elsevier, 2010.)

shows the schematic diagram of the polymeric nanoparticles along with an example of excised tumor treated with saline and paclitaxel-loaded HGC nanoparticles (PTX-CNP).

Chances of clinical translation of polymeric nanoparticles encapsulating photosensitizers are high, as photosensitizers on their own have progressed to initial clinical trials for esophageal and skin cancer.[13,14] Coupled with favorable bio-degradation and safety profiles of carrier polymeric nanoparticles, we may potentially see clinical applications in the next decade.

1.2 Carbon nanotubes

Carbon nanotubes (CNTs) have unique physical and chemical properties and have been widely explored in the fields of drug delivery, biological imaging, and thermal ablation.[15] CNTs are allotropes of carbon with cylindrical nanostructures. These cylindrical carbon molecules are categorized as single-walled nanotubes (SWNTs) and multi-walled nanotubes (MWNTs). SWNTs are made up of a single rolled up layer of graphene, while MWNTs consist of multiple tubes of graphene sheets. CNTs are able to load targeting ligands and chemotherapy agents.[16–18]

Dai and co-workers have studied the exposure of SWNTs to laser light excitation in the biological optical window, i.e. in the range of 700 nm to 900 nm wavelengths. This excitation wavelengths were could induce thermal ablation of cancer cells both *in vitro* as well as *in vivo*.[19] They functionalized SWNTs by PEGylated phospholipids with non-toxic and long-circulating properties. In order to demonstrate *in vivo* measurements, 4T1 murine breast tumor cells were injected subcutaneously in 20 BALB/c mice right shoulders to obtain tumor-bearing mice. Five days after the inoculation, 10 mice were injected with 200 μL of a 2 μmol/L solution of SWNTs through intravenous tail injection. Figure 1.2(a) shows the schematic diagram of SWNTs with the coatings. These nanotubes were coated with a 50% DSPE-mPEG/50% C_{18}-PMH-mPEG. The fluorescence of nanotubes was monitored *in vivo* for 48 hours using a 2DInGaAs detector. The other 10 mice remained untreated as controls. Three days after injection, all mice were irradiated with an 808 nm NIR laser for 5 minutes at 0.6 W/cm^2 with a laser spot size of 4.4 cm^2. The temperature was monitored continuously with thermal imaging. The images showed that the tumors on non-SWNT injected mice heated up very slowly and reached an average temperature of 43°C. However, the tumors of SWNTs injected mice heated up very fast and after 5 minutes of irradiation the temperature reached up to 52.9°C. All tumors in the control NIR radiation group without SWNTs injection did not show any sign of treatment, and all mice were considered as "non-survival" by the 36th day after thermal therapy. However, the tumors shrank in all SWNT injected and laser treated mice on average about two weeks after the treatment. All 10 mice in this group were considered survivors by ~6 months post-treatment without any tumor relapse.

Figures 1.2(b) and 1.2(c) depict optical imaging and SWNTs photoluminescence imaging of BALB/c mice with breast tumors on both

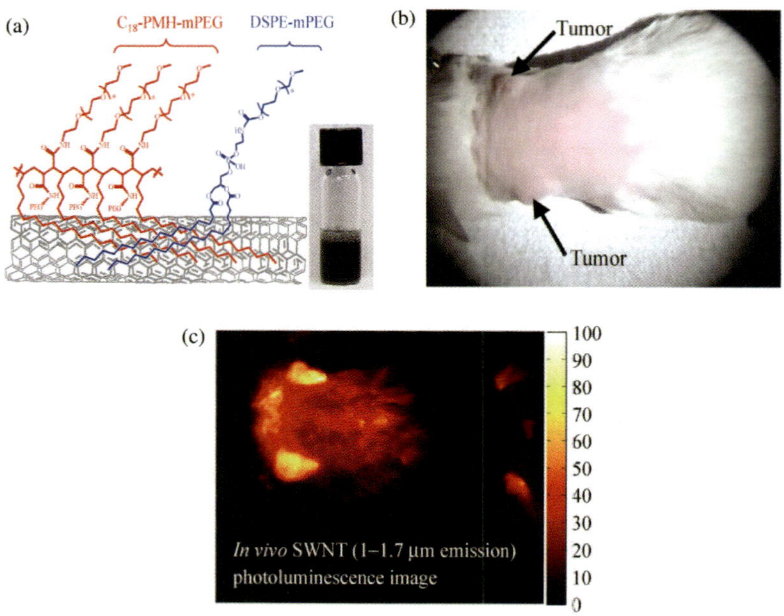

Figure 1.2 (a) Schematic diagram of SWNTs coated with DSPE-mPEG and C_{18}-PMH-mPEG, (b) Optical image of a BALB/c mouse with two 4T1 tumors (indicated by arrows), (c) NIR photo-luminesce image taken 48 hours post-injection, (d) thermal images of a tumor-bearing mouse injected with solubilized SWNTs (top image) and non-injected mice (bottom image), and (e) optical image of the BALB/m mouse immediately before laser treatment with SWNT injection (top image) and non-injected SWNT (bottom image). (Reproduced with permission from Ref. 19, copyright © Tsinghua University Press and Springer-Verlag Berlin Heidelberg, 2010.)

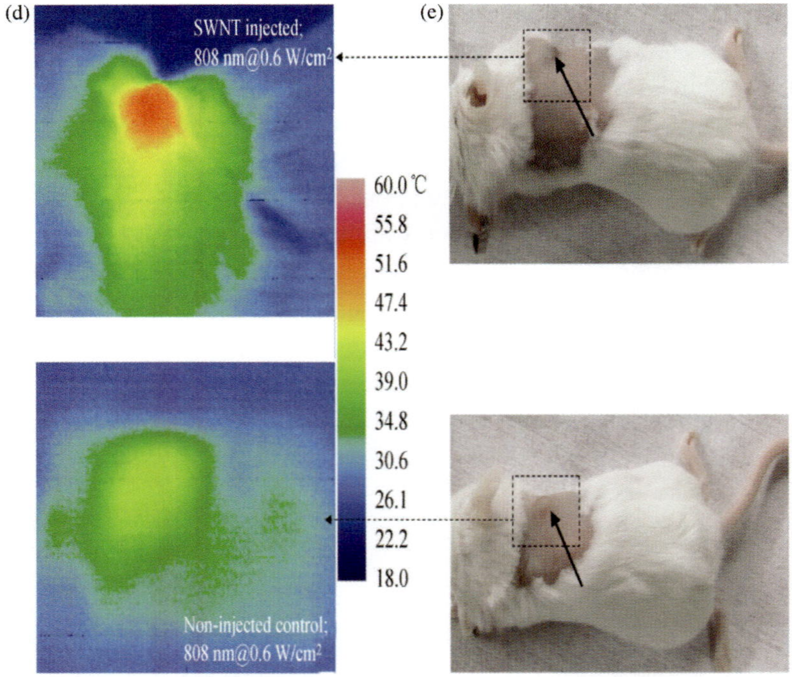

Figure 1.2 (*Continued*)

shoulders. Figure 2(d) depicts thermal imaging of an SWNT-injected mouse taken 4.5 minutes into NIR laser irradiation (top image) and a control mouse with the same irradiation (bottom image). Figure 1.2(e) shows the optical image of the BALB/c mouse injected with SWNTs (top image) and non-injected mouse (bottom image) immediately before NIR laser irradiation.

In addition to tumor cell treatment with SWNTs, Dai and co-workers have reported photothermal ablation of vascular inflammation using these nanoparticles.[15] Aqueous SWNT suspension was functionalized with Cy5.5 fluorescent dye and molecular and cellular imaging was possible with fluorescence imaging *via* NIR light irradiation. They have shown that SWNTs were taken up by macrophages both in cell culture and in mice carotid artery ligation. Cy5.5-SWNTs were injected intravenously *via* the tail vein, and fluorescence imaging was taken 24 and 48 hours after the injection using irradiation light of 808 nm wavelength. Photothermal destruction of mouse macrophage cells was also performed with SWNTs exposed to NIR light of 808 nm wavelength to treat atherosclerosis. Light exposure on macrophage-targeted SWNTs induced 93 ± 3% cell death.

While promising in pre-clinical studies, CNTs might be constrained for clinical translation because of uncertain clearance mechanisms, and unexplored safety and toxicity mechanisms. Several studies on the cellular toxicity of carbon based nanomaterials have been reported that the toxicity of these particles is attributed to various factors including length, concentration, surface chemistry, aspect ratio, type of functionalization, duration of exposure, and method of distribution.[20,21]

In the class of carbon based theranostic agents, nanographenes have been used in photothermal and photodynamic therapies. Liu and co-workers were the first to study the behaviors of PEGylated nano-graphene oxide in mice bearing cancer tumor.[22] The nGO-PEG was injected intravenously in mice as a photothermal agent followed by irradiation of tumors with an 808 nm laser for 5 minutes. They showed that tumors were cured effectively after treatment. In another work, Shen and co-workers have reported the application of graphene oxide as a photosensitive anti-cancer drug.[23] A novel GO-hypocrellin A (GO-HA) hybrid was prepared, and its photodynamic activity was studied *in vitro*. HeLa cancer cell lines were incubated, and GO-HA were added to the cells followed by laser irradiation for 15 minutes. The drug (GO-HA) and laser irradiation significantly changed cell morphology and activity.

1.3 Silica nanoparticles

In the class of theranostic nanoparticles, silica-based nanoparticles have emerged as one of the most promising agents for diagnostic and therapeutic applications. These particles are robust, stable, and easy to control in size and morphology, with known safety and toxicity properties. These attributes make silica-based nanoparticles good candidates for biomedical applications.[24,25]

Prasad and co-workers developed photosensitizer cargo loaded silica nanoparticles for PDT applications.[26] An effective hydrophobic photosensitizer 2-devnyl-2-(1-hexyloxyethyl) pyropheophorbide (HPPH) was encapsulated in hydrated ceramic-based nanoparticles. The morphology and size of the nanoparticles was shown to be spherical with an average size of 30 nm *via* Transmission Electron Microscopy (TEM). HPPH-SNTs nanoparticles were studied *in vitro* for the uptake, viability, and bio-imaging of the particles using two kinds of cancer cells, i.e. HeLa and UCI-107 cells. In order to determine whether HPPH was taken up by tumor cells, the cells were imaged with confocal microscopy after irradiation with ~530 nm wavelength light. The fluorescence images of HeLa and UCI-107 cells showed significant intracellular uptake of HPPH-SNTs in the cytoplasm of the cancer cells, and significant damage to tumor cells was observed upon the irradiation with NIR light of ~650 nm.

In another application of theranostic SNTs, *in vivo* studies of male athymic BALB/c mice with subcutaneously implanted HeLa cells were performed with methylene blue-encapsulated phosphonate-terminated silica nanoparticles (MB-encapsulated PSiNPs).[27] Intratumoral injection of MB-encapsulated PSiNPs was performed after the acceptable growth of the tumors. The tumors were exposed to 635 nm NIR light for 5 minutes with the approximate irradiation density of 500 mW/cm^2. The luminescence fluorescence imaging of MB-encapsulated PSiNPs were imaged with an optical *in vivo* imaging system, consisting of a liquid crystal tunable filter (LCTF) with a scanning wavelength in the range of 680–800 nm, an optical coupler, and a scientific grade monochrome CCD camera. The tumor response post-treatment was also tracked with optical imaging and measured by necrotic area of the tumor. The tumors treated with joint MB-encapsulated PSiNPs and NIR light irradiation shrank gradually. However, the tumors treated with just MB-encapsulated PSiNPs or just NIR light irradiation still kept growing. Figure 1.3 shows an example of subcutaneously tumor-bearing mice. In Figure 1.3(a), MB-encapsulated PSiNPs were injected intratumorally, and the tumor was exposed to NIR light for photodynamic therapy. Figures 1.3(b) and 1.3(c) show control images of the model tumor with just nanoparticle injection or just laser light exposure respectively.

Among various silica-based nanoparticles, mesoporous silica particles have been investigated the most for drug delivery and bio-imaging applications. For example, Brinker and collaborators have reported a class of nanoparticles that combine the features of mesoporous silica and liposomes.[28] In this study, nanoporous silica particles were first

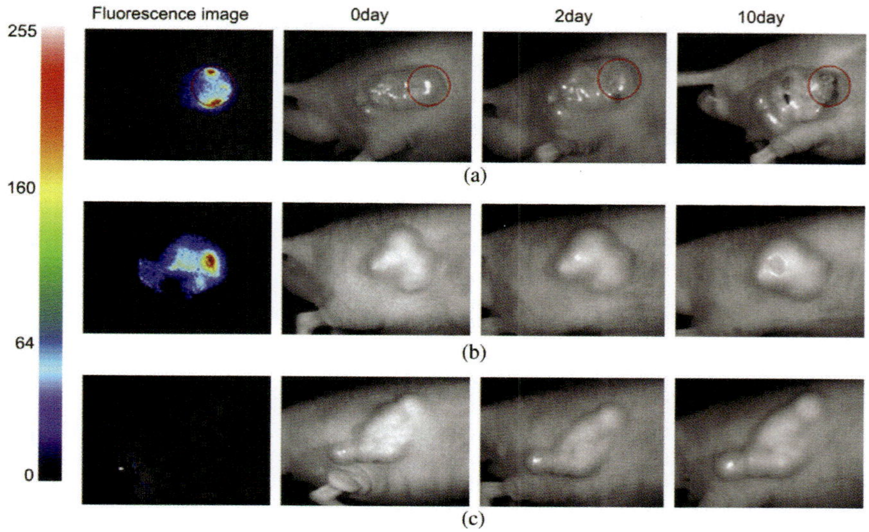

Figure 1.3 Fluorescence luminescence *in vivo* imaging and PDT of HeLa-tumor mice after different treatments: (a) 100 μl 44 mg/ml MB-encapsulated PSiNPs injection and NIR light exposure, (b) 100 μl 44 mg/ml MB-encapsulated PSiNPs injection, (c) NIR light exposure (Reproduced with permission from Ref. 27, copyright © Elsevier, 2009.)

synthesized and particles larger than ~150 nm in diameter were removed by differential centrifugation. Then, protocells were formed by fusing liposomes with a length of ~120 nm in diameter to the nanoporous nanoparticles. Finally, the particles were modified with multiple copies of a targeting peptide, a fusogenic peptide, and PEG. Nanoparticles in the range of 100 nm to 150 nm in diameter were employed as a cargo of therapeutic (drugs, small interfering RNA and toxins) and diagnostic (quantum dots) agents.

1.4 Gold nanoparticles

Gold-based nanomaterials have received the greatest attention as theranostics mainly due to their long term *in vivo* safety and clearance profiles. Moreover, gold-based nanostructures exhibit unique and tunable surface plasmon resonances that can be used for field enhancement of the irradiation light around them to increase the excitation efficiency of photosensitizers.[29–31] Colloidal gold nanoparticles absorb electromagnetic radiation in the visible wavelength range, while other gold nanoparticle geometries *viz*, nanoshells, nanorods, nanostars, nanocages, or hollow shells can be designed to resonate preferentially with NIR radiation.[32,33]

El-Sayed and co-workers have reported an extensive study on three classes of gold nanoparticles, i.e. gold nanospheres, gold nanoshells, and gold nanorods.[34] The optical properties of these particles exhibit the dependency of the resonance wavelength, the extinction cross-section, and the ratio scattering to absorption, on the nanoparticle dimensions. For example, the typical dimension of gold nanospheres in the range of 20 nm to 80 nm limits the maximum resonance wavelength to be in ~520 nm to 550 nm. Therefore, these nanoparticles have limited *in vivo* applications for deep tissue applications. The nanoshell size or the ratio of core-to-shell size is used to tune the resonance wavelength in NIR region.[35] Similarly, the effective size or the aspect ratio of the nanorods linearly controls their optical properties across the range of near-infrared wavelengths.[36,37]

Drezek and coworkers were the first to report *in vitro* and *in vivo* demonstration of targeted gold nanoshells for integrated human breast cancer imaging and therapy.[38] First, the gold nanoshells were synthesized by fabrication of a 10 nm thickness gold sells on a 120 nm diameter silica particles. Either anti-HER2 or a non-specific antibody (anti-IgG) was attached to a PEG linker, and the linker was then attached to the gold nanoshell. The nanoshells were incubated within HER2-positive SKBr3 breast adenocarcinoma cells. Cells were imaged with a black/white CCD camera under a dark-field microscope sensitive to scattered light. Following the imaging, NIR light (820 nm with a power density of 0.008 W/m^2) was irradiated to the cells for 7 minutes. Cytotoxicity was observed in cells targeted with anti-HER2 gold nanoshells that were treated with a NIR-emitting laser only. However, gold nanoshells were only indirectly imaged by employing their high absorption and/or scattering cross-sections, which limited the *in vivo* and deep tissue visualization of nanoparticle transport through tissue. In recent work, Bardhan *et al.* and Chen *et al.* have demonstrated the modifications of silica-coated gold nanoshells for simultaneous enhancement of NIR fluorescence and T2-weighted MR contrast, and shown molecular targeting of breast, ovarian, and pancreatic cancers in mouse models (Figure 1.4).[39–41]

El-Sayed and co-workers have reported *in vitro* demonstration of gold nanorods as contrast agents for both molecular imaging and photothermal cancer treatment.[42] Gold nanorods were conjugated to anti-epidermal growth factor receptor (anti-EGFR) antibodies and then incubated with epithelial cell lines of both malignant and non-malignant nature: HaCaT (human keratinocytes), and two malignant cell lines, HOC 313 clone 8 and HSC 3 (human oral squamous cell carcinoma). After incubation of the cells with antibody-conjugated nanorods for 30 minutes, the cells were illuminated with a continuous NIR laser light of 800 nm at power values of 40, 80, 120, and 160 mW. The sign of photothermal destruction of the cells *via* exposure to the laser light irradiation was observed on both cell types. However, the death of malignant cells was observed at 50% laser power compared to nonmalignant cells. Moreover, the gold nanorods incubated accumulated within the malignant cells efficiently scattered light in red wavelengths and as a result the cells were clearly visualized with a laboratory microscope.

Rege and co-authors have showed spatiotemporal temperature distribution of gold nanorods with NIR laser irradiation on cancer cells and evaluated the mathematical model for nanorod mediated therapy.[43] They synthesized gold nanorods via seed growth methods, with the absorbance maximum at 800 nm. The gold nanorods were conjugated with thiolatedpoly(ethylene glycol), resulting in mPEG-GNR. In vitro study was performed with human prostate cancer cell line (PC3-PSMA) to validate theoretical models of spatiotemporal temperature distribution. Cells were cultured in 96 well plates and PEGylated nanorods were added to them, followed by 800 nm laser light irradiation. Cell viability was assayed 24 hours after light treatment using the 3-(4,5-dimethylthiazol-2-yl)-2,5-diphenylterrazolium bromide (MTT) as a reporter. The cell viability was measured using a plate reader (Bio-Tek Synergy 2) at 570 nm, and the absorbance readouts were normalized to the live (untreated) and dead controls. They showed theoretical agreement with experimental measurements of photothermal cell death.

In another example, Zharov and coworkers have reported the potential of conjugated TNF-gold nanoshells for cancer theranostics.[44] They

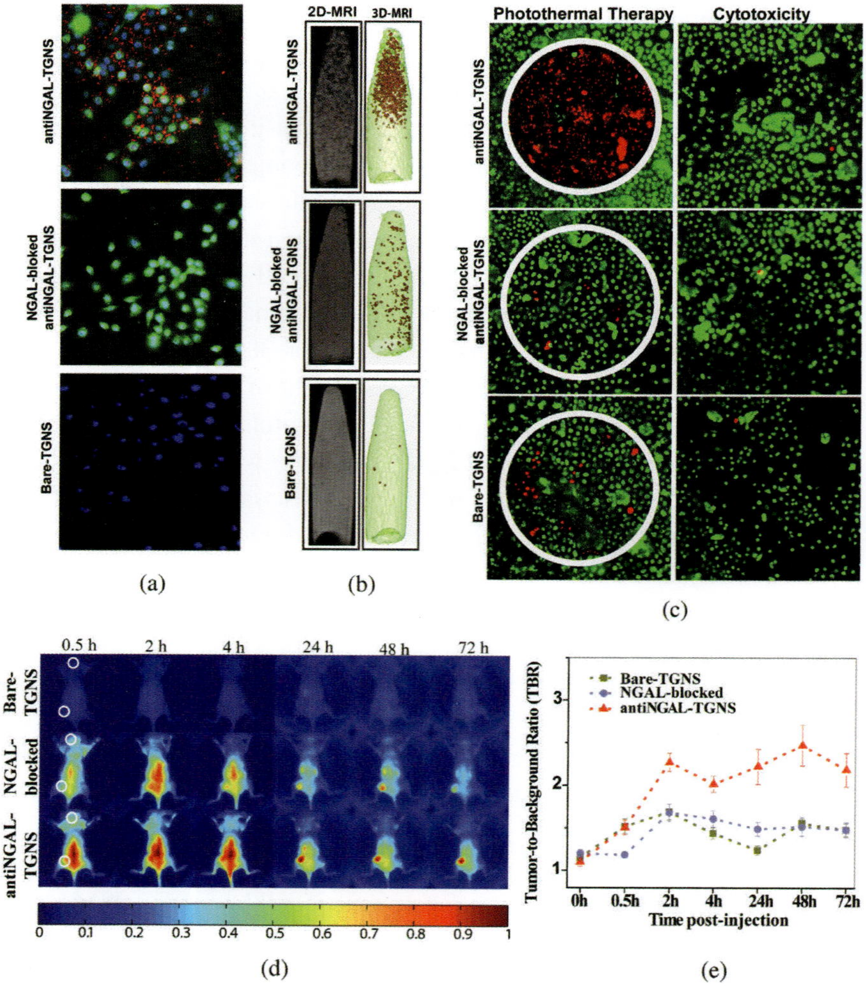

Figure 1.4 Combined NIR fluorescence and MR imaging with magneto-fluorescent theranostic gold nanoshells (TGNS). (a) Fluorescence imaging of neutrophilegelatinase associated lipocalin (NGAL)-expressing AsPC-1 cells showing nuclei stained with DAPI (blue), immunolocalization of antibody (antiNGAL) stained with a secondary antibody-Alexa-fluor 488 (green) and NIR fluorescence of ICG doped inside the silica layer of the antiNGAL-TGNS (red). Cells were incubated with antiNGAL-TGNS, NGAL was blocked with antiNGAL before adding antiNGAL-TGNS, cells were incubated with unconjugated-TGNS respectively. Specific binding was observed on the AsPC-1 cell surface for the antiNGAL-TGNS. (b) 2-D MR images of AsPC-1 cells suspended in 0.5% agarose. 3-D MRI maximum intensity projections of thresholded T2 maps of the images corresponding

Figure 1.4 (*Continued*) to 2-D MRI. (c) Photothermal ablation and live/dead stain of AsPC-1 cells. Live cells were stained green with calcein and dead cells were stained red with propidium iodide. (d) Delivery of nanocomplex *in vivo* and imaging *via* NIR-FOI in mice with overexpressing NGAL xenografts. (e) Tumor to background ratio (TBR) for each group of mice across the imaging time points (Reproduced with permission from Ref. 40, copyright © Future Medicine, 2013.)

have explored whether biochemical drugs in nanoparticle-drug conjugates can be activated by the thermal effect of laser induction. In this work, gold nanospheres with a diameter of ~33 nm were coated with PEG and then necrosis factor-alpha (TNF) was conjugated to its surface. TEM image of Au-TNF revealed a relatively large fraction of ellipsoidal nanoparticles in the sample. Therefore, the plasmon absorption of the gold nanoparticles was reported to be at 525 nm and 690 nm. Figure 1.5(e) shows the schematic of Au-TNF nanodrug and its photothermal activation. For the *in vivo* experiments, 6–8 weeks old nu/nu A/J mice were used and murine mammary cancer cells (SCK) were subcutaneously injected in the hind limb of the mice. The experiments were performed in the mice after 5–7 days of cell injection, with the tumor diameter of 7–8 mm. The tumor bearing mice were randomized into seven experimental groups and Au-TNF and Au-PEG were intravenously injected in the tail. The first group was the normal saline control, the 2nd and 3rd group was the control group treated with just laser 532 nm and laser 690 nm, the 4th group through 7th group were treated with Au-TNF alone, Au-TNF followed by laser 532 nm radiation, Au-TNF followed by laser 690 nm radiation, and Au-PEG followed by laser 690 nm radiation respectively. Figure 1.5(b) shows the tumor image in each group, and Figures 1.5(c)and 1.5(d) show the relative tumor volume increase after the treatment in each group.

In addition to tumor cell treatment with gold nanoparticles, Li and co-workers have reported *in vitro* demonstration of targeted hollow gold nanoshells (HAuNS) on tumor cells.[32] Anti-epidermal growth factor receptor (anti-EGFR) antibodies were covalently conjugated to the HAuNS surface with a total diameter of 37 nm. *In vitro* cell treatment was performed by incubation of C225-HAuNS to A431 tumor cells for one hour followed by NIR laser irradiation. The continuous wave laser

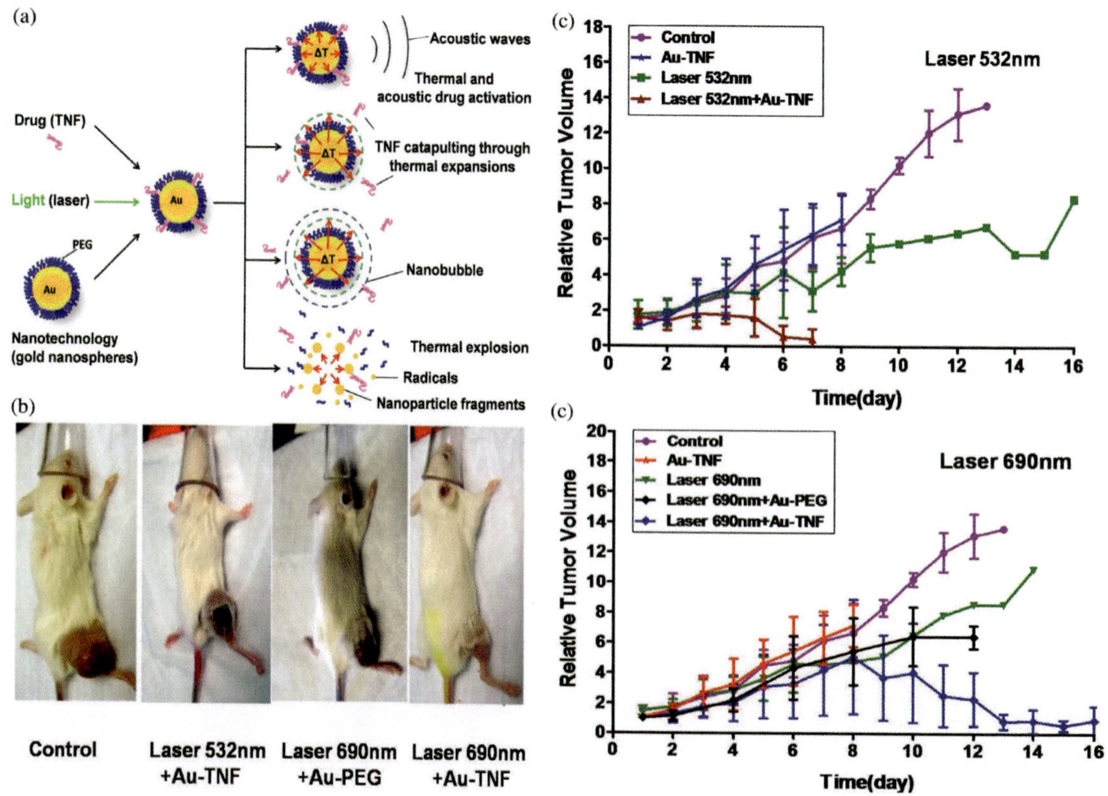

Figure 1.5 (a) Schematic of Au-TNF nanodrug and the principle photothermal activation of the drug, (b) the tumor images of different mice groups, (c) 532 nm, and (d) 690 nm laser irradiation and the relative tumor volume increase in each group. (Reproduced with permission from Ref. 44, copyright © Nature Publishing Group, 2013.)

used in this study had a center wavelength at 808 nm with an output power of 40 W/cm^2 for 5 minutes. Twenty four hours after laser treatment, cells were washed with HBSS and stained with calcein AM and EthD-1 for visualization of live and dead cells. Cell visualization was performed with a laser scanning confocal microscope system with filter sets specific for excitation/emission wavelengths at 494/517 nm for calcein and 528/617 nm for EthD-1. The results showed selective destruction of the tumor cells incubated with anti-EGFR-HAuNS followed by NIR laser.

1.5 Lipid based nanoparticles

Liposomes, which are primarily comprised of phospholipids, are one the most advanced theranostic nanocarriers. The physiochemical properties of liposomes are controllable depending on their phospholipid compositions.[45] For example, the melting temperature of the liposome, which is the temperature at which the phospholipids undergo a transition into the liquid crystalline phase, can be tuned by adding cholesterol. Optically modulated liposomes can be triggered with the heat generated by photothermal action upon illumination with a suitable laser wavelength.[46] In photothermal applications, the membrane is destabilized at the melting temperature, and the content of the liposome, frequently cancer chemotherapeutics, can be released. In addition to indirect thermal sensitivity, direct light-sensitive phospholipid compositions can be used to trigger liposomes upon exposure of light, by utilizing photochemistry techniques. Irradiation of light at a certain wavelength can lead to the photo-polymerization of the light-sensitive components resulting in the formation of chains of covalently linked lipid molecules in the bilayers. The creation of the phase boundary defects in the membrane causes a higher membrane permeability resulting in the release of the liposome content.[47, 48] By modifying the liposomes with targeting ligands, e.g. with antibodies, the liposomes are also able to accumulate specifically in the target tissue.[49] The targeting potential and the triggering are essential for the translational into clinical trials.

Puri and coworkers have reported heat-sensitive liposomes with modified HER2-affisomes (HER2$^+$affisomes).[49] The liposomes were either loaded with rhodomine-PE and calcein or with Doxil. HER2$^+$ cells and

$HER2^-$ cells were both incubated with the liposomes. Fluorescently labeled $HER2^+$-affisomes showed a 10-fold higher increase in binding to $HER2^+$ cells compared to the binding to $HER2^-$ cells. After incubation with a 45°C buffer for 20 minutes, a substantial calcein release was detected in the cytosol of the cells. Compared to control liposomes, $HER2^+$ cells showed a 2–3 fold higher accumulation of DOX in the cells therefore higher DOX-mediated cytotoxity.

In another work, light-sensitive liposomes[50] consisting of two different lipid combinations and formulations: (i) DPPC (1,2-Dipalmitoyl-*sn*-Glycero-3-phosphocloline), and (ii) $DC_{8,9}PC$ (1,2-bis (tricosa-10,12-diynoyl)-*sn*- Glycero-3-phosphocloline) were compared. Liposomes were placed in a plate and then triggered by a UV lamp at 254 nm at room temperature for 0–45 minutes. Depending on the liposome formulation, almost 80% of calcein release was observed by UV radiation. However, no calcein was released under light treated control liposomes without $DC_{8,9}PC$, demonstrating that $DC_{8,9}PC$ can be used as a light-sensitive component in drug carrier liposomes.

In another example, Lovell and co-workers have recently described the application of porphyrin-phospholipid liposome as a therapeutic agent for tumor cells.[51] PoP-liposome release was controlled by irradiation of a 658 nm NIR laser *in vitro* and *in vivo*. These nanoparticles were loaded with DOX and the particles intra-tumorally injected in nude mice bearing subcutaneous tumors grown from the KB cancer cell line, followed by laser irradiation for 12.5 minutes. Their study showed that the tumor mice were effectively cured with DOX-PoP-liposomes with laser treatment.

1.6 Up conversion nanoparticles

Up converting nanoparticles (UCNPs) is an attractive new approach to deliver light into internally deep or large tissue volumes.[52,53] Unlike conventional fluorescence effect, up-conversion refers to a non-linear optical process that absorbs low energy light in the near infrared region and emits higher energy visible light. Similar to regular theranostic nanoparticles, UCNPs can be used as the activation agent of a photosensitizer or as imageable drug carriers. For the application of UCNPs in photodynamic therapy, cytotoxic singlet oxygen is generated through the excitation of

UCNPs and activation of photosensitizer *via* the energy transfer. In addition to photodynamic therapy, UCNPs have been used as energy donors in other photo-responsive drug release systems.[54]

Liu and coworkers have reported the first *in vivo* demonstration of UCNP based photodynamic therapy.[55] UCNPs were synthesized by NaYF4 nanocrystals followed by PEG coating. Then, the nanoparticle was loaded with a photosensitizer molecule, i.e. chlorine e6, forming UCNP-Ce6 supramolecular complex. The nanoparticles were studied for PDT efficacy in *in vitro* studies on HeLa human cervical cell lines, and for *in vivo* studies on breast tumor bearing mice with the 4T1 tumor model. The *in vivo* phototherapy experiments were performed by irradiation with 980 nm NIR light for 30 minutes at 0.5 W/cm^2. Tumor regression effect was observed for UCNP-Ce6 injected tumors, while, neither UCNP-Ce6 injection alone nor laser illumination alone was able to delay growth tumor. Detailed summary of this study is shown in Figure 1.6. Figure 1.6(a) shows examples of treated mice in each group at the 6th day of post-treatment. Figure 1.6(b) shows the relative size of tumors after treatment in each group of mice to the size of tumor right before starting the treatment, and Figure 1.6(c) shows the survival curves of mice after treatment in each group. Further, they observed that UCNPs after PDT therapy gradually cleared out from mice organs.

In another example of UCNPs, Zhang and coworkers have reported the use of $NaYF_4$ nanocrystals in photodynamic therapy.[56] First, they synthesized $NaYF_4$ nanocrystals, and then coated mesoporous silica on the particles. After preparation of UCNPs, zinc phthalocyanine (ZnPc) photosensitizer was loaded into the pores of the mesoporous silica. The nanoparticles were incubated with MB49 bladder cancer cells. The live cells were imaged using a confocal microscope that was coupled to a 980 nm NIR laser light source. Exposure of the nanoparticle to NIR light irradiated visible light that activated the photosensitizer. The emission peak of the UCNP $NaYF_4$ nanocrystals overlaps approximately the absorption peak of ZnPc photosensitizer; therefore the emitted light is able to release singlet oxygen species, resulting in local cell death. UCNP based triggered photodynamic therapy is in its infancy and further *in vivo* experiments on bio-distribution and acute and chronic toxicity and safety studies will be required for further translation of these promising technologies.

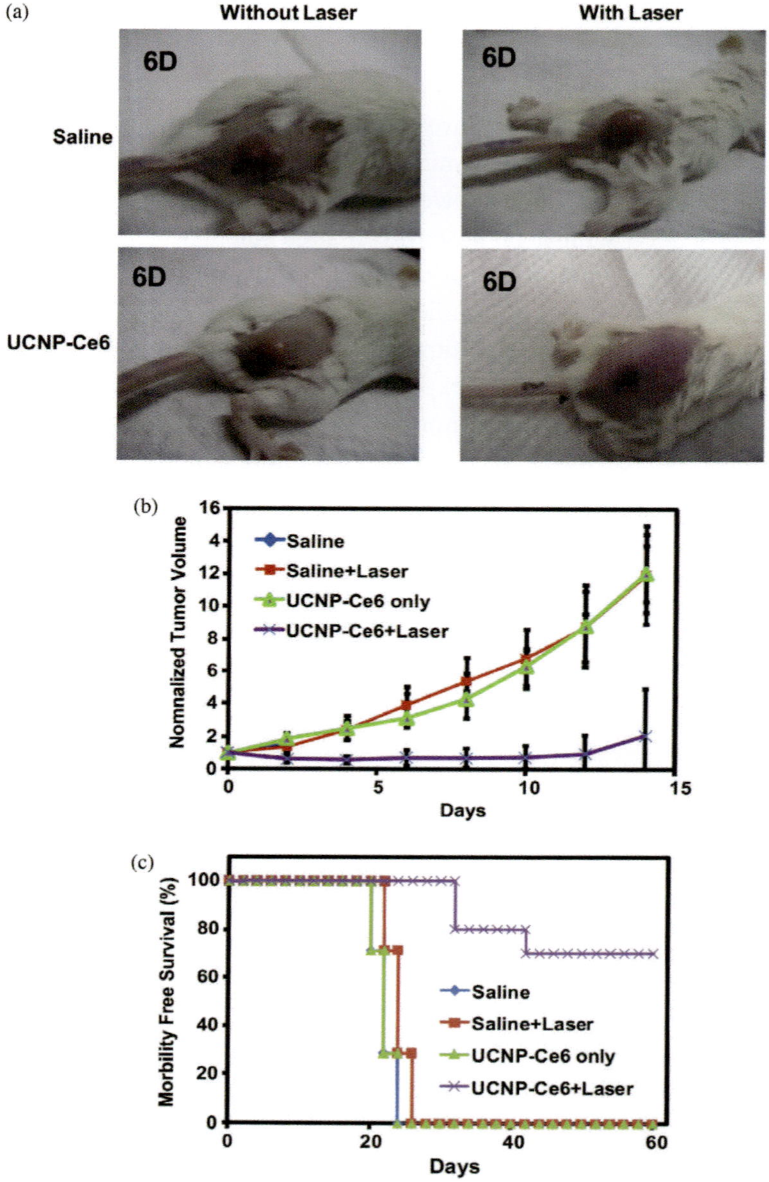

Figure 1.6 (a) Photos of tumor bearing mice in each group at the 6th day of post-treatment, (b) the growth of 4T1 tumor size on different mice treated groups. The tumor size was normalized to the size of tumor right before the treatment, and (c) the survival curves of mice in 60 days post-treatment (Reproduced with permission from Ref. 55, copyright © Elsevier, 2011.)

1.7 Conclusions

While tremendous progress has been made in physio-chemical synthesis/characterization, and pre-clinical validation of optically modulated theranostic nanoparticles discussed in this chapter, clinical translation of these technologies is constrained by two major factors. The first issue relates with the scale-up of synthesis for producing large batches with reproducible optical and other physio-chemical properties. By their very nature, most theranostic nanoparticles introduce multiple components, resulting in difficult standardization and a convoluted FDA clearance pathway. Even limited human trials require orders of magnitude more nanoparticle amounts than large nude mice based pre-clinical studies, necessitating large investments in synthesis and scale-up of these novel materials. A second related issue with theranostics is the absence of commonly agreed upon efficacy metrics and the ill-defined nature of image guided therapy in the clinical arena. Multiple imaging and therapeutic modalities are attractive in small animal models, but the therapy timing and concrete applications in human context are in their infancy. Some of the challenges for translation of nanoparticle based imaging are described by Chapman *et al.*[57] Continued progress in pre-clinical studies and accumulation of safety and toxicity data for theranostic nanoparticles will likely attract attention and investments from the larger biotechnology and pharmaceutical industries and propel these therapies to clinic.

References

1. Brigger I, Dubernet C, Couvreur P, Nanoparticles in cancer therapy and diagnosis. *Adv Drug Deliv Rev*, 2002; 54(5): 631–651.
2. Urban C, *et al.*, Externally modulated theranostic nanoparticles. T*ransl Cancer Res*, 2013; 2(4): 292–308.
3. Ntziachristos V, *et al.*, Fluorescence molecular tomography resolves protease activity *in vivo*. *Nat Med*, 2002; 8(7): 757–760.
4. Choi KY, *et al.*, Theranostic nanoplatforms for simultaneous cancer imaging and therapy: current approaches and future perspectives. *Nanoscale*, 2012; 4(2): 330–342.
5. Peer D, *et al.*, Nanocarriers as an emerging platform for cancer therapy. *Nat Nanotechnol*, 2007; 2(12): 751–760.

6. Janib SM, Moses AS, MacKay JA, Imaging and drug delivery using theranostic nanoparticles. *Adv Drug Deliv Rev*, 2010; 62(11): 1052–1063.
7. McCarthy JR, *et al.*, Polymeric nanoparticle preparation that eradicates tumors. *Nano Lett*, 2005; 5(12): 2552–2556.
8. Lee SJ, *et al.*, Tumor specificity and therapeutic efficacy of photosensitizer-encapsulated glycol chitosan-based nanoparticles in tumor-bearing mice. *Biomaterials*, 2009; 30(15): 2929–2939.
9. Heneghan HM, *et al.*, MicroRNAs as novel biomarkers for breast cancer. *J Oncol*, 2009; 950201.
10. Lovell JF, *et al.*, Activatable photosensitizers for imaging and therapy. *Chem Rev*, 2010; 110(5): 2839–2857.
11. Kim, K, *et al.*, Tumor-homing multifunctional nanoparticles for cancer theragnosis: Simultaneous diagnosis, drug delivery, and therapeutic monitoring. *J Control Release*, 2010; 146(2): 219–227.
12. Kim, JH, *et al.*, Hydrophobically modified glycol chitosan nanoparticles as carriers for paclitaxel. *J Control Release*, 2006; 111(1–2): 228–234.
13. MacCormack MA, Photodynamic therapy in dermatology: an update on applications and outcomes. *Semin Cutan Med Surg*, 2008; 27(1): 52–62.
14. Brown SB, Brown EA, Walker I, The present and future role of photodynamic therapy in cancer treatment. *Lancet Oncol*, 2004; 5(8): 497–508.
15. Kosuge H, *et al.*, Near infrared imaging and photothermal ablation of vascular inflammation using single-walled carbon nanotubes. *J Am Heart Assoc*, 2012; 1(6): e002568.
16. Liu Z, *et al.*, *In vivo* biodistribution and highly efficient tumour targeting of carbon nanotubes in mice. *Nat Nanotechnol*, 2007; 2(1): 47–52.
17. Liu Z, *et al.*, Drug delivery with carbon nanotubes for *in vivo* cancer treatment. *Cancer Res*, 2008; 68(16): 6652–6660.
18. Bhirde AA, *et al.*, Combining portable Raman probes with nanotubes for theranostic applications. *Theranostics*, 2011; 1: 310–321.
19. Robinson, J.T., *et al.*, *High Performance In Vivo Near-IR (>1 mum) Imaging and Photothermal Cancer Therapy with Carbon Nanotubes.* Nano Res, 2010; 3(11): p. 779–793.
20. Firme CP 3rd, Bandaru PR, Toxicity issues in the application of carbon nanotubes to biological systems. *Nanomedicine*, 2010; 6(2): 245–256.
21. Magrez A, *et al.*, Cellular toxicity of carbon-based nanomaterials. *Nano Lett*, 2006; 6(6): 1121–1125.

22. Yang K, *et al.*, Nano-graphene in biomedicine: theranostic applications. *Chem Soc Rev*, 2013; 42(2): 530–547.
23. Zhou L, *et al.*, Graphene oxide noncovalent photosensitizer and its anticancer activity *in vitro. Chemistry*, 2011; 17(43): 12084–12091.
24. Vallet-Regi M, Balas F, Arcos D, Mesoporous materials for drug delivery. *Angew Chem Int Ed Engl*, 2007; 46(40): 7548–7558.
25. Liong M, *et al.*, Multifunctional inorganic nanoparticles for imaging, targeting, and drug delivery. *ACS Nano*, 2008; 2(5): 889–896.
26. Roy I, *et al.*, Ceramic-based nanoparticles entrapping water-insoluble photosensitizing anticancer drugs: a novel drug-carrier system for photodynamic therapy. *J Am Chem Soc*, 2003; 125(26): 7860–7865.
27. He X, *et al.*, Methylene blue-encapsulated phosphonate-terminated silica nanoparticles for simultaneous *in vivo* imaging and photodynamic therapy. *Biomaterials*, 2009; 30(29): 5601–5609.
28. Ashley CE, *et al.*, The targeted delivery of multicomponent cargos to cancer cells by nanoporous particle-supported lipid bilayers. *Nat Mater*, 2011; 10(5): 389–397.
29. Eustis S, el-Sayed MA, Why gold nanoparticles are more precious than pretty gold: noble metal surface plasmon resonance and its enhancement of the radiative and nonradiative properties of nanocrystals of different shapes. *Chem Soc Rev*, 2006; 35(3): 209–217.
30. Huang X, *et al.*, Gold nanoparticles: interesting optical properties and recent applications in cancer diagnostics and therapy. *Nanomedicine* (*Lond*), 2007; 2(5): 681–693.
31. Boisselier E, Astruc D, Gold nanoparticles in nanomedicine: preparations, imaging, diagnostics, therapies and toxicity. *Chem Soc Rev*, 2009; 38(6): 1759–1782.
32. Melancon MP, *et al.*, *In vitro* and *in vivo* targeting of hollow gold nanoshells directed at epidermal growth factor receptor for photothermal ablation therapy. *Mol Cancer Ther*, 2008; 7(6): 1730–1739.
33. Chen J, *et al.*, Gold nanocages as photothermal transducers for cancer treatment. *Small*, 2010; 6(7): 811–817.
34. Jain PK, *et al.*, Calculated absorption and scattering properties of gold nanoparticles of different size, shape, and composition: applications in biological imaging and biomedicine. *J Phys Chem B*, 2006; 110(14): 7238–7248.

35. Oldenburg SJ, *et al.*, Nanoengineering of optical resonances. *Chem Phys Lett*, 1998; 288: 243–247.
36. Jana NR, Gearheart L, Murphy CJ, Seed-mediated growth approach for shape-controlled synthesis of spheroidal and rod-like gold nanopariticles using a surfactant template. *Adv. Mater*, 2001; 13: 1389–1393.
37. Link S, El-Sayed MA, Spectral properties and relaxation dynamics of surface plasmon electronic oscillations in gold and silver nanodots and nanorods. *J Phys Chem B*, 1999; 103: 8410–8426.
38. Loo C, *et al.*, Immunotargeted nanoshells for integrated cancer imaging and therapy. *Nano Lett*, 2005; 5(4): 709–711.
39. Bardhan R, *et al.*, Tracking of multimodal therapeutic nanocomplexes targeting breast cancer *in vivo. Nano Lett*, 2010; 10(12): 4920–4928.
40. Chen W, *et al.*, Targeting pancreatic cancer with magneto-fluorescent theranostic gold nanoshells. *Nanomedicine (Lond)*, 2013.
41. Chen W, *et al.*, A molecularly targeted theranostic probe for ovarian cancer. *Mol Cancer Ther*, 2010; 9(4): 1028–1038.
42. Huang X, *et al.*, Cancer cell imaging and photothermal therapy in the near-infrared region by using gold nanorods. *J Am Chem Soc*, 2006; 128(6): 2115–2120.
43. Huang HC, Rege K, Heys JJ, Spatiotemporal temperature distribution and cancer cell death in response to extracellular hyperthermia induced by gold nanorods. *ACS Nano*, 2010; 4(5): 2892–2900.
44. Shao J, *et al.*, Photothermal nanodrugs: potential of TNF-gold nanospheres for cancer theranostics. *Sci Rep*, 2013; 3: 1293.
45. Al-Jamal WT, Kostarelos K, Liposomes: from a clinically established drug delivery system to a nanoparticle platform for theranostic nanomedicine. *Acc Chem Res*, 2011; 44(10): 1094–1104.
46. Yavlovich A, *et al.*, Light-sensitive lipid-based nanoparticles for drug delivery: design principles and future considerations for biological applications. *Mol Membr Biol*, 2010; 27(7): 364–381.
47. Needham D, Dewhirst MW, The development and testing of a new temperature-sensitive drug delivery system for the treatment of solid tumors. *Adv Drug Deliv Rev*, 2001; 53(3): 285–305.
48. Shum P, Kim JM, Thompson DH, Phototriggering of liposomal drug delivery systems. *Adv Drug Deliv Rev*, 2001; 53(3): 273–284.

49. Puri A, *et al.*, Lipid-based nanoparticles as pharmaceutical drug carriers: from concepts to clinic. *Crit Rev Ther Drug Carrier Syst*, 2009; 26(6): 523–580.
50. Yavlovich A, *et al.*, Design of liposomes containing photopolymerizable phospholipids for triggered release of contents. *J Therm Anal Calorim*, 2009; 98(1): 97–104.
51. Carter KA, *et al.*, Porphyrin-phospholipid liposomes permeabilized by near-infrared light. *Nat Commun*, 2014; 5: 3546.
52. Chen G, *et al.*, (alpha-NaYbF4:Tm(3+))/CaF2 core/shell nanoparticles with efficient near-infrared to near-infrared upconversion for high-contrast deep tissue bioimaging. *ACS Nano*, 2012; 6(9): 8280–8287.
53. Wang J, *et al.*, Single-band upconversion emission in lanthanide-doped KMnF3 nanocrystals. *Angew Chem Int Ed Engl*, 2011; 50(44): 10369–10372.
54. Yang Y, *et al.*, *In vitro* and *in vivo* uncaging and bioluminescence imaging by using photocaged upconversion nanoparticles. *Angew Chem Int Ed Engl*, 2012; 51(13): 3125–3129.
55. Wang C, *et al.*, Near-infrared light induced in vivo photodynamic therapy of cancer based on upconversion nanoparticles. *Biomaterials*, 2011; 32(26): 6145–6154.
56. Qian HS, *et al.*, Mesoporous-silica-coated up-conversion fluorescent nanoparticles for photodynamic therapy. *Small*, 2009; 5(20): 2285–2290.
57. Chapman S, *et al.*, Nanoparticles for cancer imaging: The good, the bad, and the promise. *Nano Today*, 2013; 8(5): 454–460.

Chapter 2

GENE THERAPY TREATMENTS FOR CANCER

Jacob Elmer

Department of Chemical Engineering,
College of Engineering, Villanova University, Villanova, PA, USA
Jacob.Elmer@Villanova.edu

Abstract

Gene therapy has been investigated as a cancer treatment in over 1,100 clinical trials (as of 2013). This chapter provides an overview of those trials, including both viral and non-viral gene delivery methods. The delivery of tumor suppressor genes (e.g. P53, Rb, BRCA 1 and 2, MDA-7, E1A, and FUS-1) is highlighted as a common approach, along with the development of DNA vaccines that reprogram the immune system to target cancer-specific antigens (e.g. PSA, PSMA). Strategies to optimize the effectiveness of each of these techniques are also discussed. Alternative approaches such as suicide gene therapy (e.g. delivery of cytosine deaminase), oncogene silencing with RNAi (e.g. VEGF), and oncolytic virotherapy (e.g. Oncorine) are also discussed, but chimeric antigen receptor T-cell (CAR-T) therapy is not included within the scope of this chapter.

2.1 Introduction

Gene therapy is a promising technique that could potentially cure hundreds of genetic disorders by replacing or supplementing mutated genes. Indeed, as of 2013, more than 1800 clinical trials involving gene therapy had been approved for a variety of different diseases. Cancer is the most prominent target for gene therapy, accounting for more than half (64.4%) of all clinical trials.[1] These trials have shown that gene therapy can be used to halt or reverse tumor growth and increase survival rates for cancer patients. However, well-designed and rigorous clinical trials are required before these treatments can be approved for clinical use. This chapter focuses on the most effective gene therapy strategies that have been developed thus far for cancer, including delivery of tumor suppressor genes to promote cell cycle arrest and/or apoptosis, activation of the immune system against tumor cells with DNA vaccines, and a brief commentary on cytotoxic suicide genes, oncogene silencing with RNAi, and oncolytic viruses.

2.1.1 *Overview of gene delivery methods*

Successful delivery and expression of therapeutic genes in mammalian cells is a particularly difficult task, since eukaryotic cells have a myriad of defenses that repel foreign DNA. For example, foreign DNA must

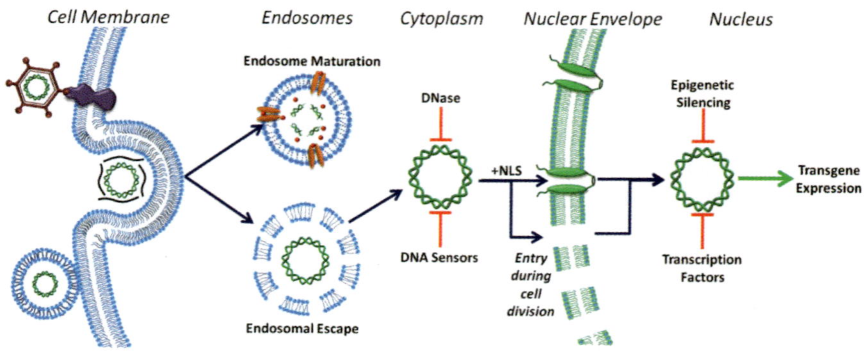

Figure 2.1 Barriers to Gene Delivery. Transgenes require a vehicle (e.g. virus, lipid, or polymer) to enter most cells. Once the transgene is endocytosed, it must escape the endosome before it matures and acidifies. It must then travel through the cytoplasm, in the presence of DNases and inhibitory DNA sensors to reach the nucleus. Transgenes are too large to passively enter the nuclear pore complex by themselves, so they must either bind to a protein with a nuclear localization sequence (NLS) or wait until the nuclear envelope breaks down during mitosis. The transgene may then be expressed in the nucleus, unless it is repressed by epigenetic mechanisms or host transcription factors.

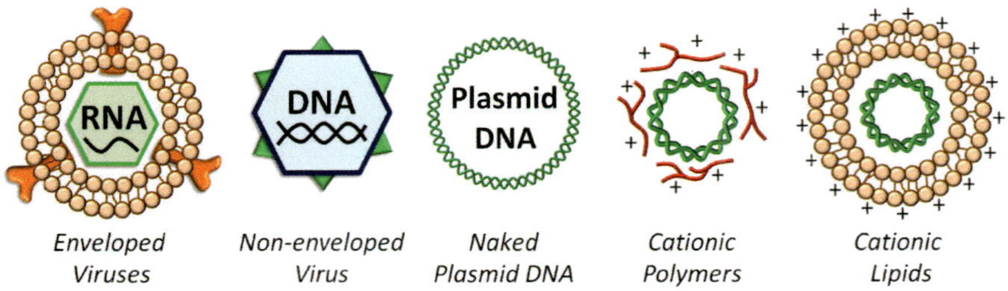

Figure 2.2 Delivery vehicles for transgenes. From left to right: enveloped viruses (e.g. lentivirus) consist of a plasmid membrane (with or without viral proteins) and a protein capsid containing RNA. Non-enveloped viruses only have a capsid. Naked plasmid DNA from bacteria is a common gene therapy agent, but it may also be complexed with cationic polymers to make polyplexes or enveloped in cationic lipids to make liposomes.

overcome three physical barriers to reach the nucleus — the cell membrane, endosomal compartment, and nuclear envelope (see Figure 2.1).[2] The cell also has several biochemical defenses against foreign DNA, ranging from cell surface receptors[3] to DNase enzymes[4] and cytosolic DNA sensors that activate cellular defenses.[5,6] Once inside the nucleus, transgene expression may also be repressed by transcription factors and histone proteins that condense foreign DNA.[7,8] Nonetheless, several effective delivery vehicles based on viral[9,10] and non-viral[11–15] methodologies have been developed to safely deliver genetic cargo to the nucleus (see Figure 2.2).

2.1.1.1 *Viral gene delivery methods*

Viruses have evolved over countless millennia to efficiently transfer nucleic acids into the nuclei of eukaryotic cells. Therefore, viruses were a natural starting point for gene delivery research[16] and are still widely used today. Viral gene therapy has been used to treat several genetic disorders, including some types of blindness (Leber's Congenital Amaurosis[17]), immunodeficiency,[18] hemophilia,[19] Parkinson's disease,[20] and several types of cancers.[21] Many different types of viruses are used, including adenovirus,[22] retrovirus,[23] vaccinia virus,[24] poxvirus,[25] adeno-associated virus (AAV),[206] herpes simplex virus (HSV),[27] and lentivirus.[28] These viruses are modified for gene therapy by removing genes for viral replication and replacing them with therapeutic genes to make them non-pathogenic.

Consequently, these replication-deficient viruses must be manufactured in cells that are simultaneously infected with "helper viruses" or in transgenic packaging cell lines that express the missing viral replication genes.[29]

Viruses may infiltrate host cells *via* membrane fusion or endocytosis. Membrane fusion is observed only in enveloped viruses (e.g. HSV, pox, vaccinia, and retroviruses), which are surrounded by a phospholipid bilayer containing glycoproteins that bind specific receptors on host cells. For example, HSV envelopes have four glycoproteins (gB, gD, gH, and gL)[30] that facilitate fusion of the envelope and cell membrane to release the viral DNA directly into the cytoplasm. Viruses without envelopes rely on capsid proteins to target receptors on specific host cells (e.g. adenoviruses bind to the CAR receptor) and trigger endocytosis. The virus is then taken up into an endosome that will eventually acidify and degrade the viral genome, unless the virus escapes into the cytoplasm. Viruses facilitate endosomal escape with fusogenic peptides that change conformation and disrupt the endosomes as their pH decreases.[31] For example, at pH 5.0, the hemagglutinin protein of influenza viruses undergoes a conformational change that exposes an HA2 domain and causes destabilization of the endosomal membrane.[31] Similar proteins have also been observed in HIV-1 (gp160[32] and TAT [33]) and herpes virus (Vp22).[34]

Regardless of how a virus enters the cell, it must eventually transport its genome into the nucleus. This is a daunting task, since the pores of the nuclear membrane are quite small (9–40 nm).[35] Some retroviruses can only cross the nuclear envelope when it breaks down during mitosis.[36] Therefore, these viruses are unable to transfect non-dividing quiescent cells (e.g. stem or brain cells). Other viruses have special DNA sequences (i.e. "enhancers") in their genomes which bind peptides with nuclear localization signals (NLS) that interact with importin α in the nuclear pore complex to facilitate transport through nuclear pores.[37] For example, the simian virus 40 (SV40) genome contains an enhancer sequence that binds the SV40 large T antigen, which contains an NLS. Incorporation of this SV40 enhancer into plasmids rapidly increases nuclear accumulation and transgene expression up to 100-fold in a variety of cell lines.[38,39] Consequently, the SV40 enhancer is commonly used in both viral and non-viral gene therapy.

When the viral genomes reach the nucleus, they must either integrate into host chromosomes or reside in the nucleus as extrachromosomal elements (i.e. episomes). Genomic integration is typically associated with retroviruses and lentiviruses which have ssRNA genomes.[40,41] Immediately after cellular invasion, these viruses express an integrase enzyme in the cytoplasm that binds to long terminal repeats flanking the viral genome. The integrase-RNA complex then travels to the nucleus and randomly integrates itself into host chromosomes.[42] This integration strategy provides stable long-term expression (even after multiple rounds of cell division). Unfortunately, integration of viral genes can be oncogenic if they also influence the expression of other host cell genes (e.g. silencing tumor suppressor genes or enhancing oncogenes). For example, although viral gene therapy was used to successfully treat 90% of patients with X-linked SCID in a clinical trial, two of the patients eventually developed leukemia. Further analysis of these patients revealed that the retroviral genes had integrated near the promoter region of the LMO2 oncogene, leading to its overexpression and a subsequent increase in hematopoiesis and cell division.[43,44] Therefore, while integrating viruses are especially effective tools for long-term gene therapy, the possibility of insertional mutagenesis is a serious safety risk. However, this risk might eventually be abrogated by specifically targeting viral integration to safe regions in the genome with sequence-specific integrase or recombinase enzymes.[45,46]

Non-integrating viruses (e.g. HSV1, vaccinia, and adenoviruses) avoid the risks associated with insertional mutagenesis, but they are typically limited to only transient gene expression since viral episomes may be lost during cell division. However, transgene expression can be significantly extended by incorporating origins of replication[47] and scaffold/matrix attachment regions (S/MARs)[48] into episomes. Some viral episomes have been observed to persist in human cells for up to two months.[49,50] However, it is also important to mention that transient gene expression may be sufficient for the treatment of cancer, since transgene expression may be unnecessary once the patient enters remission.

Altogether, viruses have been shown to be highly efficient gene delivery vehicles, but they do have some limitations. Aside from the insertional mutagenesis observed with retroviruses, viral gene delivery may also be intercepted by the immune system, especially after multiple doses of the same viral vector.[51,52] The dimensions of viral capsids can also limit the size and number of genes that can be delivered. Smaller adeno-associated viruses have a maximum payload capacity of 4.5 kb, while larger herpes virus capsids can accommodate constructs above 30 kb.[53] Lastly, the manufacturing costs associated with viruses can also be higher than the non-viral alternatives discussed in the following sections. Despite these drawbacks, viruses are still the most commonly used gene delivery vehicles, accounting for more than two thirds of all clinical trials.[1]

2.1.1.2 *Non-viral gene delivery methods*

The immunogenic and oncogenic events observed with viral gene delivery have motivated the development of many different non-viral gene delivery methods involving lipids, polymers, proteins, polymers, and/or other polycations (examples shown in Figure 2.3). In general, non-viral gene delivery vehicles have a net positive charge that allows them to simultaneously bind the negative charges on DNA and the cell surface. Therapeutic genes are simply cloned into plasmids and produced in bacteria, then mixed with a polycation to form "polyplexes" or "lipoplexes" that are structurally similar to viral capsids, but much less efficient. In most cases, plasmids delivered non-virally are maintained as extrachromosomal elements and do not integrate into the host genome.

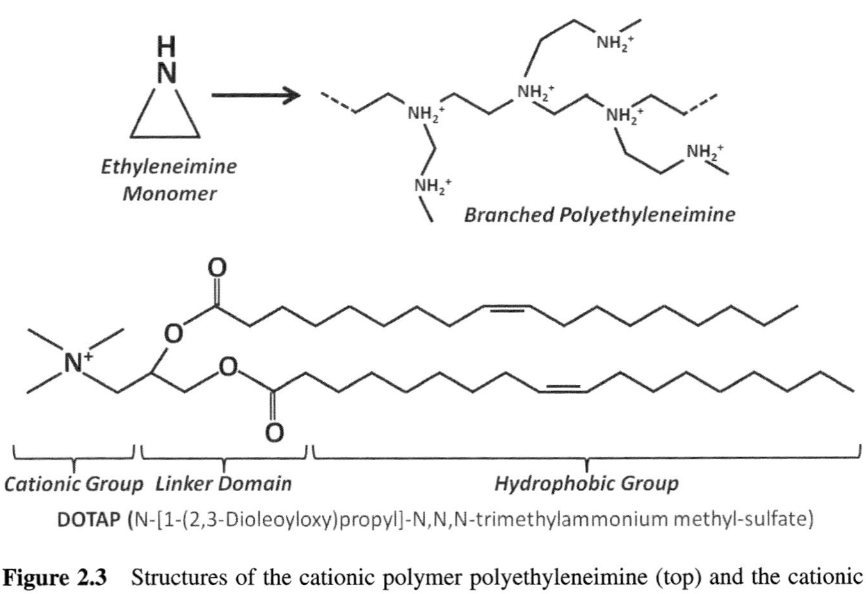

Figure 2.3 Structures of the cationic polymer polyethyleneimine (top) and the cationic lipid DOTAP (bottom).

Cationic lipids are the most commonly used non-viral gene delivery vehicles in clinical trials (6% of total).[1] Many different cationic lipids have been developed, but they generally consist of three variable parts — a hydrophilic headgroup, hydrophobic domain, backbone domain, and a linker that connects them.[54] The hydrophilic headgroup is responsible for binding DNA and usually contains some sort of protonable amine(s) or other nitrogenous moiety.[55] The hydrophobic domain may consist of a simple aliphatic chain or a more complex steroid group (e.g. cholesteryl) or fluorinated lipid.[56] Multiple hydrophobic domains assemble to form a liposomal structure that surrounds the DNA to form lipoplexes. The hydrophobic moiety has also been shown to influence the overall toxicity of cationic lipids.[57] Lastly, carbamate, ester, amide, or ether bonds may be used to link the hydrophobic and hydrophilic domains with varying stability (ether > carbamate > ester). Since easily biodegradable linker groups tend to have lower cytotoxicity, linker groups with pH and redox-sensitive bonds (e.g. disulfides) have also been investigated.[54]

Altogether, the diversity of components available for cationic lipid synthesis has enabled researchers to finely tune and optimize several different cationic lipids for gene therapy. One of the first cationic lipids synthesized for gene delivery was DOTMA,[58] which consists of a quarternary amine group and glycerol backbone linked to unsaturated oleoyl hydrophobic chains with an ether bond. It was later discovered that the lipoplex formation and transfection efficiency of DOTMA and other cationic lipids could be greatly enhanced by mixing them with neutral "helper lipids" like dioleoylphosphatidylethanolamine (DOPE).[59] In contrast to other neutral lipids, DOPE has the unique ability to destabilize the endosomal membrane to release DNA into the cytoplasm.[59,60] DOPE is also used as a helper lipid in the commercial transfection reagent Lipofectamine®, which uses the cationic lipid DOSPA (DOSPA is similar to DOTMA, but with the polycation spermine in place of its quarternary amine). Transfection efficiency and circulation half life of lipoplexes has also been increased by PEGylating the surface of liposomes to create "stealth polyplexes" that evade clearance by the mononuclear phagocyte system *in vivo*.[61] Both traditional and PEGylated polyplexes can be further modified with antigens to specifically target cancer cells (e.g. transferrin for colon cancer[62] and haloperidol for breast cancer[63]). Overall, although the transfection efficiency of cationic lipids is still somewhat lower than viruses, their versatility and safety profiles make them interesting alternatives to viral gene therapy.

In addition to lipids, several cationic polymers have also been developed for non-viral gene therapy.[64] The synthesis of these polymers typically involves chain reactions between one or two monomers, making them easier to prepare than lipids. Instead of forming liposomes, cationic polymers simply form electrostatic bonds with DNA to form nanoparticles called "polyplexes" with a net positive surface charge that is readily endocytosed by cells. The most commonly used polymer for gene delivery is polyethyleneimine (PEI,[65] see Figure 2.3), which is commercially available in several molecular weights (5–500 kDa) and in different isoforms (branched or linear polymers). In general, high MW branched PEI polymers tend to have high transfection efficiency and toxicity, while lower MW linear PEI polymers have lower transfection efficiencies and toxicity.[66,67] Likewise, carbohydrate and peptide polymers (e.g. chitosan[68] and poly-L-lysine[69]) with higher biocompatibility and lower toxicity also tend to have low transfection efficiencies. These shortcomings have inspired the search for novel polymers with high transfection efficiency and low toxicity. Indeed, several promising polymers with higher transfection efficiency than PEI have been identified in high-throughput screens of combinatorial polymer libraries, including poly(diglycidyl ethers),[70] diamine-end-modified poly(β-amino) esters,[71] and polyaminoglycosides.[72] In addition, polymer-lipid hybrids have recently been shown to provide exceptionally high transfection efficiencies.[73–75] Any of these polymers may also be conjugated to antigens to enhance delivery to specific cell types (e.g. antibodies,[76] sugars,[77] peptides,[78] or vitamins[79]). Therefore, despite their humble beginnings, cationic polymers are emerging as potent gene delivery vehicles that are comparable to viruses and lipids.

Several other interesting non-viral vechicles have also been developed, including dendrimers, nanoparticles, and histone peptides. Dendrimers share many of the same characteristics as polymers, but their size and composition is much easier to control.[80] Gold nanorods (coated with polymer) have also been shown to be useful gene delivery vehicles with interesting spectral properties.[81] In addition, magnetic nanoparticles have been used to form complexes with viruses, lipids, and polymers that physically force genes into a variety of cell lines using a magnetic field.[82] Finally, genes have also been delivered into cells with histones, the nuclear proteins which condense chromosomes inside the nucleus. The presence of several positively charged lysine and arginine residues on histone N-terminal tail domains, along with their nuclear localization signals, makes histonefection an effective and highly biocompatible non-viral gene delivery method.[83]

2.1.1.2.1 Naked plasmid DNA

In some cases, "naked" plasmid DNA can be used to transfect cells without any type of delivery vehicle, although the effects are usually non-specific and at low levels.[84] Nonetheless, naked plasmids are the most commonly used non-viral gene delivery technique, accounting for 18.3% of all clinical gene therapy trials.[1] Following intravascular injection, plasmid DNA quickly and specifically accumulates in hepatocyte cells in the liver (endothelial and Kupffer cells are unaffected).[85] Transfections with naked plasmid DNA were shown to induce high-level expression of clotting factor IX (0.5–2.0 μg/mL) for up to 1.5 years in mice.[86] The exact mechanisms of this phenomenon remain unclear, but observations that plasmid DNA is not taken up by hepatocytes *in vitro* suggests that tissue architecture may somehow facilitate transfection by naked DNA[84,87] Naked plasmids may also be injected intramuscularly or intradermally to efficiently transfect non-dividing muscle or skin tissue.[84] Transfection efficiencies in these tissues may be further enhanced using local electroporation.[88] Intramuscular injection of naked plasmids (with or without electroporation) is commonly used to deliver DNA vaccines, as described in Section 2.3.[89]

2.2 Delivery of tumor suppressor genes into tumor cells

The advent of high-throughput DNA sequencing and proteomic analysis technologies have revolutionized cancer diagnosis and treatment. For example, now that it is possible to quickly sequence the entire genome of a cancer patient, physicians may identify the specific mutation(s) responsible for their cancer. This personalized healthcare approach allows doctors to immediately prescribe the most effective drug(s) and treatments for each patient, instead of wasting precious time and resources screening multiple chemotherapeutic drugs. In addition, genomic sequencing can also be used to identify target genes for gene therapy. This chapter specifically discusses the delivery of tumor suppressor genes (e.g. p53, BRCA-1, and others) that are down-regulated in cancer cells. Hypothetically, this type of treatment should effectively inhibit or reverse tumor growth with minimal side effects by directly treating the root of the disease.

2.2.1 *Genetic dysregulation in cancer cells*

Decades of research have identified mutations in thousands of different genes that can lead to cancer.[90] (These mutations have been organized into the Catalogue of Somatic Mutations in Cancer (COSMIC) database, which is available online[91]) Broadly speaking, genetic mutations may cause cancer by either activating/overexpressing oncogenes or downregulating/mutating tumor suppressor genes. A recent census of gene expression patterns in a variety of different cancers identified more than 70 different genes that were overexpressed in cancer cells, including AKT2, ERBB2/HER2, MYC, MYCL1, MYCN, REL, and others.[92] Mutations in these genes may initiate/accelerate the cell cycle or inhibit apoptosis, thereby allowing cancer cells to rapidly proliferate. One of the best known oncogenes is HER2 (aka ERBB2, Neu, or CD340), a receptor tyrosine protein kinase that is associated with breast and ovarian cancers. HER2 is one of several HER receptors (HER1, 3, and 4) that control initiation of the cell cycle. While most of these HER receptors must be activated by growth factors, HER2 is constitutively active and down-regulated in most cell types. Therefore, overexpression of HER2 in cancer cells can cause unregulated cell growth.[93] Many different drugs have been developed to target and inhibit HER2, including the monoclonal antibody Trastuzumab (aka Herceptin®).[94] HER2 has also been successfully targeted with gene silencing,[95] but shRNA delivery is outside the scope of this chapter.

2.2.2 *Delivery of tumor suppressor genes*

In contrast to oncogenes, tumor suppressor genes are expressed to prevent cancer in healthy cells.[96] They are typically divided into two

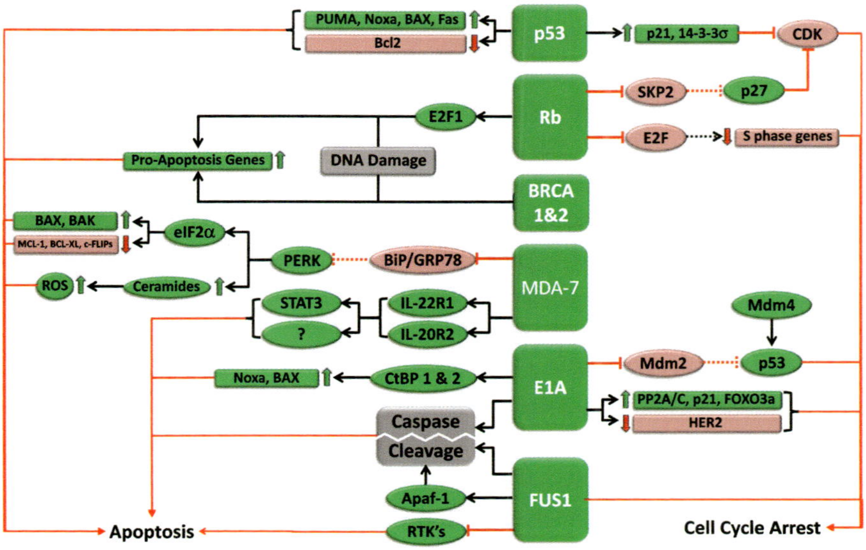

Figure 2.4 Pathways of apoptosis induction (left) and cell cycle arrest (right) by tumor suppressor genes. Genes are shown in rectangles and enzymes in ovals, while activated enzymes are shown in green and inhibited enzymes are red. For example, expression of p53 upregulates the p21 and 14-3-3s genes, which inhibit cyclin dependent kinase (CDK) to arrest the cell cycle. Likewise, MDA-7 binds to BiP/GRP78 to prevent it from inhibiting PERK, allowing PERK to activate eIF2d to transcribe the pro-apoptotic genes BAX and BAK.

categories — gatekeepers and caretakers (see Figure 2.4 for an overview of tumor suppressor gene activities). Gatekeeper genes (e.g. p53, p73, PTEN, Fhit, and Rb) suppress tumor formation by regulating the initiation and progression of the cell cycle. Meanwhile, caretaker genes (e.g. ATM, ATR, BRCA1, and BRCA2) are expressed to detect and repair harmful genetic mutations.[97] If the mutation is irreparable, caretakers may also induce cell death or apoptosis to prevent the spread of these mutations. Since many of these tumor suppressor genes may be mutated or lost in cancer cells, they are attractive targets for gene therapy. The following section discusses several different tumor suppressor genes that have been used in clinical trials to effectively inhibit or reverse tumor growth, including p53, retinoblastoma protein (Rb), BRCA-1 & 2, MDA-7, E1A, and FUS1.

2.2.2.1 *Delivery of the gatekeeper p53 gene*

The vast majority of clinical gene therapy trials have focused on the tumor suppressor protein p53 (aka TP53). It is a transcription factor that is activated in response to cellular stresses (including DNA damage[98]) to regulate the expression of over 150 different genes that prevent cell division and promote apoptosis, if necessary.[99] For example, p53 activates the p21[100] and 14-3-3σ[101] genes, which inhibit cyclin dependent kinases and arrest the cell cycle in the G1 or G2 phases. In response to DNA damage, P53 can also induce apoptosis by activating a variety of pro-apoptotic genes (e.g. Bax, Noxa, p53AIP1, PUMA or Fas; see Figure 2.4) or down-regulating anti-apoptosis genes (e.g. Bcl2).[102–105] These functions allow p53 to prevent the proliferation of unhealthy or cancerous cells. However, mutations in p53 have been observed in the majority of human cancers, thereby allowing cancer cells to proliferate.[106]

Hypothetically, restoration of wild type p53 expression in cancer cells with p53 mutation(s) should inhibit cell division and/or induce apoptosis. This theory has been the driving force for decades of p53 gene delivery research that have yielded some promising results. Preclinical studies showed that p53 could be effectively delivered to a myriad of different cancers (e.g. colorectal, head and neck, and gliomas) using adenovirus,[107–109] retrovirus,[110] and liposomal vectors.[111,112] Several studies also showed that radiotherapy[113–115] and chemotherapy[116,117] could be synergistcally enhanced by p53 gene delivery, due to its ability to induce apoptosis in response to DNA damage. These results motivated a series of Phase I clinical trials in the 1990's that used adenoviruses to deliver p53 to lung cancer cells. These trials showed successful gene delivery in at least 25–50% of the patients, most of which experienced tumor stabilization or regression.[118,119] In another Phase I trial, the combination of p53 adenoviruses with the DNA-damaging chemotherapeutic cisplatin proved to be even more effective, with 17 out of 23 patients showing disease stabilization for up to two months.[120]

Additional clinical trials with adenoviral p53 gene delivery have shown further promise. One Phase II trial showed that simultaneous administration of radiation (60 Gy) and p53 gene delivery significantly enhanced non-small cell lung cancer patient outcome, with a one year progression-free survival of 45.5%.[121] Similar results were also observed with ovarian cancer patients, which showed that multiple doses of adenoviral p53 were necessary to ensure long-term survival.[122] Advexin, an adenoviral p53 therapy developed by Introgen therapeutics, showed promising results in Phase I/II trials, but stagnated during Phase III trials.[123–125] In contrast, a similar adenoviral treatment called Gendicine (SiBiono GeneTech, Shenzhen, China) flourished throughout its clinical trials for head and neck squamous cell carcinoma (HNSCC), showing complete regression in 64% of patients and partial regression in 29% of patients following weekly doses of virus. Gendicine was officially approved for clinical use with HNSCC in China in 2003, but has not yet been approved in the United States or other countries.[126,127] Instead, non-viral p53 gene delivery vehicles have gained momentum in the USA, since they do not elicit the immune responses seen with adenoviruses. One liposomal formulation (SGT-53) with an anti-transferrin receptor single chain antibody fragment has also been shown to selectively target tumor cells to maximize therapeutic effects. Recent phase I clinical trials of SGT-53 have shown impressive selectivity for tumor cells, with partial responses in some patients.[128,129]

2.2.2.2 *Delivery of the gatekeeper retinoblastoma protein (Rb) gene*

Another heavily studied tumor suppressor is the retinoblastoma protein (Rb). Although Rb mutations were originally discovered in ocular tumors (i.e. retinoblastomas),[130,131] they have also been observed in a wide variety of other cancer types (e.g. small and non-small cell lung cancers).[132] In healthy cells, Rb plays a crucial role in regulating cell cycle arrest and apoptosis. For example, Rb prevents entry into the S phase of the cell cycle by binding and inhibiting E2F transcription factors that would otherwise activate genes required for DNA synthesis and cell division. However, this effect can be reversed by cyclin-dependent kinases (CDK), which phosphorylate Rb to prevent E2F binding in proliferating cells.[133] Interestingly, Rb can also inhibit CDK by inactivating SKP2 (S phase kinase-associated protein 2), an antagonist of the CDK inhibitor p27. Therefore, inhibition of SKP2 by Rb allows p27 to accumulate and inhibit

CDK.[134] In addition, recent studies have shown that Rb can also use E2F proteins as adaptors to make sequence-specific epigenetic modifications (including histone deacetylation, methylation (H4K20me3), and DNA methylation)[135,136] that manipulate gene expression to control cell cycle progression. The Rb-E2F1 complexes may also inhibit cell growth by activating apoptotic pathways. Indeed, it has been shown that DNA damage activates the ATM (ataxiatelangiectasia mutated) and CHK2 (checkpoint kinase 2) kinases, which phosphorylate the Rb-E2F1 complex, leading to the activation of several pro-apoptotic genes.[137,138]

Mutation or loss of the gatekeeper Rb can lead to rapid cell proliferation and oncogenesis due to loss of control at the G1/S cell cycle checkpoint. In addition, loss of Rb also contributes to genomic instability, since constitutively active E2F transcription factors induce aberrant gene expression patterns that lead to faulty chromosome replication, DNA damage, abnormal chromosome segregation during mitosis, and aneuploidy (an irregular number of chromosomes in cells).[139,140] While these symptoms would normally activate apoptosis in healthy cells, Rb mutations also prevent apoptosis and allow cancerous cells to survive. Hypothetically, delivery of a functional Rb gene to Rb-deficient tumors could restore cell cycle control and apoptosis induced by DNA damage to slow or reverse tumor growth. Indeed, animal studies have shown that adenoviral delivery of the Rb gene to Rb-deficient pituitary tumors in mice resulted in stabilization or regression of the tumors.[141] A truncated form of the Rb gene (Rb94) has also been shown to decrease tumor growth and induce apoptosis in mouse tumors.[142] These promising results have motivated a Phase I clinical trial using liposomes to deliver the Rb94 gene into cancer patients, but the results of this study are not yet available (Clinicaltrials.gov Identifier: NCT01517464).

2.2.2.3 *Delivery of the caretaker BRCA-1 and BRCA-2 genes*

The "caretakers" BRCA-1 and BRCA-2 are tumor suppressor genes that encode large nuclear proteins involved in cell cycle progression, transcriptional regulation, and induction of apoptosis. However, their main function is to detect and repair DNA damage *via* error-free non-homologous end joining.[143–145] DNA repair may still occur without BRCA-1 and

BRCA-2, but through other mechanisms (e.g. single-stranded annealing) that are more error-prone, thereby making cells more susceptible to oncogenic mutations. Indeed, mutations in these genes can increase a patient's risk of breast cancer from 12% (general population's risk) to 57% and 49% for BRCA-1 and BRCA-2 mutations, respectively.[146,147] BRCA-1 and BRCA-2 mutations have also been implicated in ovarian cancer,[147] leukemia, and lymphoma.[148]

Although no clinical trials have been conducted with BRCA-2 gene delivery, BRCA-1 gene delivery showed early promise in clinical trials. In a Phase 1 trial with ovarian cancer patients, retroviral delivery of BRCA-1 stabilized tumors in 8 of 12 patients, while 1 patient showed signs of tumor remission.[149] Unfortunately, a similar Phase II trial failed to show any positive effects, most likely due to immune responses to the retroviral vector used in the study.[150] The same group later designed a new viral vector that had lower immunogenicity *in vitro*, but no clinical trials were attempted.[151] Due to the lack of progress in BRCA-1 gene delivery, it is now much more common to treat breast cancer patients with chemotherapeutic drugs that intentionally damage DNA (e.g. the DNA cross-linker cisplatin) and selectively kill cells that are unable to repair their DNA due to BRCA mutations.[152]

2.2.2.4 *Delivery of the cytokine MDA-7/IL-24 gene*

The melanoma differentiation-associated gene 7 (MDA-7) is a novel type of tumor suppressor gene that is downregulated in melanoma and several other cancers.[153,154] In addition to its intracellular effects, MDA-7 is also secreted to activate responses in distant cells, much like a cytokine. Consequently, MDA-7 has been recognized as part of the IL-10 family and given the alternative designation MDA-7/IL-24.[155] The ability of MDA-7/IL-24 to amplify its effects in neighboring or distant cells makes it an especially attractive candidate for cancer gene therapy, since it is often impossible to successfully transfect each of the cells inside a tumor.

MDA-7 has been shown to activate apoptosis in cancer cells in three distinct ways — (1) modulation of protein translation, (2) generation of reactive oxygen species (ROS), and (3) activation of apoptotic pathways in distant cells by signal transduction. When MDA-7 is expressed intracellularly, it binds to the chaperonin protein BiP/GRP78 to prevent it from binding and inhibiting the protein-kinase R-like endoplasmid reticulum kinase (PERK).[156] This allows PERK to phosphorylate and activate eIF2 alpha, an inhibitor of protein translation that suppresses anti-apoptotic proteins (MCL-1, BCL-XL, and c-FLIP's).[157,158] In addition, pro-apoptotic genes (e.g. BAX and BAK) are simultaneously activated to induce apoptosis.[159,160] In addition to eIF2 alpha activation, PERK also increases ceramide concentrations, which leads to ROS generation. Elevated ROS levels trigger apoptosis in multiple ways, including activation of the JNK and p38MAPK pathways and concomitant increases in BAK and BAX expression.[161] As a secreted cytokine, MDA-7/IL-24 seeks out and binds cells with IL-22R1 and/or IL-20R2 receptors.[162] Receptor binding activates the STAT3 transcription factor and induces apoptosis, but the exact mechanism of this phenomenon is not yet fully understood.[163]

Several preclinical studies have shown that restoration of MDA-7/IL-24 expression *via* gene therapy rapidly induces apoptosis in a variety of cancer cells[164,165] while having no effect on non-cancerous cells.[154,165] In addition, MDA-7/IL-24 gene delivery was also shown to enhance the effects of chemotherapeutics (5-fluorouracil, cisplatin, mitomycin C, and etoposide)[166] and ROS generation by radiotherapy.[165,167–169] These promising results motivated clinical trials of an adenoviral vector for MDA-7/IL-24 named INGN 241. Following intratumoral injection, MDA-7/IL-24 expression was observed in 100% of patients and clinical responses were observed in 44% of patients. In addition to apoptosis around the injection site, MDA-7 and its effects were also observed in neighboring cells and distant sites.[170] Interestingly, expression of the MDA-7/IL-24 cytokine also increased the concentrations of other cytokines (IL-6, IL-10, and TNF-alpha) and enhanced the activation of immune cells ($CD8^+$ T cells).[171] These preliminary results suggest that MDA-7/IL-24 gene therapy may be a particularly effective strategy, since it is able to induce apoptosis in both transduced and non-transduced cancer cells.

2.2.2.5 *Delivery of the adenoviral E1A gene*

In addition to the human tumor suppressor genes, some viral genes have also been observed to have tumor suppressor activity. For example, the adenoviral early region 1A gene (E1A)[172] arrests cells in the S phase and modifies the expression of several host genes to enhance replication of the viral genome. In addition, host cell proteins can also recognize E1A as a pathogenic protein and initiate apoptosis to prevent viral replication.[173] Therefore, E1A may suppress tumor growth by inducing either cell cycle arrest or apoptosis.

Expression of E1A has been found to have a wide variety of effects on several host cell proteins and pathways implicated in cancer treatment. For instance, E1A suppresses cell proliferation by stabilizing its fellow tumor suppressor p53 following repression of the p53 inhibitor Mdm2 and enhanced binding between p53 and Mdm4. The p53-Mdm4 complex is still functional, but its large size prevents its export from the nucleus, thereby prolonging the nuclear residence time of p53.[174,175] In addition to p53, E1A also activates several other tumor suppressors, including PP2A/C,[176] p21,[177] and FOXO3a.[178] Interestingly, E1A has also been shown to reduce cell proliferation by inhibiting the p300/CBP transactivator complex and repression of the HER2/neu oncogene.[179] In addition, E1A binds and activates the C-terminal binding proteins CtBP1 and CtBP2 to enhance the expression of several pro-apoptotic genes (BAX, Noxa, Perp, p21, and Bik).[180] E1A also enhances apoptosis by increasing PARP cleavage and enhancing the activation of caspases 8 and 9. Consequently, E1A expression also enhances the induction of apoptosis by TRAIL (Tumor necrosis factor-related apoptosis inducing ligand), which relies on these caspases.[181]

Preclinical studies have shown that delivery of the E1A gene is an effective tumor suppressor, especially in combination with the histone deacetylase inhibitor Vorinostat[182] and chemotherapeutics (e.g. paclitaxel, cisplatin, and doxorubicin) in cancer cells that overexpress HER2/neu.[183–185] Multiple Phase I clinical trials with E1A were completed from 2001–2004, all of which used liposomes instead of adenovirus to deliver the E1A gene. These trials showed promising results, including cell cycle arrest, increased apoptosis, and decreased HER2 expression in breast and ovarian cancer patients. Some dose-limiting toxicity was observed at high concentrations of E1A liposomes, but lower doses proved safe and effective.[186–188] A Phase II clinical trial that used E1A liposomes to treat 24 head and neck cancer patients also showed a modest tumor response, as one patient showed a complete response, two had minor responses, and seven patients had stable disease.[189] Several more clinical trials with E1A are currently underway, but the current results show that E1A is an effective tumor suppressor with unique effects, including HER2 repression.

2.2.2.6 *Delivery of the gatekeeper FUS-1 gene*

One of the most common genetic alterations in lung and breast cancer cells is the deletion of a large portion of the short arm of chromosome 3p.[190,191] More specifically, a 120 kbp segment containing several putative tumor suppressor genes (CACNA2D2, PL6, 101F6, FUS1, BLU, RASSF1, NPRL2, HYAL2, AND HYAL1)[192,193] within the 3p21.3 region is commonly deleted. The most potent tumor suppressor in the group appears to be FUS1 (aka Tumor Suppressor Candidate 2, TUSC2), a gatekeeper gene that induces G1 arrest when reintroduced into FUS1-deficient lung cancer cells.[194] FUS1 can also induce apoptosis by activating the apoptotic protease-activating factor 1 (Apaf-1),[195] which activates caspases and triggers the release of cytochrome p450 from mitochondria to initiate apoptosis.[196,197] Alternatively, FUS1 can also induce apoptosis *via* inhibition of receptor tyrosine kinases (e.g. EGFR, PDGFR, c-Abl, c-Kit, and AKT), which also leads to caspase activation.[195]

Preclinical studies have shown that FUS1 expression also enhances the effects of chemotherapy (cisplatin) and radiotherapy.[198] Interestingly, co-delivery of FUS1 and p53 genes also displays synergy, since FUS1 inhibits binding of Mdm2 to p53.[197] A recent Phase I clinical trial using DOTAP-cholesterol nanoparticles to deliver a FUS1 gene to 31 lung cancer patients showed promising results, including FUS1 expression in seven out of eight tumors tested and a significant increase in tumor cell apoptosis.[199] These positive results have prompted an ongoing phase I/II study of FUS1 gene delivery with the EGFR inhibitor erlotinib to determine if there is any significant synergy between them (Clinicaltrials.gov Identifier: NCT01455389).

2.2.3 *Strategies to enhance tumor suppressor gene expression*

While all the tumor suppressor genes discussed thus far have shown promise in clinical trials, transgene delivery and expression techniques still require significant improvement to realize their full potential. For example, while adenoviruses are extremely efficient gene delivery vehicles, they tend to elicit immune responses that reduce their effectiveness when injected systemically. Therefore, viruses must be injected intratumorally to avoid immune detection, making them less effective for treatment of metastatic cancers. Non-viral delivery vehicles can be administered systemically since they do not elicit immune responses, but they are typically less effective than viruses. The main disadvantage of both viral and non-viral gene delivery techniques is that transgene expression is typically limited to a few days or weeks. Indeed, most gene therapies require several weekly treatments for maximum effectiveness. Therefore, gene therapy treatments could be substantially improved by finding ways to prolong expression of transgenes inside the cell nucleus. This section discusses the modification of plasmid/episome DNA sequences to enhance and prolong the expression of transgenes inside host cells.

2.2.3.1 *Optimization of the transgene expression cassette*

Transgene expression from both viral and non-viral plasmids is dependent upon several different genetic elements (shown in Figure 2.5). For example, the amount of protein expressed inside a cell is directly proportional to the strength of the promoter that initiates transcription of the corresponding gene. Viral promoters (e.g. SV40,[200] CMV,[201] and RSV[202]) have traditionally been used for gene therapy, since they are constitutively active in most human cell lines. However, the relative strength of these promoters has been shown to vary significantly between cell types (CMV > RSV >SV40, in general)[203,204] and they are prone to silencing in the presence of cytokines.[205–207] These shortcomings have prompted the search for alternative human promoters that are equally strong but not influenced by cytokines. The two most popular human promoters identified thus far are the elongation factor 1 alpha (EF1α) and tetracyclin response element (TRE) promoters, which provide equal or higher levels of gene expression compared to CMV in a variety of different cell lines.[208] Several other cell-specific mammalian promoters have also been used to target transgene expression to specific cell types, including a desmin promoter for muscle cells,[209] a metallothionein promoter for keratinocytes,[210] and a platelet

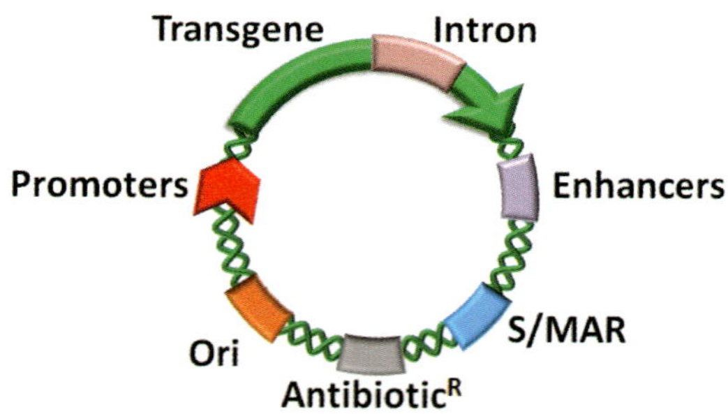

Figure 2.5 Plasmid genetic elements. Transgene expression can be driven by a variety of promoters. Transgene expression may be enhanced by optimizing the codons of the transgene or incorporating introns into its sequence. Translation may also be enhanced by adding enhancers that stabilize mRNA. The plasmid backbone must contain a bacterial origin of replication (Ori) and antibiotic resistance gene for production in *E. coli*. Finally, plasmid replication and segregation in eukaryotic cells may be enhanced by adding a scaffold/matrix attachment region (S/MAR) sequence downstream of the transgene.

derived growth factor β (PDGF-β) promoter for neurons.[211] These tissue-specific promoters are especially effective for suicide gene therapy, in which the transgene is expressed in disease cells to convert a non-toxic prodrug into its cytotoxic form.[212]

Translation of transgene mRNA can also be enhanced by including special sequences within the mRNA transcript. First of all, although ribosomal binding sequences are not strictly required for eukaryotic translation, transgene expression is significantly enhanced by adding a "Kozak" sequence (GCCGCC(A or G)CC immediately upstream of the transgene to assist in ribosomal binding.[213] Long terminal repeats at the 5′ or 3′ end of the mRNA transcript have also been shown to reduce the rate of mRNA degradation, allowing more protein to be translated. For example, addition of the SV40 polyA tail sequence to mRNA transcripts has been shown to enhance transgene expression up to 125-fold *in vivo*.[214] Lastly, the woodchuck hepatitis virus post-transcriptional regulatory element (WPRE) has also been shown to assist in nuclear export of transgene mRNA and enhance expression up to 700-fold in murine liver tissue.[215]

The sequence of the transgene itself can also be modified to optimize expression. Most importantly, the codons within the transgene must strictly adhere to the codon bias of the host organism. A codon bias indicates the codons and tRNA's that are most frequently (or rarely) used by different organisms. For example, although leucine is encoded by six different codons, the TTA and CTA codons only occur rarely (frequency = 7%) in human genes. Since the corresponding tRNA's are equally rare, including these codons in transgene mRNA can lead to ribosomal stalling and abortion of translation. However, codon optimization for human cells avoids this problem and significantly enhances transgene expression (especially for the heterologous proteins used in DNA vaccines).[216,217] Introns may also be inserted into transgenes to enhance their expression.[218] For example, intron A from the cytomegalovirus has been shown to significantly enhance transgene expression.[219] The mechanism of this enhancement is still unclear, but it may be due to increased nuclear export of spliced transcripts and the deposition of beneficial proteins onto the mRNA during splicing that prevent mRNA degradation.[218]

2.2.3.2 *Replication and maintenance of plasmids by S/MAR's*

For transgene expression to be maintained in proliferating cells, plasmids must be faithfully replicated and segregated during mitosis. This may be accomplished by inserting a scaffold/matrix attachment region (S/MAR) into the plasmid immediately downstream of the transgene. The AT-rich composition of S/MARs makes them relatively unstable, especially when exposed to the supercoiling strain created by transcription of the upstream transgene. Destabilization of the S/MAR creates a region of single stranded DNA that is bound by scaffold attachment factor A (SAFA) and connected to the nuclear matrix. Association with the nuclear matrix has several benefits, including increased gene expression and replication.[220] In addition, since chromosomes are also attached to the nuclear matrix, plasmids with S/MARs are able to "hitch-hike" with them as they segregate during mitosis.[221] Indeed, an episomal plasmid with a S/MAR from the human interferon β gene (i.e. pEPI-1) was shown to replicate inside a variety of different cell lines (copy number = 5–10) and segregate alongside chromosomes into both daughter cells during mitosis.[222,223] It is worth mentioning that viral replication origins from the SV40 and Epstein-Barr viruses also induce plasmid replication, but they require the expression of additional viral proteins and tend to be oncogenic.[224]

2.2.3.3 *Plasmid DNA sequences can also decrease transgene expression*

Sequences within plasmid DNA can have both positive and negative effects on transport inside the cell. For example, unmethylated cytosine-guanosine bases (i.e. CpGs) in bacterial and viral DNA are immediately recognized as pathogenic molecules by TLR9 receptors that coat the inside of endosomes.[225,226] Activation of TLR9 by unmethylated CpGs initiates an innate immune response that triggers expression of several cytokines that inhibit transgene expression.[3,227] However, bacterial plasmids without CpGs avoid detection by TLR9 and significantly enhance and prolong transgene expression, which may be up to 56 days in lung tissue.[228]

In addition to unmethylated CpGs, the bacterial sequences required for plasmid maintenance in *E. coli* (e.g. origins of replication and antibiotic resistance genes) can also silence transgene expression in eukaryotic cells.[229] Chen *et al.* devised a way to solve this problem by flanking a transgene expression cassette with attB and attP recombination sites that split the plasmid into two "minicircles" when exposed to ØC31 integrase.[230] This process provides transgene minicircles that are devoid of bacterial sequences with significantly enhanced and prolonged transgene expression.[230–232] These findings show that it is important to optimize both the transgene and the rest of the plasmid backbone to avoid negative responses to bacterial elements (e.g. unmethylated CpGs and bacterial genes) that might hinder gene therapy.

2.2.3.4 *Epigenetic effects on transgene expression*

Humans are able to maintain an immense genome (~six billion bp[233]) by condensing their DNA with histone proteins that tightly bind DNA to form chromatin and chromosomes.[234] However, while DNA condensation allows cells to maintain thousands of genes, it can also sterically inhibit gene transcription. For example, genes within tightly packed chromatin (i.e.

heterochromatin) tend to be repressed, while relaxed chromatin (i.e. euchromatin) usually contains active genes. Cells can alter local chromatin structure to activate or repress gene expression with several different enzymes that make distinct site-specific modifications to the N-terminal tails of histones. For example, genes may be activated by histone acetyl transferases (HATs) that remodel chromatin by neutralizing positive charges on histone lysines to reduce histone-DNA binding.[235] In contrast, histone deacetylases (HDACs) remove acetyl groups to restore heterochromatin and silence genes.[236,237] The study of the effect of histone acetylation and many other modifications (e.g. methylation.[238–240] phosphorylation.[241,242] ubiquitinylation.[243] and many others[244–248]) on gene expression is collectively called epigenetics.

Epigenetic phenomena have also been observed to influence transgene expression. For example, latent (inactive) viral DNA is often found within tightly condensed regions of centromeric heterochromatin. However, viruses can be brought out of latency with HDAC inhibitors (e.g. trichostatin A and MS-275), which relax heterochromatin to activate viral genes.[7,249–254] Bacterial plasmids are also prone to epigenetic silencing, since histones have been shown to tightly bind plasmid DNA and form nucleosomes *in vitro*.[255,256] Therefore, epigenetic regulation of transgene expression is obviously an important factor to consider in both viral and non-viral gene therapy.

2.3 DNA vaccines reprogram the immune system to attack cancer cells

DNA vaccines are an attractive alternative to tumor suppressors for cancer gene therapy. Instead of delivering genes that induce cell cycle arrest or apoptosis in cancer cells, DNA vaccines use genes for cancer-specific antigens to reprogram immune cells to seek out and destroy cancer cells. This strategy can be much more effective than tumor suppressor gene therapy, since low levels of transgene expression are amplified by immune cells into a prolonged and systemic immune response. Indeed, several clinical trials with DNA vaccines have shown considerable potential, as discussed in this section.

2.3.1 *Overview of the immune system*

Before we begin our discussion on DNA vaccines, let us first review the intricacies of the human immune system. First of all, humans possess two types of immunity — innate and adaptive. The innate immune system is comprised of several different types of cells (e.g. neutrophils, eosinophils, basophils, natural killer cells, monocytes, macrophages, and dendritic cells) that provide an immediate, yet non-specific response to foreign pathogens. For example, neutrophils circulate in the bloodstream until they are recruited to sites of injury/infection, where they phagocytose microbes with pathogen-associated molecular patterns (PAMPs) such as lipopolysaccharide and other bacteria-specific molecules.[257] In addition, antibodies produced by the adaptive immune system (see below) can also bind to pathogens and trigger phagocytosis by neutrophils and natural killer cells.[258]

While the innate immune system efficiently eliminates extracellular microbes, it is often unable to neutralize viruses or other pathogens that reside within cells. These types of infection are controlled by the adaptive immune system, which consists of three main types of cells ($CD4^+$ T helper cells, $CD8^+$ T cytotoxic cells, and B cells) that provide humoral (antibody-mediated) and/or cellular immune responses. For example, when a virus infects a cell, some of the viral proteins are inevitably degraded into peptides within proteasomes. These viral peptides are routed to the endoplasmic reticulum (ER), where they are bound by the major histocompatibility complex I (MHC1). The MHC1-peptide complex then leaves the ER and is presented on the surface of the infected cell. Presentation of multiple MHC1-peptide complexes recruits $CD8^+$ T cytotoxic (Tc) cells with T cell receptors that specifically bind MHC1 and the peptide antigen. The binding sites within T cell receptors are randomly mutated during T cell differentiation, providing a family of 10^8 unique receptors capable of binding almost any pathogenic peptide.[259] The activated T cell injects the infected cell with "granzymes" that induce apoptosis to prevent pathogen proliferation. This process also releases cytokines (e.g. interleukin 2) that trigger Tc proliferation to enhance immune surveillance throughout the body.[258]

The adaptive immune system may also be activated by extracellular viral and bacterial proteins. In this scenario, pathogenic proteins are randomly endocytosed by antigen presenting cells (APCs, e.g. dendritic cells) or specifically absorbed by antibodies on the surface of B cells. B cell antibodies are randomly mutated like T cell receptors during differentiation, but they are much more diverse (B cells may present ~10^{10} different antibodies[259]). In either case, the foreign protein is digested into peptides within a mature endosome that eventually fuses with vesicles containing the major histocompatibility complex II (MHC2). The activated MHC2-peptide complex is then presented on the surface of the cell like a MHC1 complex, but the MHC2-peptide complexes are bound by T cell receptors on $CD4^+$ T helper (Th) cells instead of Tc cells. MHC2 binding activates the Th cell to release several interleukins that induce B cell recruitment, proliferation, and maturation. Mature B cells then secrete antibodies that bind the pathogen and trigger their destruction by innate immune cells.[258] In some special cases, antigen binding by B cells can directly trigger antibody secretion without Th cells, but this type of response is relatively short-lived.[260]

2.3.1.1 *Cancer cells evade natural immune responses*

In addition to viruses and bacteria, tumor cells can also be recognized as pathogenic cells by the immune system. Indeed, several cases of spontaneous remission in cancer patients have been attributed to immune responses.[261,262] Unfortunately, the vast majority of cancer cells develop mutations that allow them to evade detection by the immune system. For example, most cancer cells defend themselves from T cells by decreasing their MHC1 expression levels.[263] Cancer cells may also aggressively inhibit T cells by overexpressing inoleamine 2,3-dioxygenase to deplete tryptophan (a vital nutrient required by T cells)[264] or the PD-L1/B7-H1 surface receptor, which inhibits T cell activation.[265] Cancer cells can also hijack parts of the immune system by secreting cytokines that convert macrophages to tumor-associated macrophages (TAMs), which actually enhance tumor growth, angiogenesis, and metastasis.[266]

2.3.2 *Development of DNA vaccines for cancer treatment*

The primary goal of DNA vaccine therapy is to deliver genes for cancer-specific antigens to antigen presenting cells (APCs) that activate an adaptive immune response against tumor cells (see Figure 2.5). Many different cancer-specific antigen targets have been identified for DNA vaccines, including tyrosinase (skin cancer), the androgen receptor (prostate cancer), and the HER2 receptor (breast cancer).[267–270] If the transgene is delivered to an APC, it may be transported to the proteasome for MHC1 presentation and Tc activation or to endosomes for MHC2 presentation and Th activation. Alternatively, transfected somatic cells may also present the antigen in MHC1 complexes or release the antigen to activate a humoral B cell response.[271,272] By simultaneously activating both T and B cells, DNA vaccines provide potent humoral and cellular immune responses against tumors. Indeed, DNA vaccines have been proven effective in several different clinical trials.

2.3.2.1 *Clinical trials of DNA vaccines*

DNA vaccines are the most common type of gene therapy, accounting for 20.5% of all clinical trials.[1] Naked plasmids and viruses have both shown significant promise in treating a variety of different cancers. For example, when naked plasmid DNA was used to deliver human and mouse tyrosinase genes to advanced melanoma patients, 41% (7/17) developed T cell responses to the tyrosinase antigen and the median survival of these patients increased from 36 to 42 months.[273] An even higher rate of T cell activation (71% of patients) was observed in a similar trial that simultaneously delivered multiple epitopes for common melanoma antigens (tyrosinase, melan-A, MAGE-1, MAGE-3, and NY-ESO) with naked plasmid DNA and a viral booster.[274,275] Several clinical trials have also shown that DNA vaccines for PSA (prostate-specific antigen),[276] PSMA (prostate-specific membrane antigen),[277,278] and PAP (prostatic acid phosphatase)[279,280] can effectively treat prostate cancer. For example, monthly injections of a DNA vaccine for PSA elicited long-term immune responses against PSA in four out of eight patients. Immune-stimulating adjuvants (granulocyte/macrophage colony stimulating factor and IL-2) were also used in this study to stimulate the immune response.[276] A similar trial using electroporation to enhance the delivery of DNA vaccines for PSMA obtained even higher levels of immune activation, with 29/30 patients exhibiting a CD4+ Th cell response and 16/30 patients exhibiting a $CD8^+$ Tc cell activation. Significantly longer PSA doubling times (a measure of prostate tumor progression) were also observed in more than half (14/24) of the patients.[277,278]

Clinical trials with naked plasmid DNA vaccines have shown promise, but they are not completely effective. Therefore, several clinical trials have utilized recombinant viruses to enhance DNA vaccine delivery.[281] Viral vectors also enhance DNA vaccination by stimulating the innate immune pathways that attract T cells.[282] One of the most promising viral DNA vaccines is PROSTVAC, a recombinant vaccinia virus designed to deliver a DNA vaccine for PSA. It was proven safe in Phase I clinical trials and shown to increase median survival by 8.5 months while decreasing mortality rates 44% compared to a control group in phase II trials.[283–286] As of 2014, a phase III clinical trial of PROSTVAC is recruiting patients (Clinicaltrials.gov Identifier: NCT01322490). A genetically modified form of the vaccinia virus (MVA, Modified Vaccinia Ankara) carrying a vaccine for mucin 1 (MUC-1) has also been shown to enhance survival in stage III/IV lung cancer patients when administered with chemotherapy.[287] MVA has also been used to effectively treat cervical[288] and colorectal[289] cancers in phase II trials, providing motivation for phase III trials that are currently underway. Despite these promising results, it is important to mention that an MVA DNA vaccine targeting tyrosinase in melanoma patients failed to induce an immune response.[290] Other types of viruses (e.g. canary pox or adenovirus) also appear to be significantly less effective than vaccinia viruses, probably due to pre-existing immunities for these viruses.[271] Therefore, vaccinia viruses appear to be the best method for DNA vaccine therapy.

2.3.3 *Strategies to enhance the activity of DNA vaccines*

While DNA vaccines have shown considerable promise, they are still not 100% effective. However, it may be possible to improve DNA vaccine

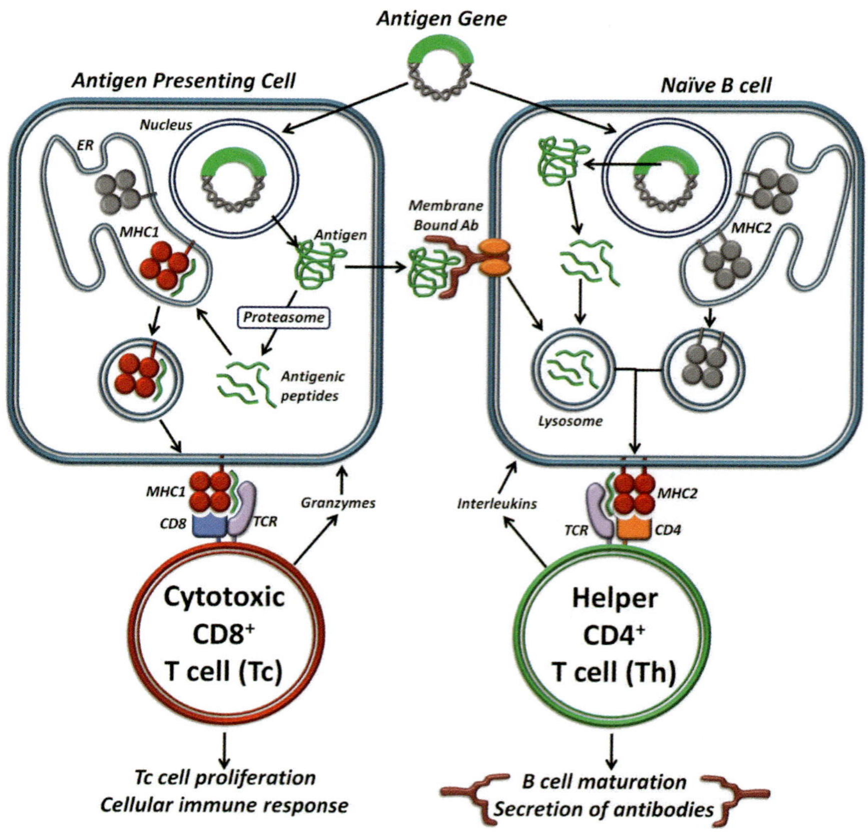

Figure 2.6 Immune activation by DNA vaccines. Antigenic transgenes may activate the immune system in two ways. (1) If an APC is transfected, the antigen may be degraded by the proteasome and transported to the ER, where it binds an MHC1 molecule. The MHC1-peptide complex is then shuttled to the cell surface, where it may be bound by CD8-TCR complexes on $CD8^+$ cytotoxic T cells to trigger a cellular immune response. The antigen may also be released prior to degradation within APCs and bound by antibodies on the surface of naïve B cells. The antigen is then endocytosed, degraded, and fused with vesicles containing MHC2 molecules. (2) Alternatively, antigen expressed inside B cells may also be targeted to lysosomes for MHC2 presentation. Either way, the MHC2-peptide complex is then bound by $CD4^+$ T helper cells that secrete interleukins to promote B cell maturation and antibody secretion for a humoral immune response.

efficiency by increasing antigen expression, processing, and presentation. This section focuses only on antigen processing and presentation (as summarized in Figure 2.6), but any of the methods described in Section 2.2.3 may be applied to increase antigen expression.

2.2.3.1 *Antigen processing*

The first step in immune activation by DNA vaccines is degradation of the transgenic antigen into smaller peptides within the proteasome. This process may be enhanced by adding an ubiquitin sequence to the transgene, since ubiquitinylated proteins are quickly shuttled to the proteasome. Indeed, several studies have shown that addition of a single ubiquitin or polyubiquitin tag to different antigens significantly enhances T cell responses *in vivo*.[291–293] Insertion of an arginine residue between the ubiquitin tag and the antigen sequence decreased antigen half life from 10 hours to 15 minutes, leading to a stronger T cell response in mice.[294] Aside from ubiquitin, antigen degradation can also be enhanced by adding PEST tags (sequences rich in proline, glutamic acid, serine, and threonine), which also target proteins to the proteasome.[295] Finally, intentional misfolding of antigens may also be used to increase their degradation and MHC1 presentation.[296]

2.3.2.2 *Antigen presentation with MHC1 or MHC2*

Once an intracellular antigen is degraded, it is usually transported to the ER to bind MHC1 molecules, which are then shuttled to the cell surface to activate Tc cells (see Figure 2.5). Antigen transport to the ER may be enhanced by adding an ER translocation signal (e.g. adenoviral E3/19K) to the N-terminus, but not the C-terminus, of the antigen gene.[297,298] Heat shock proteins (HSPs) that rapidly transport antigen peptides to the ER may also be co-expressed with the antigen to enhance peptide transport and MHC1 presentation.[299] In addition, expression of the HSPs on the cell surface (e.g. gp96 and HSP70) significantly enhance the activation of T cells and dendritic cells.[300,301] Alternatively, MHC2 expression and Th cell activation may be induced by targeting antigen peptides to endosomes and/or lysosomes with a lysosome-associated membrane protein I (LAMP-1) tag. Indeed, adding a LAMP-1 tag to the human papillomavirus antigen E7 significantly enhanced Th cell activation and slowed cervical tumor growth *in vivo*.[302,303]

2.4 Additional genetic strategies for cancer treatment

In addition to tumor suppressor gene therapy and DNA vaccines, several alternative genetic strategies have been developed to fight cancer. This section will briefly describe a few of the most successful strategies, including suicide gene therapy, oncolytic viruses, and gene silencing with RNA interference (RNAi).

2.4.1 *Suicide gene therapy*

The primary goal of suicide gene therapy is to deliver transgenes that induce apoptosis in cancer cells. The toxicity of these suicidal transgenes may be restricted to cancer cells by utilizing delivery vehicles that target cancer-specific membrane receptors (e.g. VEGFR, EGFR, folate receptor, transferring receptor, and many others) and using promoters that are only active in cancer cells (e.g. HER2/neu, EGFR, transferrin receptor, and telomerase-hTERT promoters).[212] Many different suicide transgenes have been developed,[212,304] including antibodies that inhibit antioxidant enzymes to increase oxidative stress[305] and inhibit constitutively active caspases[306,307] and DNase[308] enzymes to trigger apoptosis. However, the most effective suicide genes encode enzymes that metabolize harmless prodrugs into cytotoxic drugs selectively inside cancer cells. This section will discuss the thymidine kinase suicide gene from herpes simplex virus (HSV-tk) and the cytosine deaminase (CD) suicide gene from *E. coli*, both of which have been the subject of multiple clinical trials.

The HSV-tk enzyme converts the anti-viral prodrug ganciclovir (GCV) to ganciclovir monophosphate (GCV-1P). GCV-1P is then converted by host cell enzymes to ganciclovir triphosphate (GCV-3P), a structural analog of the nucleoside guanosine triphosphate (GTP). Consequently, GCV-3P is frequently substituted for GTP during DNA synthesis, leading to chain termination and cell cycle arrest[309] or apoptosis.[310] Preclinical experiments using a variety of vehicles to deliver the HSV-tk suicide gene to bladder,[311] liver[312] and colon[313] cancer cells showed increased apoptosis in tumor cells and motivated several clinical trials. In a Phase I trial, liposomal delivery of HSV-tk with GCV in patients with recurrent glioblastoma multiforme significantly reduced tumor size in 25% (2/8) patients and was well tolerated.[314] Another phase I trial with hormone-refractory prostate cancer was promising, where five out of eight patients showed a significant increase in PSA doubling time from 2.9 to 6 months.[315] However, later phase III trials showed no clinical benefit from HSV-tk suicide gene therapy compared to a control group in glioblastoma patients. The authors of the study attributed this failure to low levels of gene delivery that were insufficient for a significant response.[316]

The cytosine deaminase (CD) suicide gene from *E. coli* is another attractive option for suicide gene therapy. This enzyme catalyzes the hydrolytic deamination of non-toxic 5-fluorocytosine (5-FC) into 5-fluorouracil (5-FU), which is then converted by host cell enzymes to cytotoxic pyrimidine antimetabolites (5-FdUMP, 5-FdUTP, and 5-FUTP). These antimetabolites induce apoptosis[317] by interfering with a variety of vital processes within the cell, including RNA and DNA synthesis.[318] In a phase I clinical trial, a CD gene under the control of the breast cancer-specific HER2/neu promoter was shown to provide highly targeted transgene expression in 90% of patients.[319] A combination of the CD and HSV-tk suicide genes in an adenoviral vector was shown to be effective at decreasing PSA levels in another phase I trial for prostate cancer.[320] A retroviral CD therapy is also in progress for brain cancer (Clinicaltrials.gov Identifier: NCT01470794).[321]

2.4.2 *Silencing of oncogenes with RNAi*

While most gene therapy research focuses on delivering genes, an interesting alternative strategy is to deliver short interfering RNAs (siRNAs) that silence oncogene expression. Gene silencing by siRNAs (aka RNA interference or RNAi) is a phenomenon that was discovered decades ago in plants, in which double stranded RNA molecules activate a host defense mechanism that destroys any complementary mRNA to silence expression of that gene. RNAi may be induced by directly delivering siRNA to the cytoplasm or by delivering a plasmid that expresses shRNA (short hairpin RNA) in the nucleus, which is then processed by the enzyme Dicer to generate siRNA fragments that are approximately 22 base pairs long. In either case, siRNAs are then bound by the RNA induced silencing complex (RISC), which uses the RNA fragments to find complementary single stranded mRNA transcripts and degrade them. This process effectively prevents translation of any genes with sequences complementary to the siRNA sequence.[322] For example, delivery of siRNA containing sequences from the HER2 gene would reduce HER2 expression by inducing the degradation of HER2 mRNA.

Although RNAi has been extensively used in preclinical experiments to silence cancer genes (e.g. in colon[323] and gastric[324] cancer models), only one clinical trial has been performed with RNAi. In this phase I trial, liposomes were used to deliver siRNA designed to silence vascular endothelial growth factor (VEGF) and kinesin spindle proteins (KSP). VEGF is overexpressed in many cancer types to stimulate angiogenesis,[325,326] while KSP is required for cell division.[327] Therefore, silencing of these genes could substantially reduce tumor growth. Indeed, delivery of these siRNA's elicited anti-tumor activity in 16% (4/24) of patients. One patient with endometrial cancer even experienced complete regression of liver metastases.[328] These results suggest that RNAi therapy may be an effective strategy that warrants further clinical trials.

2.4.3 *Oncolytic virotherapy*

Although most of the viruses used for gene therapy are modified to prevent their replication *in vivo*, the viruses used for oncolytic virotherapy are engineered to enhance their replication inside cancer cells. Much like suicide gene expression, viral replication may be restricted to cancer cells by targeting cancer-specific receptors for infection[329] and promoters for viral gene expression.[330] In addition to direct lysis of tumor cells, oncolytic viral infection also activates a systemic Tc cell response against other cancer cells that enhances tumor regression.[331] More than a dozen different oncolytic viruses have been developed and clinically tested thus far,[332] two of which (OncoVEX and Oncorine/H101) have reached phase III clinical trials.

OncoVEX (aka talimogene laherparepvec) is an oncolytic herpes simplex virus that also expresses granulocyte/macrophage colony stimulating factor (G/MCSF), a cytokine that amplifies replication of innate immune cells to enhance immune responses. In a phase II trial, OncoVEX provided complete regression in 16% (8/50) of patients with unresectable metastatic melanoma.[333] These promising results have motivated a phase III trial for patients with stage III/IV melanoma that is currently underway.[334] Similar results have also been observed with Oncorine (aka H101), an adenovirus that is missing its E1B-55kDa gene (a p53 inhibitor). This deletion restricts viral replication to cancer cells that lack p53, while the p53 in non-cancerous cells responds to viral infection by inducing cell cycle arrest or apoptosis. Indeed, a phase III clinical trial showed that a combination of chemotherapy (cisplatin and 5-Fluoruracil) and Oncorine provided a clinical response in twice as many patients (41/52) as chemotherapy alone (21/53).[335] Oncorine was subsequently approved for clinical use in China in 2006 and is slated to begin clinical trials in the United States and Europe.[336]

2.5 Concluding remarks

Several different gene therapy strategies have shown promise for cancer treatment. Indeed, two gene therapy treatments have already been approved for clinical use in China (Gendicine and Oncorine), although no gene therapy treatments have been approved elsewhere. Therefore, much work remains to be done before the full potential of gene therapy for cancer (and many other diseases) can be realized. For example, the development of improved viral or non-viral delivery vehicles with higher transfection efficiencies could significantly enhance all of the gene therapy strategies mentioned in this chapter. Specifically, limiting the immunogenicity of viral vehicles can substantially enhance gene therapy by allowing them to be delivered intravenously instead of intratumorally. Inhibition of transgene expression by host cell mechanisms (e.g. epigenetic regulation and DNA sensors) must also be addressed to enhance the magnitude and duration of transgene expression. Once these problems are solved, gene therapy may become a particularly powerful treatment option for a variety of cancers. Progress towards this goal is monitored by the *Journal of Gene Medicine*, which provides a regularly updated online database of ongoing clinical trials for cancer and other genetic disorders at www.abedia.com/wiley/.

References

1. Ginn SL, Alexander IE, Edelstein ML, Abedi MR, Wixon J, Gene therapy clinical trials worldwide to 2012 — an update. *J Gene Med,* 2013; 15: 65–77.
2. Wiethoff CM, Middaugh CR, Barriers to nonviral gene delivery. *J Pharm Sci,* 2003; 92: 203–217.
3. Kumagai Y, Takeuchi O, Akira S, TLR9 as a key receptor for the recognition of DNA. *Adv Drug Deliv Rev* 2008; 60: 795–804.
4. Evans CJ, Aguilera RJ, DNase II: genes, enzymes and function. *Gene,* 2003; 322: 1–15.
5. Barber GN, Cytoplasmic DNA innate immune pathways. *Immunol Rev,* 2011; 243: 99–108.
6. Paludan SR, Bowie AG, Immune sensing of DNA. *Immunity,* 2013; 38: 870–880.
7. Poleshko A, Palagin I, Zhang R, Boimel P, Castagna C, Adams PD, Skalka AM, Katz R, A. Identification of cellular proteins that maintain retroviral epigenetic silencing: evidence for an antiviral response. *J Virol,* 2008; 82: 2313–2323.
8. Raja P, Sanville BC, Buchmann RC, Bisaro DM, Viral genome methylation as an epigenetic defense against geminiviruses. *J Virol,* 2008; 82: 8997–9007.
9. Thomas CE, Ehrhardt A, Kay MA, Progress and problems with the use of viral vectors for gene therapy. *Nat Rev Genet,* 2003; 4: 346–358.
10. Mancheño-Corvo P, Martín-Duque P, Viral gene therapy. *Clin Transl Oncol,* 2006; 8: 858–867.
11. Wong SY, Pelet JM, Putnam D, Polymer systems for gene delivery — past, present, and future. *Prog Polym Sci,* 2007; 32: 799–837.
12. Thomas M, Klibanov AM, Non-viral gene therapy: polycation-mediated DNA delivery. *Appl Microbiol Biotechnol,* 2003; 62: 27–34.
13. Wang W, Li W, Ma N, Steinhoff G, Non-viral gene delivery methods. *Curr Pharm Biotechnol,* 2013; 14: 46–60.

14. Davis ME, Non-viral gene delivery systems. *Curr Opin Biotechnol,* 2002; 13: 128–131.
15. Junquera E, Aicart E, Cationic lipids as transfecting agents of DNA in gene therapy. *Curr Top Med Chem,* 2014; 14: 649–663.
16. Blaese RM, Culver KW, Miller AD, Carter CS, *et al.* T lymphocyte-directed gene therapy for ADA-SCID: initial trial results after 4 years. *Science,* 1995; 270: 475–480.
17. Simonelli F, Maguire AM, Testa F, Pierce EA, *et al.* Gene therapy for Leber's congenital amaurosis is safe and effective through 1.5 years after vector administration. *Mol Ther* 2010; 18: 643–650.
18. Cavazzana-Calvo M, Gene therapy of human severe combined immunodeficiency (SCID)-X1 disease. *Science (80-),* 2000; 288: 669–672.
19. High KH, Nathwani A, Spencer T, Lillicrap D, Current status of haemophilia gene therapy. *Haemophilia,* 20 Suppl 2014; 4: 43–49.
20. Palfi S, Gurruchaga JM, Ralph GS, Lepetit H, *et al.* Long-term safety and tolerability of ProSavin, a lentiviral vector-based gene therapy for Parkinson's disease: a dose escalation, open-label, phase 1/2 trial. *Lancet,* 2014; 383: 1138–1146.
21. Cross D, Burmester JK, Gene therapy for cancer treatment: past, present and future. *Clin Med Res,* 2006; 4: 218–227.
22. Fukazawa T, Matsuoka J, Yamatsuji T, Maeda Y, Durbin ML, Naomoto Y, Adenovirus-mediated cancer gene therapy and virotherapy. *Int J Mol Med,* 2010; 25: 3–10.
23. Deregowski V, Canalis E, Gene delivery by retroviruses. *Methods, Mol Biol* 2008; 455: 157–162.
24. Guo ZS, Bartlett DL, Vaccinia as a vector for gene delivery. *Expert Opin Biol Ther,* 2004; 4: 901–917.
25. Gomez C, Najera J, Krupa M, Esteban M, The poxvirus vectors MVA and NYVAC as gene delivery systems for vaccination against infectious diseases and cancer. *Curr Gene Ther,* 2008; 8: 97–120.
26. Lu Y, Recombinant adeno-associated virus as delivery vector for gene therapy — a review. *Stem Cells Dev,* 2004; 13: 133–145.
27. Goins WF, Huang S, Cohen JB, Glorioso JC, Engineering HSV-1 vectors for gene therapy. *Methods Mol Biol,* 2014; 1144: 63–79.
28. Cockrell AS, Kafri T, Gene delivery by lentivirus vectors. *Mol Biotechnol* 2007; 36: 184–204.

29. Rodrigues A, Alves PM, Coroadinha A, in *Viral Gene Therapy* (ed. Xu K,) (Intech, 2011). DOI:978-953-307-539-6.
30. Reske A, Pollara G, Krummenacher C, Chain BM, Katz DR, Understanding HSV-1 entry glycoproteins. *Rev Med Virol,* 17, 205–215.
31. Plank C, Zauner W, Wagner E, Application of membrane-active peptides for drug and gene delivery across cellular membranes. *Adv Drug Deliv Rev,* 1998; 34: 21–35.
32. Freed EO, Myers DJ, Risser R, Characterization of the fusion domain of the human immunodeficiency virus type 1 envelope glycoprotein gp41. *Proc Natl Acad Sci USA,* 1990; 87: 4650–4654.
33. Biology M, Baltimore M, Cellular uptake of the tat protein from human immunodeficiency virus. *Dis Markers,* 1988; 8: 33–34.
34. Elliott G, O'Hare P, Intercellular trafficking and protein delivery by a herpesvirus structural protein. *Cell,* 1997; 88: 223–233.
35. Nitin N, LaConte L, Rhee WJ, Bao G, Tat peptide is capable of importing large nanoparticles across nuclear membrane in digitonin permeabilized cells. *Ann Biomed Eng,* 2009; 37: 2018–2027.
36. Yamashita M, Emerman M, Retroviral infection of non-dividing cells: old and new perspectives. *Virology,* 2006; 344: 88–93.
37. Lange A, Mills RE, Lange CJ, Stewart M, Devine SE, Corbett AH, Classical nuclear localization signals: definition, function, and interaction with importin alpha. *J Biol Chem,* 2007; 282: 5101–5105.
38. Dean DA, Dean BS, Muller S, Smith LC, Sequence requirements for plasmid nuclear import. *Exp Cell Res* 1999; 253: 713–722.
39. Li S, MacLaughlin FC, Fewell JG, Gondo M, Wang J, Nicol F, Dean DA, Smith LC, Muscle-specific enhancement of gene expression by incorporation of SV40 enhancer in the expression plasmid. *Gene Ther,* 2001; 8: 494–497.
40. Kotin RM, Menninger JC, Ward DC, Berns KI, Mapping and direct visualization of a region-specific viral DNA integration site on chromosome 19q13-qter. *Genomics,* 1991; 10: 831–834.
41. Bushman F, Lewinski M, Ciuffi A, Barr S, Leipzig J, Hannenhalli S, Hoffmann C, Genome-wide analysis of retroviral DNA integration. *Nat Rev Microbiol,* 2005; 3: 848–858.
42. Hindmarsh P, Leis J, Retroviral DNA integration. *Microbiol Mol Biol Rev,* 1999; 63: 836–843.

43. Kohn DB, Sadelain M, Glorioso JC, Occurrence of leukaemia following gene therapy of X-linked SCID. *Nat Rev Cancer,* 2003; 3: 477–488.
44. Hacein-Bey-Abina S, Von Kalle C, Schmidt M, McCormack MP, *et al.* LMO2-associated clonal T cell proliferation in two patients after gene therapy for SCID-X1. *Science,* 2003; 302: 415–419.
45. Gersbach CA, Gaj T, Gordley RM, Mercer AC, Barbas CF, Targeted plasmid integration into the human genome by an engineered zinc-finger recombinase. *Nucleic Acids Res,* 2011; 39: 7868–7878.
46. Groth AC, Olivares EC, Thyagarajan B, Calos MP, A phage integrase directs efficient site-specific integration in human cells. *Proc Natl Acad Sci USA,* 2000; 97: 5995–6000.
47. Kirchmaier AL, Sugden B, Rep*: a viral element that can partially replace the origin of plasmid DNA synthesis of Epstein-Barr virus. *J Virol,* 1998; 72: 4657–4666.
48. Jankelevich S, Kolman JL, Bodnar JW, Miller G, A nuclear matrix attachment region organizes the Epstein-Barr viral plasmid in Raji cells Into a single DNA domain. *EMBO J,* 1992; 11: 1165–1176.
49. Milanesi G, Barbanti-Brodano G, Negrini M, Lee D, Corallini, A, Caputo, A, Grossi MP, Ricciardi RP, BK virus-plasmid expression vector that persists episomally in human cells and shuttles into *Escherichia coli*. *Mol Cell Biol,* 1984; 4: 1551–1560.
50. Cooper M, Miron S, Efficient episomal expression vector for human transitional carcinoma cells. *Hum Gene Ther,* 2008; 4: 557–566.
51. Kafri T, Morgan D, Krahl T, Sarvetnick N, Sherman L, Verma I, Cellular immune response to adenoviral vector infected cells does not require de novo viral gene expression: implications for gene therapy. *Proc Natl Acad Sci USA,* 1998; 95: 11377–11382.
52. Chirmule N, Propert K, Magosin S, Qian Y, Qian R, Wilson J, Immune responses to adenovirus and adeno-associated virus in humans. *Gene Ther,* 1999; 6: 1574–1583.
53. Thomas CE, Ehrhardt A, Kay MA, Progress and problems with the use of viral vectors for gene therapy. *Nat Rev Genet,* 2003; 4: 346–358.
54. Martin B, Sainlos M, Aissaoui A, Oudrhiri N, Hauchecorne M, Vigneron J-P, Lehn J-M, Lehn P, The design of cationic lipids for gene delivery. *Curr Pharm, Des* 2005; 11: 375–394.

55. Zhi D, Zhang S, Cui S, Zhao Y, Wang Y, Zhao D, The headgroup evolution of cationic lipids for gene delivery. *Bioconjug Chem,* 2013; 24: 487–519.
56. Zhi D, Zhang S, Wang B, Zhao Y, Yang B, Yu S, Transfection efficiency of cationic lipids with different hydrophobic domains in gene delivery. *Bioconjug Chem* 2010; 21: 563–577.
57. Lv H, Zhang S, Wang B, Cui S, Yan J, Toxicity of cationic lipids and cationic polymers in gene delivery. *J Control Release,* 2006; 114: 100–109.
58. Felgner PL, Gadek TR, Holm M, Roman R, Chan HW, Wenz M, Northrop JP, Ringold GM, Danielsen M, Lipofection: a highly efficient, lipid-mediated DNA-transfection procedure. *Proc Natl Acad Sci USA,* 1987; 84: 7413–7417.
59. Farhood H, Serbina N, Huang L, The role of dioleoyl phosphatidylethanolamine in cationic liposome mediated gene transfer. *Biochim Biophys Acta — Biomembr,* 1995; 1235: 289–295.
60. Zhou X, Huang L, DNA transfection mediated by cationic liposomes containing lipopolylysine: characterization and mechanism of action. *Biochim Biophys Acta — Biomembr,* 1994; 1189: 195–203.
61. Immordino ML, Dosio F, Cattel L, Stealth liposomes: review of the basic science, rationale, and clinical applications, existing and potential. *Int J Nanomedicine,* 2006; 1: 297–315.
62. Ishida O, Maruyama K, Tanahashi H, Iwatsuru M, Sasaki K, Eriguchi M, Yanagie H, Liposomes bearing polyethyleneglycol-coupled transferrin with intracellular targeting property to the solid tumors *in vivo*. *Pharm Res,* 2001; 18: 1042–1048.
63. Mukherjee A, Prasad TK, Rao NM, Banerjee R, Haloperidol-associated stealth liposomes: a potent carrier for delivering genes to human breast cancer cells. *J Biol Chem,* 2005; 280: 15619–1527.
64. Aied A, Greiser U, Pandit A, Wang W, Polymer gene delivery: overcoming the obstacles. *Drug Discov Today,* 2013; 18: 1090–1098.
65. Godbey WT, Wu KK, Mikos AG, Poly(ethylenimine) and its role in gene delivery. *J Control Release,* 1999; 60: 149–160.
66. Cai J, Yue Y, Rui D, Zhang Y, Liu S, Wu C, Effect of chain length on cytotoxicity and endocytosis of cationic polymers. *Macromolecules,* 2011; 44: 2050–2057.
67. Kunath K, von Harpe A, Fischer D, Petersen H, Bickel U, Voigt K, Kissel T, Low-molecular-weight polyethylenimine as a non-viral vector for DNA delivery: comparison of physicochemical properties, transfection

efficiency and in vivo distribution with high-molecular-weight polyethylenimine. *J Control Release,* 2003; 89: 113–125.
68. Casettari L, Vllasaliu D, Lam JKW, Soliman M, Illum L, Biomedical applications of amino acid-modified chitosans: a review. *Biomaterials,* 2012; 33: 7565–7583.
69. Kwoh DY, Coffin CC, Lollo CP, Jovenal J, Banaszczyk MG, Mullen P, Phillips A, Amini A, Fabrycki J, Bartholomew RM, Brostoff SW, Carlo DJ, Stabilization of poly-L-lysine/DNA polyplexes for *in vivo* gene delivery to the liver. *Biochim Biophys Acta,* 1999; 1444: 171–190.
70. Barua S, Joshi A, Banerjee A, Matthews D, Sharfstein ST, Cramer SM, Kane RS, Rege K, Parallel synthesis and screening of polymers for nonviral gene delivery. *Mol Pharm,* 2008; 6: 86–97.
71. Green JJ, Langer R, Anderson DG, A combinatorial polymer library approach yields insight into nonviral gene delivery. *Acc Chem Res,* 2008; 41: 749–759.
72. Potta T, Zhen Z, Grandhi TSP, Christensen MD, Ramos J, Breneman CM, Rege K, Discovery of antibiotics-derived polymers for gene delivery using combinatorial synthesis and cheminformatics modeling. *Biomaterials,* 2014; 35: 1977–1988.
73. Li J, He Y, Li W, Shen Y, Li Y, Wang Y, A novel polymer-lipid hybrid nanoparticle for efficient nonviral gene delivery. *Acta Pharmacol Sin,* 2010; 31: 509–514.
74. Han So, Mahato RI, Kim SW, Water-soluble lipopolymer for gene delivery. *Bioconjug Chem* 12, 337–345.
75. Xun M-M, Zhang X-C, Zhang J, Jiang Q-Q, Yi W-J, Zhu W, Yu X-Q, Low molecular weight PEI-based biodegradable lipopolymers as gene delivery vectors. *Org Biomol Chem,* 2013; 11: 1242–1250.
76. Chiu S-J, Ueno NT, Lee RJ, Tumor-targeted gene delivery via anti-HER2 antibody (trastuzumab, Herceptin) conjugated polyethylenimine. *J Control Release,* 2004; 97: 357–369.
77. Midoux P, Mendes C, Legrand A, Raimond J, Mayer R, Monsigny M, Roche AC, Specific gene transfer mediated by lactosylated poly-L-lysine into hepatoma cells. *Nucleic Acids Res,* 1993; 21: 871–878.
78. Kim WJ, Yockman JW, Lee M, Jeong JH, Kim Y-H, Kim SW, Soluble Flt-1 gene delivery using PEI-g-PEG-RGD conjugate for anti-angiogenesis. *J Control Release,* 2005; 106: 224–234.

79. Benns JM, Maheshwari A, Furgeson DY, Mahato RI, Kim SW, Folate-PEG-folate-graft-polyethylenimine-based gene delivery. *J Drug Target,* 2001; 9: 123–139.
80. Dufès C, Uchegbu IF, Schätzlein AG, Dendrimers in gene delivery. *Adv Drug Deliv Rev,* 2005; 57: 2177–2202.
81. Ramos J, Rege K, Transgene delivery using poly(amino ether)-gold nanorod assemblies. *Biotechnol Bioeng,* 2012; 109: 1336–1346.
82. Kami D, Takeda S, Itakura Y, Gojo S, Watanabe M, Toyoda M, Application of magnetic nanoparticles to gene delivery. *Int J Mol Sci,* 2011; 12: 3705–3722.
83. Kaouass M, Beaulieu R, Balicki D, Histonefection: Novel and potent non-viral gene delivery. *J Control Release,* 2006; 113: 245–254.
84. Herweijer H, Wolff JA, Progress and prospects: naked DNA gene transfer and therapy. *Gene Ther,* 2003; 10: 453–458.
85. Budker V, Zhang G, Knechtle S, Wolff JA, Naked DNA delivered intraportally expresses efficiently in hepatocytes. *Gene Ther,* 1996; 3: 593–598.
86. Miao CH, Thompson AR, Loeb K, Ye X, Long-term and therapeutic-level hepatic gene expression of human factor IX after naked plasmid transfer *in vivo*. *Mol Ther,* 2001; 3: 947–957.
87. Wolff JA, Budker V, The mechanism of naked DNA uptake and expression. *Adv Genet,* 2005; 54: 3–20.
88. Sardesai NY, Weiner DB, Electroporation delivery of DNA vaccines: prospects for success. *Curr Opin Immunol,* 2011; 23: 421–429.
89. Chiarella P, Massi E, De Robertis M, Sibilio A, Parrella P, Fazio VM, Signori E, Electroporation of skeletal muscle induces danger signal release and antigen-presenting cell recruitment independently of DNA vaccine administration. *Expert Opin Biol Ther,* 2008; 8: 1645–1657.
90. Futreal PA, Coin L, Marshall M, Down T, Hubbard T, Wooster R, Rahman N, Stratton MR, A census of human cancer genes. *Nat Rev Cancer,* 2004; 4: 177–183.
91. Forbes SA, Bhamra G, Bamford S, Dawson E, Kok C, Clements J, Menzies A, Teague JW, Futreal PA, Stratton MR, The catalogue of somatic mutations in cancer (COSMIC). *Curr Protoc Hum Genet,* Chapter 10, Unit 10.11 (2008).
92. Santarius T, Shipley J, Brewer D, Stratton MR, Cooper CS, A census of amplified and overexpressed human cancer genes. *Nat Rev Cancer,* 2010; 10: 59–64.

93. Ménard S, Tagliabue E, Campiglio M, Pupa SM, Role of HER2 gene overexpression in breast carcinoma. *J Cell Physiol,* 2000; 182: 150–162.
94. Tan AR, Swain SM, Ongoing adjuvant trials with trastuzumab in breast cancer. *Semin Oncol,* 2003; 30: 54–64.
95. Faltus T, Yuan J, Zimmer B, Krämer A, Loibl S, Kaufmann M, Strebhardt K, Silencing of the HER2/neu gene by siRNA inhibits proliferation and induces apoptosis in HER2/neu-overexpressing breast cancer cells. *Neoplasia,* 2004; 6: 786–795.
96. Lee EYHP, Muller WJ, Oncogenes and tumor suppressor genes. *Cold Spring Harb Perspect Biol,* 2, a003236 (2010).
97. The genetic analysis of cancer. *J Intern Med,* 1998; 243: 413–417.
98. Lakin ND, Jackson SP, Regulation of p53 in response to DNA damage. *Oncogene,* 1999; 18: 7644–7655.
99. Farnebo M, Bykov VJN, Wiman KG, The p53 tumor suppressor: a master regulator of diverse cellular processes and therapeutic target in cancer. *Biochem Biophys Res Commun,* 2010; 396: 85–89.
100. el-Deiry WS, Tokino T, Velculescu VE, Levy DB, Parsons R, Trent JM, Lin D, Mercer WE, Kinzler KW, Vogelstein, B. WAF1, a potential mediator of p53 tumor suppression. *Cell,* 1993; 75: 817–825.
101. Yang H-Y, Wen Y-Y, Chen C-H, Lozano G, Lee M-H, 14-3-3 sigma positively regulates p53 and suppresses tumor growth. *Mol Cell Biol,* 2003; 23: 7096–7107.
102. Miyashita T, Krajewski S, Krajewska M, Wang HG, Lin HK, Liebermann DA, Hoffman B, Reed JC, Tumor suppressor p53 is a regulator of bcl-2 and bax gene expression in vitro and *in vivo. Oncogene,* 1994; 9: 1799–1805.
103. Oda E, Ohki R, Murasawa H, Nemoto J, Shibue T, Yamashita T, Tokino T, Taniguchi T, Tanaka, N. Noxa, a BH3-only member of the Bcl-2 family and candidate mediator of p53-induced apoptosis. *Science,* 2000; 288: 1053–1058.
104. Nakano K, Vousden KH, PUMA, a novel proapoptotic gene, is induced by p53. *Mol Cell,* 2001; 7: 683–694.
105. Owen-Schaub LB, Zhang W, Cusack JC, Angelo LS, Santee SM, Fujiwara T, Roth JA, Deisseroth AB, Zhang WW, Kruzel E, Wild-type human p53 and a temperature-sensitive mutant induce Fas/APO-1 expression. *Mol Cell Biol,* 1995; 15: 3032–3040.
106. Soussi T, Wiman KG, Shaping genetic alterations in human cancer: the p53 mutation paradigm. *Cancer Cell,* 2007; 12: 303–312.

107. Badie B, Drazan KE, Kramar MH, Shaked A, Black KL, Adenovirus-mediated p53 gene delivery inhibits 9L glioma growth in rats. *Neurol Res,* 1995; 17: 209–216.
108. Clayman GL, el-Naggar AK, Roth JA, Zhang WW, Goepfert H, Taylor DL, Liu TJ, *In vivo* molecular therapy with p53 adenovirus for microscopic residual head and neck squamous carcinoma. *Cancer Res,*1995; 55: 1–6.
109. Spitz FR, Nguyen D, Skibber JM, Cusack J, Roth JA, Cristiano RJ, *In vivo* adenovirus-mediated p53 tumor suppressor gene therapy for colorectal cancer. *Anticancer Res* 16, 3415–3422
110. Hwang RF, Gordon EM, Anderson WF, Parekh D, Gene therapy for primary and metastatic pancreatic cancer with intraperitoneal retroviral vector bearing the wild-type p53 gene. *Surgery,* 124, 143–150; discussion 150–151 (1998).
111. Zou Y, Zong G, Ling YH, Hao MM, Lozano G, Hong WK, Perez-Soler R, Effective treatment of early endobronchial cancer with regional administration of liposome-p53 complexes. *J Natl Cancer Inst* 1998; 90: 1130–1137.
112. Lesoon-Wood LA, Kim WH, Kleinman HK, Weintraub BD, Mixson AJ, Systemic gene therapy with p53 reduces growth and metastases of a malignant human breast cancer in nude mice. *Hum Gene Ther,* 1995; 6: 395–405.
113. Spitz FR, Nguyen D, Skibber JM, Meyn RE, Cristiano RJ, Roth JA, Adenoviral-mediated wild-type p53 gene expression sensitizes colorectal cancer cells to ionizing radiation. *Clin Cancer Res,* 1996; 2: 1665–1671.
114. Badie B, Kramar MH, Lau R, Boothman DA, Economou JS, Black KL, Adenovirus-mediated p53 gene delivery potentiates the radiation-induced growth inhibition of experimental brain tumors. *J Neurooncol,* 1998; 37: 217–222.
115. Pirollo KF, Hao Z, Rait A, Jang YJ, Fee WE, Ryan P, Chiang Y, Chang EH, p53 mediated sensitization of squamous cell carcinoma of the head and neck to radiotherapy. *Oncogene,* 1997; 14: 1735–1746.
116. Nguyen DM, Spitz FR, Yen N, Cristiano RJ, Roth JA, Gene therapy for lung cancer: enhancement of tumor suppression by a combination of sequential systemic cisplatin and adenovirus-mediated p53 gene transfer. *J Thorac Cardiovasc Surg,* 112, 1372–1376; discussion 1376–1377 (1996).
117. Miyake H, Hara I, Hara S, Arakawa S, Kamidono S, Synergistic chemosensitization and inhibition of tumor growth and metastasis by adenovirus-mediated

P53 gene transfer in human bladder cancer model. *Urology,* 2000; 56: 332–336.

118. Tursz T, Cesne AL, Baldeyrou P, Gautier E, Opolon P, Schatz C, Pavirani A, Courtney M, Lamy D, Ragot T, Saulnier P, Andremont A, Monier R, Perricaudet M, Le Chevalier T, Phase I study of a recombinant adenovirus-mediated gene transfer in lung cancer patients. *J Natl Cancer Inst,* 1996; 88: 1857–1863.
119. Swisher SG, Roth JA, Nemunaitis J, Lawrence DD, *et al.,* Adenovirus-mediated p53 gene transfer in advanced non-small-cell lung cancer. *J Natl Cancer Inst,* 1999; 91: 763–771.
120. Nemunaitis J, Swisher SG, Timmons T, Connors D, *et al.* Adenovirus-mediated p53 gene transfer in sequence with cisplatin to tumors of patients with non-small-cell lung cancer. *J Clin Oncol,* 2000; 18: 609–622.
121. Swisher SG, Roth JA, Carbone DP, Genetic and immunologic therapies for lung cancer. *Semin Oncol,* 2002; 29: 95–101.
122. Buller RE, Shahin MS, Horowitz JA, Runnebaum IB, Mahavni V, Petrauskas S, Kreienberg R, Karlan B, Slamon D, Pegram M, Long term follow-up of patients with recurrent ovarian cancer after Ad p53 gene replacement with SCH 58500. *Cancer Gene Ther* 2002; 9: 567–572.
123. Yoo GH, Moon J, Leblanc M, Lonardo F, Urba S, Kim H, Hanna E, Tsue T, Valentino J, Ensley J, Wolf G, A phase 2 trial of surgery with perioperative INGN 201 (Ad5CMV-p53) gene therapy followed by chemoradiotherapy for advanced, resectable squamous cell carcinoma of the oral cavity, oropharynx, hypopharynx, and larynx: report of the Southwest Oncology Group. *Arch Otolaryngol Head Neck Surg,* 2009; 135: 869–874.
124. Nemunaitis JM, Nemunaitis J, Potential of Advexin: a p53 gene-replacement therapy in Li-Fraumeni syndrome. *Future Oncol,* 2008; 4: 759–768.
125. Gabrilovich DI, INGN 201 (Advexin): adenoviral p53 gene therapy for cancer. *Expert Opin Biol Ther,* 2006; 6: 823–832.
126. Pearson S, Jia H, Kandachi K, China approves first gene therapy. *Nat Biotechnol,* 2004; 22: 3–4.
127. Pan J, Zhang S, Chen C, Xiao S, Sun Y, Liu C, Su X, Li D, Xu G, Xu B, Lu Y, Effect of recombinant adenovirus-p53 combined with radiotherapy on long-term prognosis of advanced nasopharyngeal carcinoma. *J Clin Oncol,* 2009; 27: 799–804.

128. Senzer N, Nemunaitis J, Nemunaitis D, Bedell C, Edelman G, Barve M, Nunan R, Pirollo KF, Rait A, Chang EH, Phase I study of a systemically delivered p53 nanoparticle in advanced solid tumors. *Mol Ther,* 2013; 21: 1096–1103.
129. Camp ER, Wang C, Little EC, Watson PM, Pirollo KF, Rait A, Cole DJ, Chang EH, Watson DK, Transferrin receptor targeting nanomedicine delivering wild-type p53 gene sensitizes pancreatic cancer to gemcitabine therapy. *Cancer Gene Ther,* 2013; 20: 222–228.
130. Friend SH, Bernards R, Rogelj S, Weinberg RA, Rapaport JM, Albert DM, Dryja TP, A human DNA segment with properties of the gene that predisposes to retinoblastoma and osteosarcoma. *Nature,* 323, 643–646.
131. Dryja TP, Friend S, Weinberg RA, Genetic sequences that predispose to retinoblastoma and osteosarcoma. *Symp Fundam Cancer Res* 1986; 39: 115–119.
132. Sherr CJ, McCormick F, The RB and p53 pathways in cancer. *Cancer Cell,* 2002; 2: 103–112.
133. Weinberg RA, The retinoblastoma protein and cell cycle control. *Cell,* 1995; 81: 323–330.
134. Ji P, Jiang H, Rekhtman K, Bloom J, Ichetovkin M, Pagano M, Zhu L, An Rb-Skp2-p27 pathway mediates acute cell cycle inhibition by Rb and is retained in a partial-penetrance Rb mutant. *Mol Cell,* 2004; 16: 47–58.
135. Bourgo RJ, Thangavel C, Ertel A, Bergseid J, McClendon AK, Wilkens L, Witkiewicz AK, Wang JYJ, Knudsen ES, RB restricts DNA damage-initiated tumorigenesis through an LXCXE-dependent mechanism of transcriptional control. *Mol Cell,* 2011; 43: 663–672.
136. Gonzalo S, García-Cao M, Fraga MF, Schotta G, Peters AHFM, Cotter SE, Eguía R, Dean DC, Esteller M, Jenuwein T, Blasco MA, Role of the RB1 family in stabilizing histone methylation at constitutive heterochromatin. *Nat Cell Biol,* 2005; 7: 420–428.
137. Dick FA, Rubin SM, Molecular mechanisms underlying RB protein function. *Nat Rev Mol Cell Biol,* 2013; 14: 297–306.
138. Munro S, Carr SM, La Thangue NB, Diversity within the pRb pathway: is there a code of conduct? *Oncogene,* 2012; 31: 4343–4352.
139. Schvartzman J,-M., Duijf PHG, Sotillo R, Coker C, Benezra R, Mad2 is a critical mediator of the chromosome instability observed upon Rb and p53 pathway inhibition. *Cancer Cell,* 2011; 19: 701–714.

140. Kabeche L, Compton DA, Checkpoint-independent stabilization of kinetochore-microtubule attachments by Mad2 in human cells. *Curr Biol,* 2012; 22: 638–644.
141. Riley DJ, Nikitin AY, Lee WH, Adenovirus-mediated retinoblastoma gene therapy suppresses spontaneous pituitary melanotroph tumors in Rb+/- mice. *Nat Med,* 1996; 2: 1316–1321.
142. Pirollo KF, Rait A, Zhou Q, Zhang X, Zhou J, Kim C-S, Benedict WF, Chang EH, Tumor-targeting nanocomplex delivery of novel tumor suppressor RB94 chemosensitizes bladder carcinoma cells in vitro and in vivo. *Clin Cancer Res,* 2008; 14: 2190–2198.
143. Lee J-S, Chung JH, Diverse functions of BRCA1 in the DNA damage response. *Expert Rev Mol Med,* 2001; 3: 1–11.
144. Scully R, Xie A, Nagaraju G, Molecular functions of BRCA1 in the DNA damage response. *Cancer Biol Ther,* 2004; 3: 521–527.
145. Venkitaraman AR, Cancer Susceptibility and the Functions of BRCA1 and BRCA2. *Cell,* 2002; 108: 171–182.
146. Antoniou A, Pharoah PDP, Narod S, Risch HA, *et al.,* Average risks of breast and ovarian cancer associated with BRCA1 or BRCA2 mutations detected in case Series unselected for family history: a combined analysis of 22 studies. *Am J Hum Genet,* 2003; 72: 1117–1130.
147. Chen S, Parmigiani G, Meta-analysis of BRCA1 and BRCA2 penetrance. *J Clin Oncol,* 2007; 25: 1329–1333.
148. Friedenson B, The BRCA1/2 pathway prevents hematologic cancers in addition to breast and ovarian cancers. *BMC Cancer,* 7, 152 (2007).
149. Tait DL, Obermiller PS, Redlin-Frazier S, Jensen RA, Welcsh P, Dann J, King MC, Johnson DH, Holt JT, A phase I trial of retroviral BRCA1sv gene therapy in ovarian cancer. *Clin Cancer Res,* 1997; 3: 1959–1968.
150. Tait DL, Obermiller PS, Hatmaker AR, Redlin-Frazier S, Holt JT, Ovarian cancer BRCA1 gene therapy: Phase I and II trial differences in immune response and vector stability. *Clin Cancer Res,* 1999; 5: 1708–1714.
151. Tait DL, Obermiller PS, Holt JT, Preclinical studies of a new generation retroviral vector for ovarian cancer BRCA1 gene therapy. *Gynecol Oncol,* 2000; 79: 471–476.
152. Clark-Knowles K, V, O'Brien AM, Weberpals JI, BRCA1 as a therapeutic target in sporadic epithelial ovarian cancer. *J Oncol,* 2010, 891059 (2010).

153. Jiang H, Lin JJ, Su ZZ, Goldstein NI, Fisher PB, Subtraction hybridization identifies a novel melanoma differentiation associated gene, mda-7, modulated during human melanoma differentiation, growth and progression. *Oncogene,* 1995; 11: 2477–2486.
154. Ellerhorst JA, Loss of MDA-7 expression with progression of melanoma. *J Clin Oncol,* 2002; 20: 1069–1074.
155. Pestka S, Krause CD, Sarkar D, Walter MR, Shi Y, Fisher PB, Interleukin-10 and related cytokines and receptors. *Annu Rev Immunol,* 2004; 22: 929–979.
156. Gupta P, Walter MR, Su Z, Lebedeva IV, Emdad L, Randolph A, Valerie K, Sarkar D, Fisher PB, BiP/GRP78 is an intracellular target for MDA-7/IL-24 induction of cancer-specific apoptosis. *Cancer Res,* 2006; 66: 8182–8191.
157. Fels DR, Koumenis C, The PERK/eIF2alpha/ATF4 module of the UPR in hypoxia resistance and tumor growth. *Cancer Biol Ther,* 2006; 5: 723–728.
158. Raven JF, Koromilas AE, PERK and PKR: old kinases learn new tricks. *Cell Cycle,* 2008; 7: 1146–1150.
159. Lebedeva IV, Sarkar D, Su Z-Z, Kitada S, Dent P, Stein CA, Reed JC, Fisher PB, Bcl-2 and Bcl-x(L) differentially protect human prostate cancer cells from induction of apoptosis by melanoma differentiation associated gene-7, mda-7/IL-24. *Oncogene,* 2003; 22: 8758–8773.
160. Fisher PB, Sarkar D, Lebedeva IV, Emdad L, Gupta P, Sauane M, Su Z, Grant S, Dent P, Curiel DT, Senzer N, Nemunaitis J, Melanoma differentiation associated gene-7/interleukin-24 (mda-7/IL-24): novel gene therapeutic for metastatic melanoma. *Toxicol Appl Pharmacol,* 2007; 224: 300–307.
161. Sauane M, Su Z-Z, Dash R, Liu X, Norris JS, Sarkar D, Lee S-G, Allegood JC, Dent P, Spiegel S, Fisher PB, Ceramide plays a prominent role in MDA-7/IL-24-induced cancer-specific apoptosis. *J Cell Physiol,* 2010; 222: 546–555.
162. Parrish-Novak J, Xu W, Brender T, Yao L, Jones C, West J, Brandt C, Jelinek L, Madden K, McKernan PA, Foster DC, Jaspers S, Chandrasekher YA, Interleukins 19, 20, and 24 signal through two distinct receptor complexes. Differences in receptor-ligand interactions mediate unique biological functions. *J Biol Chem,* 2002; 277: 47517–47523.
163. Chada S, Mhashilkar AM, Ramesh R, Mumm JB, Sutton RB, Bocangel D, Zheng M, Grimm EA, Ekmekcioglu S, Bystander activity of Ad-mda7: human MDA-7 protein kills melanoma cells via an IL-20 receptor-dependent but STAT3-independent mechanism. *Mol Ther,* 2004; 10: 1085–1095.

164. Jiang H, Su ZZ, Lin JJ, Goldstein NI, Young CS, Fisher PB, The melanoma differentiation associated gene mda-7 suppresses cancer cell growth. *Proc Natl Acad Sci USA,* 1996; 93: 9160–9165.
165. Dent P, Yacoub A, Hamed HA, Park MA, *et al.* MDA-7/IL-24 as a cancer therapeutic: from bench to bedside. *Anticancer Drugs,* 2010; 21: 725–731.
166. Ma G, Kawamura K, Shan Y, Okamoto S, Li Q, Namba M, Shingyoji M, Tada Y, Tatsumi K, Hiroshima K, Shimada H, Tagawa M, Combination of adenoviruses expressing melanoma differentiation-associated gene-7 and chemotherapeutic agents produces enhanced cytotoxicity on esophageal carcinoma. *Cancer Gene Ther,* 2014; 21: 31–37.
167. Yacoub A, Mitchell C, Lebedeva IV, Sarkar D, Su Z-Z, McKinstry R, Gopalkrishnan R, V, Grant S, Fisher PB, Dent P, mda-7 (IL-24) Inhibits growth and enhances radiosensitivity of glioma cells *in vitro via* JNK signaling. *Cancer Biol Ther,* 2, 347–353.
168. Su Z-Z, Lebedeva IV, Sarkar D, Gopalkrishnan RV, Sauane M, Sigmon C, Yacoub A, Valerie K, Dent P, Fisher PB, Melanoma differentiation associated gene-7, mda-7/IL-24, selectively induces growth suppression, apoptosis and radiosensitization in malignant gliomas in a p53-independent manner. *Oncogene,* 2003; 22: 1164–1180.
169. Emdad L, Sarkar D, Lebedeva IV, Su Z-Z, Gupta P, Mahasreshti PJ, Dent P, Curiel DT, Fisher PB, Ionizing radiation enhances adenoviral vector expressing mda-7/IL-24-mediated apoptosis in human ovarian cancer. *J Cell Physiol,* 2006; 208: 298–306.
170. Cunningham CC, Chada S, Merritt JA, Tong A, Senzer N, Zhang Y, Mhashilkar A, Parker K, Vukelja S, Richards D, Hood J, Coffee K, Nemunaitis J, Clinical and local biological effects of an intratumoral injection of mda-7 (IL24; INGN 241) in patients with advanced carcinoma: a phase I study. *Mol Ther,* 2005; 11: 149–159.
171. Tong AW, Nemunaitis J, Su D, Zhang Y, Cunningham C, Senzer N, Netto G, Rich D, Mhashilkar A, Parker K, Coffee K, Ramesh R, Ekmekcioglu S, Grimm EA, van Wart Hood J, Merritt J, Chada S, Intratumoral injection of INGN 241, a nonreplicating adenovector expressing the melanoma-differentiation associated gene-7 (mda-7/IL24): biologic outcome in advanced cancer patients. *Mol Ther,* 2005; 11: 160–172.
172. Flint J, Shenk T, Adenovirus E1A protein paradigm viral transactivator. *Annu Rev Genet,* 1989; 23: 141–161.

173. Rao X-M, Tseng MT, Zheng X, Dong Y, Jamshidi-Parsian A, Thompson TC, Brenner MK, McMasters KM, Zhou HS, E1A-induced apoptosis does not prevent replication of adenoviruses with deletion of E1b in majority of infected cancer cells. *Cancer Gene Ther,* 2004; 11: 585–593.
174. Itamochi H, Kigawa J, Kanamori Y, Oishi T, Bartholomeusz C, Nahta R, Esteva FJ, Sneige N, Terakawa N, Ueno NT, Adenovirus type 5 E1A gene therapy for ovarian clear cell carcinoma: a potential treatment strategy. *Mol Cancer Ther,* 2007; 6: 227–235.
175. Li Z, Day C-P, Yang J-Y, Tsai W-B, Lozano G, Shih H,-M. Hung M,-C. Adenoviral E1A targets Mdm4 to stabilize tumor suppressor p53. *Cancer Res,* 2004; 64: 9080–9085.
176. Liao Y, Hung M-C, A new role of protein phosphatase 2a in adenoviral E1A protein-mediated sensitization to anticancer drug-induced apoptosis in human breast cancer cells. *Cancer Res,* 2004; 64: 5938–5942.
177. Najafi SMA, Li Z, Makino K, Shao R, Hung M-C, The adenoviral E1A induces p21WAF1/CIP1 expression in cancer cells. *Biochem Biophys Res Commun,* 2003; 305: 1099–1104.
178. Su J-L, Cheng X, Yamaguchi H, Chang Y-W, Hou C-F, Lee D-F, Ko H-W, Hua K-T, Wang Y-N, Hsiao M, Chen PB, Hsu J-M, Bast RC, Hortobagyi GN, Hung M-C, FOXO3a-Dependent mechanism of E1A-induced chemosensitization. *Cancer Res,* 2011; 71: 6878–6887.
179. Chen H, Yu D, Chinnadurai G, Karunagaran D, Hung MC, Mapping of adenovirus 5 E1A domains responsible for suppression of neu-mediated transformation *via* transcriptional repression of neu. *Oncogene,* 1997; 14: 1965–1971.
180. Chinnadurai G, The transcriptional corepressor CtBP: a foe of multiple tumor suppressors. *Cancer Res,* 2009; 69: 731–734.
181. Shao R, Lee D-F, Wen Y, Ding Y, Xia W, Ping B, Yagita H, Spohn B, Hung M-C, E1A sensitizes cancer cells to TRAIL-induced apoptosis through enhancement of caspase activation. *Mol Cancer Res,* 2005; 3: 219–226.
182. Yamaguchi H, Chen C-T, Chou C-K, Pal A, Bornmann W, Hortobagyi GN, Hung M-C, Adenovirus 5 E1A enhances histone deacetylase inhibitors-induced apoptosis through Egr-1-mediated Bim upregulation. *Oncogene,* 2010; 29: 5619–5629.
183. Brader KR, Wolf JK, Hung MC, Yu D, Crispens MA, van Golen KL, Price JE, Adenovirus E1A expression enhances the sensitivity of an ovarian can-

cer cell line to multiple cytotoxic agents through an apoptotic mechanism. *Clin Cancer Res,* 1997; 3: 2017–2024.

184. Liao Y, Zou Y-Y, Xia W-Y, Hung M-C, Enhanced paclitaxel cytotoxicity and prolonged animal survival rate by a nonviral-mediated systemic delivery of E1A gene in orthotopic xenograft human breast cancer. *Cancer Gene Ther,* 2004; 11: 594–602.
185. Ueno NT, Bartholomeusz C, Herrmann JL, Estrov Z, Shao R, Andreeff M, Price J, Paul RW, Anklesaria P, Yu D, Hung MC, E1A-mediated paclitaxel sensitization in HER-2/neu-overexpressing ovarian cancer SKOV3.ip1 through apoptosis involving the caspase-3 pathway. *Clin Cancer Res,* 2000; 6: 250–259.
186. Hortobagyi GN, Ueno NT, Xia W, Zhang S, *et al.,* Cationic liposome-mediated E1A gene transfer to human breast and ovarian cancer cells and its biologic effects: a phase I clinical trial. *J Clin Oncol,* 2001; 19: 3422–3433.
187. Yoo GH, Hung M-C, Lopez-Berestein G, LaFollette S, Ensley JF, Carey M, Batson E, Reynolds TC, Murray JL, Phase I trial of intratumoral liposome E1A Gene therapy in patients with recurrent breast and head and neck cancer. *Clin Cancer Res,* 2001; 7: 1237–1245.
188. Madhusudan S, Tamir A, Bates N, Flanagan E, Gore ME, Barton DPJ, Harper P, Seckl M, Thomas H, Lemoine NR, Charnock M, Habib NA, Lechler R, Nicholls J, Pignatelli M, Ganesan TS, A multicenter Phase I gene therapy clinical trial involving intraperitoneal administration of E1A-lipid complex in patients with recurrent epithelial ovarian cancer overexpressing HER-2/neu oncogene. *Clin Cancer Res,* 2004; 10: 2986–2996.
189. Villaret D, Glisson B, Kenady D, Hanna E, Carey M, Gleich L, Yoo GH, Futran N, Hung M-C, Anklesaria P, Heald AE, A multicenter phase II study of tgDCC-E1A for the intratumoral treatment of patients with recurrent head and neck squamous cell carcinoma. *Head Neck,* 2002; 24: 661–669.
190. Daly MC, Xiang RH, Buchhagen D, Hensel CH, Garcia DK, Killary AM, Minna JD, Naylor SL, A homozygous deletion on chromosome 3 in a small cell lung cancer cell line correlates with a region of tumor suppressor activity. *Oncogene,* 1993; 8: 1721–1729.
191. Prudkin L, Behrens C, Liu DD, Zhou X, Ozburn NC, Bekele BN, Minna JD, Moran C, Roth JA, Ji L, Wistuba II, Loss and reduction of FUS1 protein expression is a frequent phenomenon in the pathogenesis of lung cancer. *Clin Cancer Res,* 2008; 14: 41–47.

192. Lerman MI, Minna JD, The 630-kb lung cancer homozygous deletion region on human chromosome 3p21.3: identification and evaluation of the resident candidate tumor suppressor genes. The International Lung Cancer Chromosome 3p21.3 Tumor Suppressor Gene Consortium. *Cancer Res,* 2000; 60: 6116–6133.
193. Ji L, Minna JD, Roth JA, 3p21.3 tumor suppressor cluster: prospects for translational applications. *Future Oncol* 2005; 1: 79–92.
194. Kondo M, Ji L, Kamibayashi C, Tomizawa Y, Randle D, Sekido Y, Yokota J, Kashuba V, Zabarovsky E, Kuzmin I, Lerman M, Roth J, Minna JD, Overexpression of candidate tumor suppressor gene FUS1 isolated from the 3p21.3 homozygous deletion region leads to G1 arrest and growth inhibition of lung cancer cells. *Oncogene* 2001; 20: 6258–6262.
195. Ji L, Roth JA, Tumor suppressor FUS1 signaling pathway. *J Thorac Oncol* 2008; 3: 327–330.
196. Riedl SJ, Li W, Chao Y, Schwarzenbacher R, Shi Y, Structure of the apoptotic protease-activating factor 1 bound to ADP. *Nature,* 2005; 434: 926–933.
197. Deng W-G, Kawashima H, Wu G, Jayachandran G, Xu K, Minna JD, Roth JA, Ji L, Synergistic tumor suppression by coexpression of FUS1 and p53 is associated with down-regulation of murine double minute-2 and activation of the apoptotic protease-activating factor 1-dependent apoptotic pathway in human non-small cell lung cancer cells. *Cancer Res* 2007; 67: 709–717.
198. Deng W-G, Wu G, Ueda K, Xu K, Roth JA, Ji L, Enhancement of antitumor activity of cisplatin in human lung cancer cells by tumor suppressor FUS1. *Cancer Gene Ther* 2008; 15: 29–39.
199. Lu C, Stewart DJ, Lee JJ, Ji L, Ramesh R, Jayachandran G, Nunez MI, Wistuba II, Erasmus JJ, Hicks ME, Grimm EA, Reuben JM, Baladandayuthapani V, Templeton NS, McMannis JD, Roth JA, Phase I clinical trial of systemically administered TUSC2(FUS1)-nanoparticles mediating functional gene transfer in humans. *PLoS One,* 7, e34833 (2012).
200. Byrne BJ, Davis MS, Yamaguchi J, Bergsma DJ, Subramanian KN, Definition of the simian virus 40 early promoter region and demonstration of a host range bias in the enhancement effect of the simian virus 40 72-base pair repeat. *Proc Natl Acad Sci USA,* 1983; 80: 721–725.
201. Boshart M, Weber F, Jahn G, Dorsch-Häsler K, Fleckenstein B, Schaffner W, A very strong enhancer is located upstream of an immediate early gene of human cytomegalovirus. *Cell,* 1985; 41: 521–530.

202. Yamamoto T, Crombrugghe B, De. Identification of a functional promoter in the long terminal repeat of rous sarcoma virus. *Cell,* 1980; 22: 787–797.
203. Zarrin, AA, Malkin L, Fong I, Luk KD, Ghose, a Berinstein NL, Comparison of CMV, RSV, SV40 viral and Vlambda1 cellular promoters in B and T lymphoid and non-lymphoid cell lines. *Biochim Biophys Acta,* 1999; 1446: 135–139.
204. Zheng C, Baum BJ, Evaluation of viral and mammalian promoters for use in gene delivery to salivary glands. *Mol Ther,* 2005; 12: 528–536.
205. Harms J, Splitter GA, Interferon-gamma inhibits transgene expression driven by SV40 and CMV promoters but augments expression driven by the mammalian MHCI promoter. *Hum Gene Ther,* 1995; 6: 1291–1297.
206. Qin L, Ding Y, Pahud DR, Chang E, Imperiale MJ, Bromberg JS, Promoter attenuation in gene therapy: interferon-gamma and tumor necrosis factor-alpha inhibit transgene expression. *Hum Gene Ther,* 1997; 8: 2019–2029.
207. Gribaudo G, Ravaglia S, Caliendo A, Cavallo R, Gariglio M, Martinotti MG, Landolfo S, Interferons inhibit onset of murine cytomegalovirus immediate-early gene transcription. *Virology,* 1993; 197: 303–311.
208. Qin JY, Zhang L, Clift KL, Hulur I, Xiang AP, Ren B-Z, Lahn BT, Systematic comparison of constitutive promoters and the doxycycline-inducible promoter. *PLoS One* 5, e10611 (2010).
209. Kwissa M, von Kampen VK, Zurbriggen R, Glück R, Reimann J, Schirmbeck R, Efficient vaccination by intradermal or intramuscular inoculation of plasmid DNA expressing hepatitis B surface antigen under desmin promoter/enhancer control. *Vaccine,* 2000; 18: 2337–2344.
210. Itai K, Sawamura D, Meng X, Hashimoto I, Keratinocyte gene therapy: inducible promoters and *in vivo* control of transgene expression. *Clin Exp Dermatol,* 2001; 26: 531–535.
211. Liu BH, Wang X, Ma YX, Wang S, CMV enhancer/human PDGF-beta promoter for neuron-specific transgene expression. *Gene Ther,* 2004; 11: 52–60.
212. Manuscript A, Suicide Gene Therapy for Cancer — Current Strategies. (2013). DOI:10.4172/2157-7412.1000139.Suicide.
213. Kozak M, At least six nucleotides preceding the AUG initiator codon enhance translation in mammalian cells. *J Mol Biol,* 1987; 196: 947–950.
214. Xu ZL, Mizuguchi H, Ishii-Watabe, A, Uchida E, Mayumi T, Hayakawa T, Optimization of transcriptional regulatory elements for constructing plasmid vectors. *Gene,* 2001; 272: 149–156.

215. Xu Z-L, Mizuguchi H, Mayumi T, Hayakawa T, Woodchuck hepatitis virus post-transcriptional regulation element enhances transgene expression from adenovirus vectors. *Biochim Biophys Acta — Gen Subj,* 2003; 1621: 266–271.
216. Gustafsson C, Govindarajan S, Minshull J, Codon bias and heterologous protein expression. *Trends Biotechnol,* 2004; 22: 346–353.
217. Garmory HS, Brown KA, Titball RW, DNA vaccines: improving expression of antigens. *Genet Vaccines Ther,* 1, 2 (2003).
218. Le Hir H, Nott A, Moore MJ, How introns influence and enhance eukaryotic gene expression. *Trends Biochem Sci,* 2003; 28: 215–220.
219. Chapman BS, Thayer RM, Vincent KA, Haigwood NL, Effect of intron A from human cytomegalovirus (Towne) immediate-early gene on heterologous expression in mammalian cells. *Nucleic Acids Res,* 1991; 19: 3979–3986.
220. Girod P-A, Nguyen D-Q, Calabrese D, Puttini S, Grandjean M, Martinet D, Regamey A, Saugy D, Beckmann JS, Bucher P, Mermod N, Genome-wide prediction of matrix attachment regions that increase gene expression in mammalian cells. *Nat Methods,* 2007; 4: 747–753.
221. Jenke ACW, Stehle IM, Herrmann F, Eisenberger T, Baiker A, Bode J, Fackelmayer FO, Lipps HJ, Nuclear scaffold/matrix attached region modules linked to a transcription unit are sufficient for replication and maintenance of a mammalian episome. *Proc Natl Acad Sci USA,* 2004; 101: 11322–11327.
222. Piechaczek C, Fetzer C, Baiker A, Bode J, Lipps HJ, A vector based on the SV40 origin of replication and chromosomal S/MARs replicates episomally in CHO cells. *Nucleic Acids Res,* 1999; 27: 426–428.
223. Bode J, Kohwi Y, Dickinson L, Joh T, Klehr D, Mielke C, Kohwi-Shigematsu T, Biological significance of unwinding capability of nuclear matrix-associating DNAs. *Science,* 1992; 255: 195–197.
224. Ascenzioni F, Donini P, Lipps HJ, Mammalian artificial chromosomes — vectors for somatic gene therapy. *Cancer Lett,* 1997; 118: 135–142.
225. Hemmi H, Takeuchi O, Kawai T, Kaisho T, Sato S, Sanjo H, Matsumoto M, Hoshino K, Wagner H, Takeda K, Akira S, A toll-like receptor recognizes bacterial DNA. *Nature,* 2000; 408: 740–755.
226. Ishii KJ, Akira S, Innate immune recognition of, and regulation by DNA, *Trends Immunol,* 2006; 27: 525–532.
227. Suzuki M, Cerullo V, Bertin TK, Cela R, Clarke C, Guenther M, Brunetti-Pierri N, Lee B, MyD88-dependent silencing of transgene expression

during the innate and adaptive immune response to helper-dependent adenovirus. *Hum Gene Ther,* 2010; 21: 325–336.

228. Hyde SC, Pringle IA, Abdullah S, Lawton AE, *et al.* CpG-free plasmids confer reduced inflammation and sustained pulmonary gene expression. *Nat Biotechnol,* 2008; 26: 549–551.
229. Chen ZY, He CY, Meuse L, Kay MA, Silencing of episomal transgene expression by plasmid bacterial DNA elements *in vivo. Gene Ther,* 2004; 11: 856–864.
230. Chen Z-Y, He C-Y, Kay MA, Improved production and purification of minicircle DNA vector free of plasmid bacterial sequences and capable of persistent transgene expression *in vivo. Hum Gene Ther,* 2005; 16: 126–131.
231. Mayrhofer P, Schleef M, Jechlinger W, Use of minicircle plasmids for gene therapy. *Methods Mol Biol,* 2009; 542: 87–104.
232. Jia F, Wilson KD, Sun N, Gupta DM, Huang M, Li Z, Panetta NJ, Chen ZY, Robbins RC, Kay M, a, Longaker MT, Wu JC, A nonviral minicircle vector for deriving human iPS cells. *Nat Methods* 2010; 7: 197–199.
233. Annunziato AT, in *Nature Education* (2008). at *http://www.nature.com/scitable/topicpage/dna-packaging-nucleosomes-and-chromatin-310*
234. Wolffe AP, *Chromatin: Structure and Function.* (Academic Press, 1999).
235. Bannister AJ, Kouzarides T, Regulation of chromatin by histone modifications. *Cell Res,* 2011; 21: 381–395.
236. Yang X-J. Seto E, HATs and HDACs: from structure, function and regulation to novel strategies for therapy and prevention. *Oncogene,* 2007; 26: 5310–5318.
237. Xhemalce B, Dawson MA, Bannister A, *Histone Modifications Encyclopedia of Molecular Cell Biology and Molecular Medicine.* (Wiley-VCH Verlag GmbH Co. KGaA, 2011).
238. Hon GC, Hawkins RD, Ren B, Predictive chromatin signatures in the mammalian genome. *Hum Mol Genet,* 2009; 18: R195–201.
239. Trojer P, Reinberg D, Facultative heterochromatin: is there a distinctive molecular signature? *Mol Cell,* 2007; 28: 1–13.
240. Kouzarides T, Histone methylation in transcriptional control. *Curr Opin Genet Dev,* 2002; 12: 198–209.
241. Rossetto D, Avvakumov N, Côté J, Histone phosphorylation: a chromatin modification involved in diverse nuclear events. *Epigenetics,* 2012; 7: 1098–108.

242. Cheung P, Allis CD, Sassone-Corsi P, Signaling to chromatin through histone modifications. *Cell,* 2000; 103: 263–271.
243. Wang H, Wang L, Erdjument-Bromage H, Vidal M, Tempst P, Jones RS, Zhang Y, Role of histone H2A ubiquitination in Polycomb silencing. *Nature,* 2004; 431: 873–878.
244. Shiio Y, Eisenman RN, Histone sumoylation is associated with transcriptional repression. *Proc Natl Acad Sci USA,* 2003; 100: 13225–13230.
245. Hassa PO, Haenni SS, Elser M, Hottiger MO, Nuclear ADP-ribosylation reactions in mammalian cells: where are we today and where are we going? *Microbiol Mol Biol Rev,* 2006; 70: 789–829.
246. Sakabe K, Wang Z, Hart GW, Beta-N-acetylglucosamine (O-GlcNAc) is part of the histone code. *Proc Natl Acad Sci USA,* 2010; 107: 19915–19920.
247. Kobza K, Camporeale G, Rueckert B, Kueh A, Griffin JB, Sarath G, Zempleni J, K4, K9 and K18 in human histone H3 are targets for biotinylation by biotinidase. *FEBS J,* 2005; 272: 4249–4259.
248. Wang Y, Wysocka J, Sayegh J, Lee Y,-H., Perlin JR, Leonelli L, Sonbuchner LS, McDonald CH, Cook RG, Dou Y, Roeder RG, Clarke S, Stallcup MR, Allis CD, Coonrod SA, Human PAD4 regulates histone arginine methylation levels *via* demethylimination. *Science,* 2004; 306: 279–83.
249. Bishop CL, Ramalho M, Nadkarni N, Kong WM, Higgins CF, Krauzewicz N, Role for centromeric heterochromatin and pml nuclear bodies in the cellular response to foreign DNA. *Mol Cell Biol,* 2006; 26: 2583–2594.
250. Archin NM, Espeseth A, Parker D, Cheema M, Hazuda D, Margolis DM, Expression of latent HIV induced by the potent HDAC inhibitor suberoylanilide hydroxamic acid. *AIDS Res Hum Retroviruses,* 2009; 25: 207–212.
251. Danaher RJ, Jacob RJ, Steiner MR, Allen WR, Hill JM, Miller CS, Histone deacetylase inhibitors induce reactivation of herpes simplex virus type 1 in a latency-associated transcript-independent manner in neuronal cells. *J Neurovirol,* 2005; 11: 306–317.
252. Duque-Afonso J, Yalcin A, Berg T, Abdelkarim M, Heidenreich O, Lübbert M, The HDAC class I-specific inhibitor entinostat (MS-275) effectively relieves epigenetic silencing of the LAT2 gene mediated by AML1/ETO. *Oncogene,* 2011; 30: 3062–3072.
253. Kasman L, Onicescu G, Voelkel-Johnson C, Histone deacetylase inhibitors restore cell surface expression of the coxsackie adenovirus receptor and

enhance CMV promoter activity in castration-resistant prostate cancer cells. *Prostate Cancer,* 2012; 137–163.

254. Kasman L, Lu P, Voelkel-Johnson C, The histone deacetylase inhibitors depsipeptide and MS-275, enhance TRAIL gene therapy of LNCaP prostate cancer cells without adverse effects in normal prostate epithelial cells. *Cancer Gene Ther,* 2007; 14: 327–334.
255. Yaneva J, Schroth GP, van Holde KE, Zlatanova J, High-affinity binding sites for histone H1 in plasmid DNA. *Proc Natl Acad Sci USA,* 1995; 92: 7060–7064.
256. Nakagawa T, Bulger M, Muramatsu M, Ito T, Multistep chromatin assembly on supercoiled plasmid DNA by nucleosome assembly protein-1 and ATP-utilizing chromatin assembly and remodeling factor. *J Biol Chem,* 2001; 276: 27384–27391.
257. Kumar H, Kawai T, Akira S, Pathogen recognition by the innate immune system. *Int Rev Immunol,* 2011; 30: 16–34.
258. Parkin J, Cohen B, An overview of the immune system. *Lancet,* 2001; 357: 1777–1789.
259. Arstila TP, Casrouge A, Baron V, Even J, Kanellopoulos J, Kourilsky P, A direct estimate of the human alphabeta T cell receptor diversity. *Science,* 1999; 286: 958–961.
260. Mond JJ, Lees A, Snapper CM, T cell-independent antigens type 2. *Annu Rev Immunol,* 1995; 13: 655–692.
261. Challis GB, Stam HJ, The spontaneous regression of cancer. A review of cases from 1900 to 1987. *Acta Oncol,* 1990; 29: 545–550.
262. Chodorowski Z, Anand JS, Wiśniewski M, Madaliński M, Wierzba K, Wiśniewski J, Spontaneous regression of cancer--review of cases from 1988 to 2006. *Przegląd Lek,* 2007; 64: 380–382.
263. Marincola FM, Jaffee EM, Hicklin DJ, Ferrone S, Escape of human solid tumors from T-cell recognition: molecular mechanisms and functional significance. *Adv Immunol,* 2000; 74: 181–273.
264. Godin-Ethier J, Hanafi L-A, Piccirillo CA, Lapointe R, Indoleamine 2,3-dioxygenase expression in human cancers: clinical and immunologic perspectives. *Clin Cancer Res,* 2011; 17: 6985–6991.
265. Nagaraj S, Gupta K, Pisarev V, Kinarsky L, Sherman S, Kang L, Herber DL, Schneck J, Gabrilovich DI, Altered recognition of antigen is a mechanism of CD8+ T cell tolerance in cancer. *Nat Med,* 2007; 13: 828–835.

266. Quatromoni JG, Eruslanov E, Tumor-associated macrophages: function, phenotype, and link to prognosis in human lung cancer. *Am J Transl Res* 2012; 4: 376–389.
267. Olson BM, McNeel DG, CD8+ T cells specific for the androgen receptor are common in patients with prostate cancer and are able to lyse prostate tumor cells. *Cancer Immunol Immunother* 2011; 60: 781–92.
268. Disis ML, Knutson KL, Schiffman K, Rinn K, McNeel DG, Pre-existent immunity to the HER-2/neu oncogenic protein in patients with HER-2/neu overexpressing breast and ovarian cancer. *Breast Cancer Res Treat* 2000; 62: 245–252.
269. Chen YT, Stockert E, Tsang S, Coplan KA, Old LJ, Immunophenotyping of melanomas for tyrosinase: implications for vaccine development. *Proc Natl Acad Sci USA,* 1995; 92: 8125–8129.
270. Boon T, Cerottini JC, Van den Eynde B, van der Bruggen P, Van Pel A, Tumor antigens recognized by T lymphocytes. *Annu Rev Immunol,* 1994; 12: 337–365.
271. Anderson RJ, Schneider J, Plasmid DNA and viral vector-based vaccines for the treatment of cancer. *Vaccine,* 2007; 25 Suppl 2, B24–B34.
272. Kutzler MA, Weiner DB, DNA vaccines: ready for prime time? *Nat Rev Genet,* 2008; 9: 776–788.
273. Wolchok JD, Yuan J, Houghton AN, Gallardo HF, Rasalan TS, Wang J, Zhang Y, Ranganathan R, Chapman PB, Krown SE, Livingston PO, Heywood M, Riviere I, Panageas KS, Terzulli SL, Perales MA, Safety and immunogenicity of tyrosinase DNA vaccines in patients with melanoma. *Mol Ther*, 2007; 15: 2044–50.
274. Dangoor A, Lorigan P, Keilholz U, Schadendorf D, Harris A, Ottensmeier C, Smyth J, Hoffmann K, Anderson R, Cripps M, Schneider J, Hawkins R, Clinical and immunological responses in metastatic melanoma patients vaccinated with a high-dose poly-epitope vaccine. *Cancer Immunol Immunother,* 2010; 59: 863–873.
275. Smith CL, Dunbar PR, Mirza F, Palmowski MJ, Shepherd D, Gilbert SC, Coulie P, Schneider J, Hoffman E, Hawkins R, Harris AL, Cerundolo V, Recombinant modified vaccinia Ankara primes functionally activated CTL specific for a melanoma tumor antigen epitope in melanoma patients with a high risk of disease recurrence. *Int J Cancer,* 2005; 113: 259–266.

276. Pavlenko M, Roos A-K, Lundqvist A, Palmborg A, Miller AM, Ozenci V, Bergman B, Egevad L, Hellström M, Kiessling R, Masucci G, Wersäll P, Nilsson S, Pisa P, A phase I trial of DNA vaccination with a plasmid expressing prostate-specific antigen in patients with hormone-refractory prostate cancer. *Br J Cancer,* 2004; 91: 688–694.
277. Low L, Mander A, McCann K, Dearnaley D, Tjelle T, Mathiesen I, Stevenson F, Ottensmeier CH, DNA vaccination with electroporation induces increased antibody responses in patients with prostate cancer. *Hum Gene Ther* 2009; 20: 1269–78.
278. Chudley L, McCann K, Mander A, Tjelle T, Campos-Perez J, Godeseth R, Creak A, Dobbyn J, Johnson B, Bass P, Heath C, Kerr P, Mathiesen I, Dearnaley D, Stevenson F, Ottensmeier C, DNA fusion-gene vaccination in patients with prostate cancer induces high-frequency CD8(+) T-cell responses and increases PSA doubling time. *Cancer Immunol Immunother,* 2012; 61: 2161–2170.
279. Becker JT, Olson BM, Johnson LE, Davies JG, Dunphy EJ, McNeel DG, DNA vaccine encoding prostatic acid phosphatase (PAP) elicits long-term T-cell responses in patients with recurrent prostate cancer. *J Immunother,* 33, 639–647.
280. McNeel DG, Dunphy EJ, Davies JG, Frye TP, Johnson LE, Staab MJ, Horvath DL, Straus J, Alberti D, Marnocha R, Liu G, Eickhoff JC, Wilding G, Safety and immunological efficacy of a DNA vaccine encoding prostatic acid phosphatase in patients with stage D0 prostate cancer. *J Clin Oncol,* 2009; 27: 4047–4054.
281. Harrop R, John J, Carroll MW, Recombinant viral vectors: cancer vaccines. *Adv Drug Deliv Rev,* 2006; 58: 931–947.
282. Zhu J, Martinez J, Huang X, Yang Y, Innate immunity against vaccinia virus is mediated by TLR2 and requires TLR-independent production of IFN-beta. *Blood* 2007; 109: 619–625.
283. DiPaola RS, Plante M, Kaufman H, Petrylak DP, Israeli R, Lattime E, Manson K, Schuetz T, A phase I trial of pox PSA vaccines (PROSTVAC-VF) with B7-1, ICAM-1, and LFA-3 co-stimulatory molecules (TRICOM) in patients with prostate cancer. *J Transl Med,* 2006; 4, 1.
284. Arlen PM, Skarupa L, Pazdur M, Seetharam M, Tsang KY, Grosenbach DW, Feldman J, Poole DJ, Litzinger M, Steinberg SM, Jones E, Chen C, Marte J,

Parnes H, Wright J, Dahut W, Schlom J, Gulley JL, Clinical safety of a viral vector based prostate cancer vaccine strategy. *J Urol,* 2007; 178: 1515–1520.

285. Gulley JL, Arlen PM, Madan RA, Tsang K-Y, *et al.,* Immunologic and prognostic factors associated with overall survival employing a poxviral-based PSA vaccine in metastatic castrate-resistant prostate cancer. *Cancer Immunol Immunother* 2010; 59: 663–674.
286. Kantoff PW, Schuetz TJ, Blumenstein BA, Glode LM, Bilhartz DL, Wyand M, Manson K, Panicali DL, Laus R, Schlom J, Dahut WL, Arlen PM, Gulley JL, Godfrey WR, Overall survival analysis of a phase II randomized controlled trial of a Poxviral-based PSA-targeted immunotherapy in metastatic castration-resistant prostate cancer. *J Clin Oncol,* 2010; 28: 1099–1105.
287. Ramlau R, Quoix E, Rolski J, Pless M, Lena H, Lévy E, Krzakowski M, Hess D, Tartour E, Chenard M-P, Limacher J,-M., Bizouarne N, Acres B, Halluard C, Velu T, A phase II study of Tg4010 (Mva-Muc1-Il2) in association with chemotherapy in patients with stage III/IV Non-small cell lung cancer. *J Thorac Oncol,* 2008; 3: 735–744.
288. García-Hernández E, González-Sánchez JL, Andrade-Manzano A, Contreras ML, Padilla S, Guzmán CC, Jiménez R, Reyes L, Morosoli G, Verde ML, Rosales R, Regression of papilloma high-grade lesions (CIN 2 and CIN 3) is stimulated by therapeutic vaccination with MVA E2 recombinant vaccine. *Cancer Gene Ther,* 2006; 13: 592–597.
289. Harrop R, Connolly N, Redchenko I, Valle J, Saunders M, Ryan MG, Myers KA, Drury N, Kingsman SM, Hawkins RE, Carroll MW, Vaccination of colorectal cancer patients with modified vaccinia Ankara delivering the tumor antigen 5T4 (TroVax) induces immune responses which correlate with disease control: a phase I/II trial. *Clin Cancer Res,* 2006; 12: 3416–3424.
290. Meyer RG, Britten CM, Siepmann U, Petzold B, *et al.,* A phase I vaccination study with tyrosinase in patients with stage II melanoma using recombinant modified vaccinia virus Ankara (MVA-hTyr). *Cancer Immunol Immunother,* 2005; 54: 453–467.
291. Leachman SA, Shylankevich M, Slade MD, Levine D, Sundaram RK, Xiao W, Bryan M, Zelterman D, Tiegelaar RE, Brandsma JL, Ubiquitin-fused and/or multiple early genes from cottontail rabbit papillomavirus as DNA vaccines. *J Virol* 2002; 76: 7616–7624.
292. Xiang R, Lode HN, Chao TH, Ruehlmann JM, Dolman CS, Rodriguez F, Whitton JL, Overwijk WW, Restifo NP, Reisfeld RA, An autologous oral

DNA vaccine protects against murine melanoma. *Proc Natl Acad Sci USA,* 2000; 97: 5492–5497.

293. Rodriguez F, An LL, Harkins S, Zhang J, Yokoyama M, Widera G, Fuller JT, Kincaid C, Campbell IL, Whitton JL, DNA immunization with minigenes: low frequency of memory cytotoxic T lymphocytes and inefficient antiviral protection are rectified by ubiquitination. *J Virol,* 1998; 72: 5174–5181.
294. Tobery T, Siliciano RF, Cutting edge: induction of enhanced CTL-dependent protective immunity in vivo by N-end rule targeting of a model tumor antigen. *J Immunol,* 1999; 162: 639–642.
295. Sewell DA, Shahabi V, Gunn GR, Pan Z-K, Dominiecki ME, Paterson Y, Recombinant Listeria vaccines containing PEST sequences are potent immune adjuvants for the tumor-associated antigen human papillomavirus-16 E7. *Cancer Res,* 2004; 64: 8821–8825.
296. Antón LC, Schubert U, Bacík I, Princiotta MF, Wearsch PA, Gibbs J, Day PM, Realini C, Rechsteiner MC, Bennink JR, Yewdell JW, Intracellular localization of proteasomal degradation of a viral antigen. *J Cell Biol,* 1999; 146: 113–124.
297. Anderson K, Cresswell P, Gammon M, Hermes J, Williamson A, Zweerink H, Endogenously synthesized peptide with an endoplasmic reticulum signal sequence sensitizes antigen processing mutant cells to class I-restricted cell-mediated lysis. *J Exp Med* 1991; 174: 489–92.
298. Bacik I, Cox JH, Anderson R, Yewdell JW, Bennink JR, TAP (transporter associated with antigen processing)-independent presentation of endogenously synthesized peptides is enhanced by endoplasmic reticulum insertion sequences located at the amino- but not carboxyl-terminus of the peptide. *J Immunol,* 1994; 152: 381–387.
299. Tsan M,-F. Gao B, Heat shock proteins and immune system. *J Leukoc Biol,* 2009; 85: 905–910.
300. Chen X, Tao Q, Yu H, Zhang L, Cao X, Tumor cell membrane-bound heat shock protein 70 elicits antitumor immunity. *Immunol Lett,* 2002; 84: 81–87.
301. Liu B, Dai J, Zheng H, Stoilova D, Sun S, Li Z, Cell surface expression of an endoplasmic reticulum resident heat shock protein gp96 triggers MyD88-dependent systemic autoimmune diseases. *Proc Natl Acad Sci, USA* 2003; 100: 15824–15829.
302. Ji H, Wang TL, Chen CH, Pai SI, Hung CF, Lin KY, Kurman RJ, Pardoll DM, Wu TC, Targeting human papillomavirus type 16 E7 to the endo-

somal/lysosomal compartment enhances the antitumor immunity of DNA vaccines against murine human papillomavirus type 16 E7-expressing tumors. *Hum Gene Ther,* 1999; 10: 2727–2740.

303. Kim MS, Sin J-I, Both antigen optimization and lysosomal targeting are required for enhanced anti-tumour protective immunity in a human papillomavirus E7-expressing animal tumour model. *Immunology,* 2005; 116: 255–266.
304. Duarte S, Carle G, Faneca H, de Lima MCP, Pierrefite-Carle V, Suicide gene therapy in cancer: where do we stand now? *Cancer Lett,* 2012; 324: 160–170.
305. Malecki M, Malecki R, Ovarian cancer suicide gene therapy with genetically engineered, transgenically expressed, intracellular scFv antibodies against anti-oxidative enzymes. *Proc S Dak Acad Sci* 2008; 87: 249–260.
306. Carlotti F, Zaldumbide A, Martin P, Boulukos KE, Hoeben RC, Pognonec P, Development of an inducible suicide gene system based on human caspase 8. *Cancer Gene Ther,* 2005; 12: 627–639.
307. Xie X, Zhao X, Liu Y, Zhang J, Matusik RJ, Slawin KM, Spencer DM, Adenovirus-mediated tissue-targeted expression of a caspase-9-based artificial death switch for the treatment of prostate cancer. *Cancer Res,* 2001; 61: 6795–6804.
308. Malecki M, Dahlke J, Haig M, Wohlwend L, Malecki R, Eradication of human ovarian cancer cells by transgenic expression of recombinant DNASE1, DNASE1L3, DNASE2, and DFFB controlled by egfr promoter: novel strategy for targeted therapy of cancer. *J Genet Syndr Gene Ther,* 2013; 4, 152.
309. Halloran PJ, Fenton RG, Irreversible G2-M arrest and cytoskeletal reorganization induced by cytotoxic nucleoside analogues. *Cancer Res,* 1998; 58: 3855–3865.
310. Wang J, Lu X,-X., Chen D,-Z., Li S,-F. Zhang L,-S. Herpes simplex virus thymidine kinase and ganciclovir suicide gene therapy for human pancreatic cancer. *World J Gastroenterol,* 2004; 10: 400–403.
311. Tang W, He Y, Zhou S, Ma Y, Liu G, A novel Bifidobacterium infantis-mediated TK/GCV suicide gene therapy system exhibits antitumor activity in a rat model of bladder cancer. *J Exp Clin Cancer Res,* 2009; 28, 155.
312. Kakinoki K, Nakamoto Y, Kagaya T, Tsuchiyama T, Sakai Y, Nakahama T, Mukaida N, Kaneko S, Prevention of intrahepatic metastasis of liver cancer by suicide gene therapy and chemokine ligand 2/monocyte

chemoattractant protein-1 delivery in mice. *J Gene Med,* 2010; 12: 1002–1013.

313. Chen L,-S., Wang M, Ou W,-C., Fung C,-Y., Chen P,-L., Chang C,-F., Huang W,-S., Wang J,-Y., Lin PY, Chang D, Efficient gene transfer using the human JC virus-like particle that inhibits human colon adenocarcinoma growth in a nude mouse model. *Gene Ther,* 2010; 17: 1033–1041.
314. Voges J, Reszka R, Gossmann A, Dittmar C, Richter R, Garlip G, Kracht L, Coenen HH, Sturm V, Wienhard K, Heiss W,-D. Jacobs AH, Imaging-guided convection-enhanced delivery and gene therapy of glioblastoma. *Ann Neurol* 2003; 54: 479–487.
315. Nasu Y, Saika T, Ebara S, Kusaka N, Kaku H, Abarzua F, Manabe D, Thompson TC, Kumon H, Suicide gene therapy with adenoviral delivery of HSV-tK gene for patients with local recurrence of prostate cancer after hormonal therapy. *Mol Ther,* 2007; 15: 834–840.
316. Rainov NG, A phase III clinical evaluation of herpes simplex virus type 1 thymidine kinase and ganciclovir gene therapy as an adjuvant to surgical resection and radiation in adults with previously untreated glioblastoma multiforme. *Hum Gene Ther,* 2000; 11: 2389–2401.
317. Fischer U, Steffens S, Frank S, Rainov NG, Schulze-Osthoff K, Kramm CM, Mechanisms of thymidine kinase/ganciclovir and cytosine deaminase/ 5-fluorocytosine suicide gene therapy-induced cell death in glioma cells. *Oncogene,* 2005; 24: 1231–1243.
318. Springer CJ, Niculescu-Duvaz I, Gene-directed enzyme prodrug therapy (GDEPT): choice of prodrugs. *Adv Drug Deliv Rev,* 1996; 22: 351–364.
319. Pandha HS, Martin LA, Rigg A, Hurst HC, Stamp GW, Sikora K, Lemoine NR, Genetic prodrug activation therapy for breast cancer: A phase I clinical trial of erbB-2-directed suicide gene expression. *J Clin Oncol* 1999; 17: 2180–2189.
320. Freytag SO, Stricker H, Pegg J, Paielli D, Pradhan DG, Peabody J, DePeralta-Venturina M, Xia X, Brown S, Lu M, Kim JH, Phase I study of replication-competent adenovirus-mediated double-suicide gene therapy in combination with conventional-dose three-dimensional conformal radiation therapy for the treatment of newly diagnosed, intermediate- to high-risk prostate cancer. *Cancer Res,* 2003; 63: 7497–7506.
321. Huang TT, Hlavaty J, Ostertag D, Espinoza FL, Martin B, Petznek H, Rodriguez-Aguirre M, Ibañez CE, Kasahara N, Gunzburg W, Gruber HE,

Pertschuk D, Jolly DJ, Robbins JM, Toca 511 gene transfer and 5-fluorocytosine in combination with temozolomide demonstrates synergistic therapeutic efficacy in a temozolomide-sensitive glioblastoma model. *Cancer Gene Ther,* 2013; 20: 544–551.

322. Wilson RC, Doudna JA, Molecular mechanisms of RNA interference. *Annu Rev Biophys* 2013; 42: 217–239.
323. Leng A, Yang J, Liu T, Cui J, Li X-H, Zhu Y, Xiong T, Chen Y, Nanoparticle-delivered VEGF-silencing cassette and suicide gene expression cassettes inhibit colon carcinoma growth *in vitro* and *in vivo*. *Tumour Biol,* 2011; 32: 1103–1111.
324. Liu T, Ye L, He Y, Chen X, Peng J, Zhang X, Yi H, Peng F, Leng A, Combination gene therapy using VEGF-shRNA and fusion suicide gene yCDglyTK inhibits gastric carcinoma growth. *Exp Mol Pathol,* 2011; 91: 745–752.
325. Veikkola T, Karkkainen M, Claesson-Welsh L, Alitalo K, Regulation of angiogenesis *via* vascular endothelial growth factor receptors. *Cancer Res,* 2000; 60: 203–212.
326. Hervé M,-A., Buteau-Lozano H, Vassy R, Bieche I, Velasco G, Pla M, Perret G, Mourah S, Perrot-Applanat M, Overexpression of vascular endothelial growth factor 189 in breast cancer cells leads to delayed tumor uptake with dilated intratumoral vessels. *Am J Pathol,* 2008; 172: 167–178.
327. Sarli V, Giannis A, Targeting the kinesin spindle protein: basic principles and clinical implications. *Clin Cancer Res,* 2008; 14: 7583–7587.
328. Tabernero J, Shapiro GI, LoRusso PM, Cervantes A, *et al.*, First-in-humans trial of an RNA interference therapeutic targeting VEGF and KSP in cancer patients with liver involvement. *Cancer Discov,* 2013; 3: 406–417.
329. Cattaneo R, Miest T, Shashkova E, V Barry MA, Reprogrammed viruses as cancer therapeutics: targeted, armed and shielded. *Nat Rev Microbiol,* 2008; 6: 529–540.
330. Dorer DE, Nettelbeck DM, Targeting cancer by transcriptional control in cancer gene therapy and viral oncolysis. *Adv Drug Deliv Rev,* 2009; 61: 554–571.
331. Naik S, Nace R, Federspiel MJ, Barber GN, Peng K-W, Russell SJ, Curative one-shot systemic virotherapy in murine myeloma. *Leukemia,* 2012; 26: 1870–1878.

332. Russell SJ, Peng K-W, Bell JC, Oncolytic virotherapy. *Nat Biotechnol,* 2012; 30: 658–670.
333. Senzer NN, Kaufman HL, Amatruda T, Nemunaitis M, Reid T, Daniels G, Gonzalez R, Glaspy J, Whitman E, Harrington K, Goldsweig H, Marshall T, Love C, Coffin R, Nemunaitis JJ, Phase II clinical trial of a granulocyte-macrophage colony-stimulating factor-encoding, second-generation oncolytic herpesvirus in patients with unresectable metastatic melanoma. *J Clin Oncol,* 2009; 27: 5763–5771.
334. Kaufman HL, Bines SD, OPTIM trial: a phase III trial of an oncolytic herpes virus encoding GM-CSF for unresectable stage III or IV melanoma. *Future Oncol,* 2010; 6: 941–949.
335. Xia Z,-J., Chang J-H, Zhang L, Jiang W-Q, Guan Z-Z, Liu J-W, Zhang Y, Hu X-H, Wu G-H, Wang H-Q, Chen Z-C, Chen J-C, Zhou Q-H, Lu J-W, Fan Q-X, Huang J-J, Zheng X, phase III randomized clinical trial of intratumoral injection of E1B gene-deleted adenovirus (H101) combined with cisplatin-based chemotherapy in treating squamous cell cancer of head and neck or esophagus. *Ai Zheng,* 2004; 23: 1666–70.
336. Garber K, China approves world's first oncolytic virus therapy for cancer treatment. *J Natl Cancer Inst,* 2006; 98: 298–300.

Chapter 3

NANOCARRIER BASED PULMONARY GENE DELIVERY FOR LUNG CANCER: THERAPEUTIC AND IMAGING APPROACHES

Nishant S. Gandhi[*,†], Micah Glasgow[†],*
and Mahavir B. Chougule[*,†,‡,§]

*Department of Pharmaceutical Sciences,
Department of Pharmaceutics and Drug Delivery,
Research of Institute of Pharmaceutical Sciences,
School of Pharmacy, University of Mississippi,
University, MS 38677, USA
[†]Department of Pharmaceutical Sciences,
The Daniel K Inouye College of pharmacy,
University of Hawaii at Hilo, 200 W. Kawili Street,
Hilo, Hawaii 96720, USA
[‡]National Center for Natural Products Research,
Research Institute of Pharmaceutical Sciences,
University of Mississippi, University, MS 38677, USA.
[§]chougule@olemiss.edu & mahavirchougule@gmail.com*

Abstract

Lung cancer constitutes a group of aggressive malignant tumors and causes the most cancer deaths than any other form of cancer. Lung cancer treatment is extremely complicated due to various adverse effects associated with conventional chemotherapy as it targets both normal and tumor cells with equal intensity. Development of cancer cell resistance, low accumulation of therapeutic drug in the lungs, and severe adverse treatment side effects represent main obstacles to efficient chemotherapy of lung cancer. Regional chemotherapy was first used for lung cancer 30 years ago. Local drug delivery to the lungs offers several substantial advantages such as maximum therapeutic efficacy with minimal adverse side effects. Nucleic acid (siRNA/miRNA/DNA) based strategies appear to be an exciting new approach for the treatment of lung cancer. However, due to their instability in biological system, efficient delivery of nucleic acids to the target site remains a challenging task in the treatment if lung cancer. Nanotechnology has been an emerging field in drug-delivery research and has been explored on a wider scale to deliver the nucleic acids. Efforts invested in biomedical research are devoted to improve the sensitivity and accuracy of cancer diagnosis by incorporating early detection methods. This will help to improve therapeutic efficacy of the cancer treatment methods by enabling faster detection.

3.1 Introduction

Lung cancer is responsible for around 23% of the total cancer deaths around the world.[1] Cytoreductive surgery is not very effective for this type of cancer because of its size and distribution pattern, thus chemotherapy and/or radiation are the treatment of choice.[2] The patient survival rate has been as low as 18% in most countries.[2] Immense efforts have been invested in the field of pulmonary medicine with a view of finding suitable therapeutic treatment for various lung diseases, including cancer.[3] Post-transcriptional gene silencing approaches using various nucleic acid-based molecules such as antisense oligodeoxynucleotides (AS-ODNs), ribozymes, and small interfering RNAs (siRNAs) have drawn much attention because of their therapeutic potential for treating various diseases, including cancer.[4] Different approaches have been reported in the literature to deliver genes, proteins, etc., to lungs *via* inhalation; however most of them are invasive and not feasible. The two most widely used non-invasive carriers used are viral and non-viral.[5] Viral vectors have relatively high transfection efficiency compared to the non-viral vectors.[6] However, the use of viral vectors has been limited due to their potential of causing immune response and difficulties in manufacturing large scale batches.[7] On the other hand the non-viral vectors are devoid of such disadvantages along with affording several advantages such as low cost, easy manipulation, etc.[8] Aerosol-mediated gene delivery is effective and represents a noninvasive alternative for the targeting of genes to the lungs.[9] This chapter discusses the delivery of various aerosolized nucleic acid-based molecules formulated as nanocarriers.

3.2 Lung cancer, therapies, gene delivery

3.2.1 *Lung cancer*

Lung cancer is the leading cause of cancer deaths worldwide in both men and women and according to WHO accounted for around 1.59 million deaths in 2012.[10] There are two major types of lung cancer, non-small cell lung cancer (NSCLC) and small cell lung cancer (SCLC).[11] NSCLC accounts for about 85% of lung cancers and is sub-divided into adenocarcinoma, squamous cell carcinoma and large cell carcinoma. The SCLC accounts for the remaining 15% of the lung cancers, it grows rapidly and spreads to other parts of the body earlier than non-small cell lung cancer.[12]

Only about 14% of the patients suffering from lung cancer survive five years and majority dies within 2 years of diagnosis.[13] Due to limited efficiency of traditional treatment there are poor survival rates among patients with lung cancer.[14] Since, lung cancer is widely spread among various lung tissues, surgery is not a feasible option and chemotherapy and/or radiotherapy is the only treatment of choice.[12] Chemotherapy is the most widely used and important treatment for lung cancer and combination drug chemotherapy is considered as the mainstay for the treatment. The first line cytotoxic combination chemotherapy is with platinum (cisplatin or carboplatinum) and paclitaxel, gemcitabine, docetaxel, vinorelbine, irinotecan and pemetrexed.[15] Epidermal growth factor receptor (EGFR) tyrosine kinase inhibitors are also the first line of treatment options available for patients with EGFR mutations.[15] The drugs are usually administered through oral and parenteral drug administration.[16] However, conventional chemotherapeutic agents affect both cancerous and normal cells as they are distributed non-specifically and thus resulting in suboptimal treatment due to excessive toxicities and limited dose achievable within tumors.[17] Thus to avoid adverse side effects of conventional chemotherapeutic agents and to enable higher therapeutic efficiency, aerosol delivery of nucleic acids *via* inhalation route can be used for the treatment of lung cancer. Lung cancer, as with any other form of cancer, is known to metastasize from the primary site, therefore a form of administration, such as aerosol delivery *via* inhalation route, might have the additional benefit of local disease control by blocking metastasis.[18]

3.2.2 *Inhalation delivery*

A local drug delivery, including inhalation delivery of the genes to lungs, offers several substantial advantages, including simplified administration protocol and enhanced gene delivery to anatomical target site with maximum therapeutic efficacy but minimal adverse side effect.[19,20] It also allows drugs to avoid first pass metabolism and thus prevent them from degradation. However, various challenges are associated with the formulation and local delivery of therapeutic genes to the lungs *via* oral inhalation (OI).[21] The most important extracellular barriers to the delivery of therapeutic genes to the respiratory tract are mucus layer in the airways and the lung surfactant in the alveolar region.[22–25] The size of delivered particles is also an important barrier. For efficient deposition at the lower respiratory tract the optimal particle size is found to be between 1 and 5 μm.[26] The deposition increases further as the particle size decreases towards the nanoscale range due to the increasing diffusional mobility.[27] Nanoparticles that are less than 100 nm appear to settle effectively at the alveolar region with a fractional deposition of around 50%.[28]

For polymeric carriers, the physico-chemical properties of the polymer-gene complexes (polyplexes), including their size, surface charge, and therapeutic gene loading greatly affects the transfection efficiency.[29,30] In addition, the uptake, cell internalization and release of therapeutic genes inside the cell is greatly influenced by those parameters.[31] Aerosol delivery distributes material uniformly and represents a noninvasive alternative for targeting genes to the lung. Various groups have demonstrated that viral as well as nonviral carrier-mediated gene delivery *via* inhalation route may provide a means of treatment for a wide range of pulmonary disorders including cancer and offer numerous advantages over invasive modes of delivery.[32–34] Therapeutics agents can be delivered to lungs using various devices such as nebulizers, pressured metered dose inhalers (pMDIs) and dry powder inhalers (DPIs).

3.2.2.1 *Nebulizers*

Nebulizers are the oldest form of aerosol generation and are commonly used for over 25 years having a consistently same basic design and performance over this time.[35] They are used to convert liquids into fine droplets and are useful to deliver large volumes of drug solutions and suspensions that can be inhaled into the lower respiratory tract.[36] Nebulizers are the choice of devices to deliver drugs which are available only in solution form. There are mainly two types of nebulizers *viz.* jet nebulizer and ultrasonic nebulizer. The jet nebulizers use the compressed air to aerosolize the drug solutions. Jet nebulizers are connected by tubing to a compressor, that causes compressed air or oxygen to flow at high velocity through a drug solution to turn it into an aerosol, which is then inhaled by the patient (Figure 3.1). The biggest drawbacks associated with jet nebulizers are the noise and its heavy weight.[37]

Ultrasonic nebulizers use a piezoelectric transducer to produce ultrasonic waves that pass through the solution and aerosolize it at the surface of the solution (Figure 3.2). An ultrasonic nebulizer creates particle sizes of about 1–6 μm MMAD with a volume output of about 1–6 mL/minute, depending on the manufacturer of the device. Compared to jet nebulizers, they are silent and are light in weight.[39]

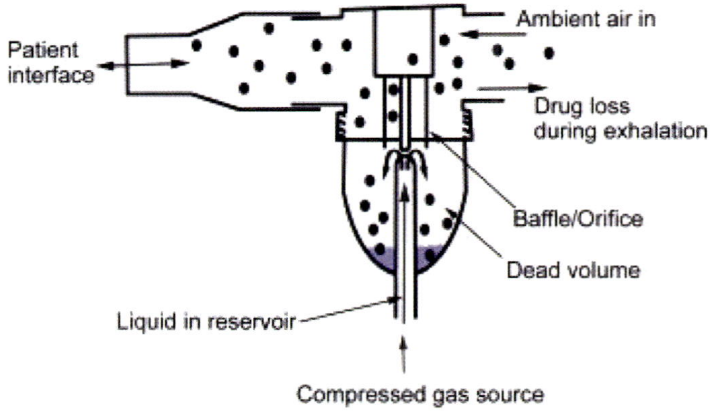

Figure 3.1 Functioning of a pneumatic jet nebulizer. Reproduced from Matah S.[38]

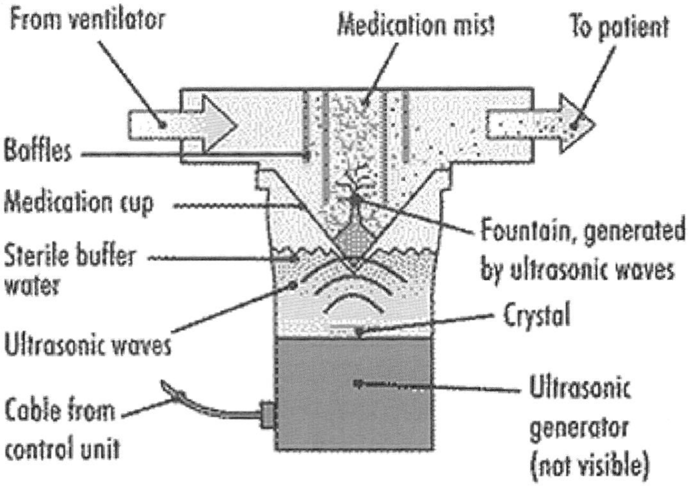

Figure 3.2 The ultrasonic nebulizer. Reproduced from Matah S.[38]

3.2.2.2 *Pressurized meter dose inhalers*

The pressurized metered dose inhaler (MDI) is the most widely used device for aerosol therapy with over 70 million patients in the world use one alone or with a spacer fitted.[40] They are designed in such a way so as to deliver a number of drug doses in sequence. The pMDIs usually consist of an active substance dissolved or suspended in the propellant (choloro-fluorocarbon or hydrofluoroalkanes) (Figure 3.3). There are many factors that influences the size of the particle delivered by MDI, including: (1) physical and chemical properties of the propellant and of the other additives; (2) pressure inside the canister; (3) metering valve and delivery outlet design volume; (4) drug used and its concentration delivered; (5) mouthpiece and delivery outlet cleanness.[41] The advantages of MDIs are their portability and low cost. The inert conditions created by the propellant vapor, together with the hermetically sealed container, protects drug from oxidative degradation and microbiological contamination.[42]

3.2.2.3 *Dry powder inhalers (DPI)*

Powder inhalers are versatile delivery systems that may require some degree of expertise to operate.[44] They dispense a fixed metered quantity of

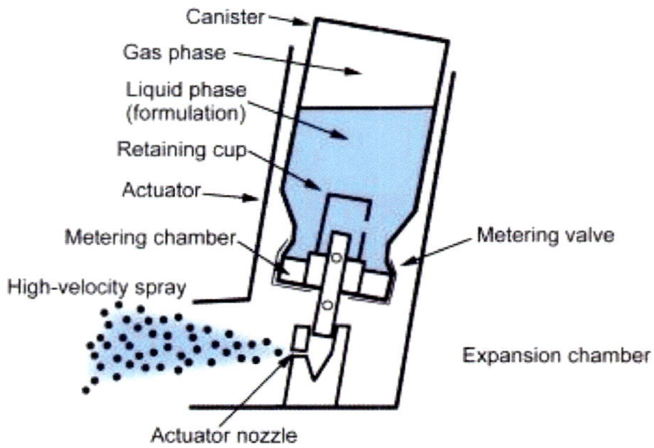

Figure 3.3 Pressurized metered dose inhaler. Reproduced from Ref. 43.

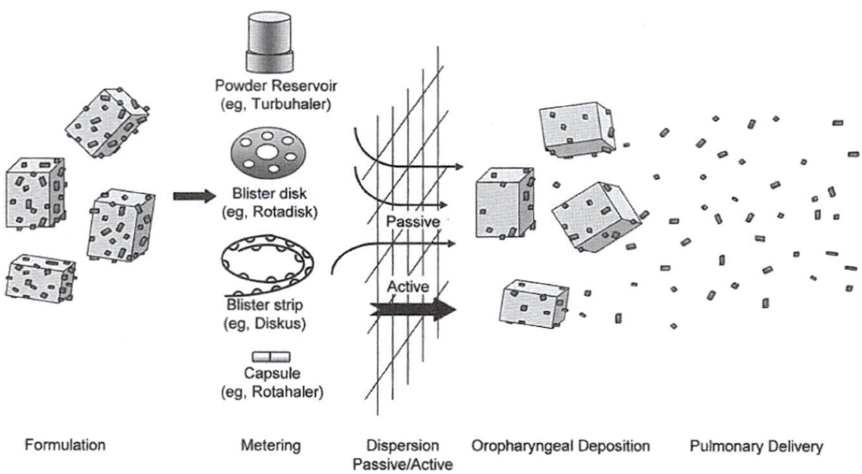

Figure 3.4 Principle of dry powder inhaler design. The formulation typically consists of micronized drug blended with larger carrier particles, dispensed by a metering system. An active or passive dispersion system entrains the particles into the patient's airways, where drug particles separate from the carrier particles and are carried into the lung. Reproduced from Telko MJ *et al.*[45]

dry powder in a stream of air through the device by the patient's own inspiration. DPIs are bolus drug delivery devices that contain solid drug, suspended or dissolved in a dry powder mix that is fluidized when the patient inhales (Figure 3.4).[45] There are two types of devices used to deliver drug to the patient, *viz.* active and passive. In case of active devices, the external energy in the form of compressed air, vibrational, piezoelectric is used to produce the aerosol cloud rather than just relying on the patient's inspiratory efforts. While in case of passive devices, the drug substance from DPI formulation is delivered to lung with the patient's inhalation. DPI offers several advantages over metered dose inhalers including environmental sustainability, stability of formulation and allowing little or no patient coordination.

3.2.3 *Therapeutic agents*

For the past 20–25 years, nucleic acids including DNA and antisense oligonucleotide have been widely studied by scientists all over the world. siRNA is the most popular candidate being studied in RNAi therapy, amongst the different types of nucleic acids that are involved in the RNAi.[46] From the delivery perspective, plasmid DNA encoding shRNA is at a disadvantage compared to siRNA. In order for DNA to be properly expressed, prior to expression it must enter the nucleus where transcription takes place, while siRNA's activity can take place in the cytoplasm. Nuclear entry is an extremely inefficient process and this is indeed considered to be one of the biggest barriers to the success of gene therapy.[28] The biggest hurdle scientific community facing is the delivery of these nucleic acids, and the same problem applies to RNAi therapy. These hydrophilic macromolecules along with being highly susceptible to nuclease degradation, are negatively charged and susceptible to cross the biological membrane on their own and thus unable to reach the site of action.[28,47] Thus in order to protect the therapeutic nucleic acids from nuclease mediated degradation, enhancing cellular uptake and facilitating release at the site of action, a delivery vector is required.[28]

3.2.3.1 *Small interfering RNA*

siRNAs are chemically synthesized duplex molecules which are 19–23 nucleotide (nt) long having 2-nt-3′ overhang comparable to that of endogenous miRNAs.[48] This allows them to be easily recognized by the enzyme known as DICER and undergo further processing.[49] siRNA induces gene specific cleavage through its complementary pairing with messenger mRNA and resulting in latter's degradation.

The potential uses of siRNA to treat cancer have been demonstrated by a number of studies.[50,51] Tumor cells usually exhibit a dysregulation of the cell cycle[52] and this protective mechanism mediated by oncogenes, tumor suppressive genes or alternative genes which can be mutated or non-mutated and are a part of list which can be targeted effectively by siRNA for the treatment of cancer.[53]

Initial targets of siRNAs in cancer pathogenesis include oncogenes p53, K-ras, BCl-2 which are primarily involved in cell cycle regulation. The initial studies were followed by several different investigations that targeted genes playing very crucial role in various different aspects related to cancer cell growth such as genes involved in therapy resistance, viral genes associated with cancer or other cancer related genes such as telomerase or Cathepsin B.[54–58] siRNA shows a tremendous potential to treat cancer, however the general obstacle is to deliver them to the target tissue at an effective dose.[53] Just like siRNA, miRNA also represents an innovative class of RNAi's that had gained considerable attention in recent years.

3.2.3.2 *Micro RNA*

miRNA are 20–24 nucleotide long, double stranded, endogenous RNA molecules which play an important role in regulating gene expression.[59–61] They are involved in mediating the post-transcriptional silencing of genes.[62] miRNA is capable of controlling the expression of more than one mRNA, a distinguishing feature from siRNA.[63]

One of the very first and well-documented oncogenic miRNAs functioning as oncogenes includes miR 17–92 cluster. The cluster targets many genes involved in the apoptotic pathway and it is more likely that the anti-apoptotic effect is mainly due to combination of suppression of many target mRNA's. It has been found that there are a wide variety of targets for miRNA members of the cluster.[64,65] Some of the member of miR 17–92 targets tumor suppressive genes PTEN and RB2, while silencing of miR 17 and miR 20a linked to B cell lymphogenesis.[66]

miRNA's functioning as tumor suppressor genes such as Let-7 family of miRNA's is a highly conserved group that are positioned at fragile sites allied with human cancers. For instance, in case of lung cancer where Let-7 negatively regulates the expression of multiple oncogenes including RAS and MYC as well as other cell cycle progression genes and acts as a tumor suppressive gene. There are other classes of miRNA's such as tumor suppressive miRNA's (ex: miR-34), which is induced by DNA damage or oncogenic stress and leads to apoptosis. In this list, miR-15a and miR-16-1 also represent key miRNA's, which negatively regulate BCL2.[67,68] The recent move in RNAi based cancer therapy is to engage them in combinatorial therapy to attain complete cancer cure. siRNA' and miRNA's have been used in combination with other chemotherapeutic agents for the treatment of cancer.[69] siRNA and miRNA have also been used as single agent for the treatment of lung cancer.[70] However, the combination of nucleic acids and chemotherapeutic is found to be more effective compared to the single agent.[71]

3.2.4 *Imaging techniques for lung tumors*

Critical factors for a favorable prognosis of cancer are its early detection and treatment. Nanotechnology is an emerging field that shows great promise in cancer diagnosis.[72] Much effort has been invested in biomedical research devoted to improving the sensitivity and accuracy of cancer diagnosis by incorporating early detection methods helping to improve efficacy and cancer treatment methods.[73] Using *in vivo* techniques, a non-invasive diagnosis of patients can be achieved. These non-invasive techniques are robust tools for assisting with the diagnosis of abnormalities in the body and monitoring any biological situations at the site of action. Several non-invasive imaging techniques include the following: computed tomography (CT), magnetic resonance imaging (MRI), nuclear imaging of positron emission tomography (PET), single-photon emission computed tomography (SPECT), and optical (or fluorescence) imaging.[74] *In vivo* images of anatomy and physiology, for both humans and animals, are acquired using these non-invasive techniques and assist scientists and clinicians in the diagnosis of certain cancers.[75–80]

Theranostics, the combination of therapeutic and imaging methods on a single level, have reached new possibilities thanks to recent advances in nanotechnology.[81,82] These nanotechnology advancements incorporate imaging probes for the enhancement of image resolution and specificity towards target tissues. Nanoparticles conjugated to active targeting ligands usually make-up these imaging probes.[83,84] Nanoparticles having no active target ligand coupled to its surface will be passive targeting, mediated by the enhanced permeability and retention effect (EPR).[85,86]

3.3 Nanocarrier systems

Nanotechnology has been an emerging field in drug-delivery research since 1970's when the cyanoacrylate based nanoparticles attached with anticancer drugs were first described.[87] Nanotechnology holds enormous promise for improving health, as well as eminence of life of people throughout the world. In recent years, there is noteworthy attention in developing biodegradable nanoparticles as a drug/gene delivery system.[88] These colloidal particles (nanoparticles) for drug delivery are referred to as drug-incorporated macromolecular soft matrixes or inorganic solid, within a size of 10–1000 nm, which is similar to that of biological macromolecules, such as proteins and DNA.[89,90] There are considerable advantages with using nanoparticles as drug delivery carriers, which includes:

- Small size
- Allowing ease of fabrication to obtain proper size, shape, surface charge and stability[91]
- Propensity to penetrate through smaller capillaries/pores
- Efficiency in resulting drug accumulation at site of action
- Modifiable surface: Surface can be modified to conjugate to a drug, protein or peptide.

Engineered nanomaterials such as the quantum dots, dendrimers, carbon nanotubes and fullerene exemplify nanomaterials that have diameters less than 100 nanometers which can be compared with the sizes of several living structures. Nanotech industries are reaching new horizons with the progression of these innovative classes of nanomaterials, extending applications to advance the excellence of healthcare commodities with superior patient compliance. Nanoparticles can be made from a variety of polymers (polymeric nanoparticles, micelles), lipids (liposomes), and organometallic compounds. Some of the classes are briefly described in the following section.

3.3.1 *Polymer-based drug carriers*

The polymers used to make drug conjugates can be divided into two groups of natural and synthetic polymers.

3.3.1.1 *Natural polymers*

Various natural polymers such as albumin,[92–94] gelatin[95,96] and chitosan[97,98] have been used to deliver genes, protein and drugs to cancer cells in the form of nanoparticles. Natural polymers have unique, intrinsic properties that render them appealing to serve as a gene carrier.[99] They are generally non-toxic, even in large doses, mucoadhesive (chitosan), biocompatible and biodegradable.[100]

3.3.1.2 *Synthetic polymers*

Various synthetic polymers poly(lactic-co-glycolic) acid (PLGA)[101-103], N-(2-hydroxypropyl)-methacrylamide copolymer (HPMA),[104] polyethylene glycol (PEG),[105,106] and poly-L-glutamic acid (PGA)[107] were used to be formulated as nanocarriers and deliver the drug load to the cancer cells. PGA was the first biodegradable polymer to be used for conjugate synthesis.[108] Several chemotherapeutic drugs have been conjugated with PGA and tested *in vitro and in vivo* and showed encouraging results. The synthetic polymer provides the ease of modification which allows targeting of drugs to the tumor cells and release of the contents inside the cell.

3.3.2 *Lipid-based drug carriers*

3.3.2.1 *Liposomes*

Liposomes are closed colloidal structures, composed of lipid bilayers having a spherical shape in which the outer hydrophobic lipid bilayer surround the inner hydrophilic aqueous space.[109] Various liposomal formulations have been used to deliver various drugs including tacrolimus through inhalation[110–112] enabling controlled and prolonged retention at the site of action. Recently a new class of liposomes called "immunoliposomes" has been investigated, which selectively delivers the drug to the site of action and may possibly be called as the next generation liposomes.[113]

3.3.2.2 *Solid lipid nanoparticles (SLN)*

The SLN usually consist of a solid lipid core which is made up of triglycerides, glyceride mixtures or waxes that are solid at room temperature and body temperature.[114] They are usually composed of biocompatible and biodegradable components and do not require the use of organic solvents for their formulation. Some of the advantages of using SLNs include control and or target drug release, excellent biocompatibility, improved stability of pharmaceuticals and enhanced drug content.

3.3.2.3 *Nanostructured lipid carriers (NLC)*

NLC are solid and liquid phase lipid based colloidal particles ranging in size from 100–500 nm.[115] They are generally solid at temperatures over 40^{0}C. Unlike SLN, the matrix of NLC is made up of amorphous structure and usually has higher drug loading with less pronounced drug release.[109] Since the recently developed SLN and NLC consist of all natural components, they are potentially attractive and marketable choices. Also since the solvent use can be avoided, they are well-positioned for large scale manufacturing.

3.3.3 *Carbon nanotubes*

Carbon nanotubes are essentially carbon cylinders composed of benzene rings.[116] They are usually insoluble in all the solvents, however chemical modification rendering them water soluble have made possible to use them in various nanoformulation preparations.[116] They have gained widespread attention owing to their distinct characteristics, such as high surface area, enhanced cellular uptake and the possibility to be easily conjugated with many therapeutics.[117]

3.4 Characterization of nanocarriers

Nanoparticles are characterized for percent drug entrapment, particle size distribution, zeta potential, percent entrapment efficiency, morphology, oxidative index, differential scanning calorimetry, gel electrophoresis, transfections, TEM and SEM.[118–122] Apart from these routine studies performed for the evaluation of nucleic acid loaded nanoparticles, several following studies are also performed for the characterization of aerosolized delivery of siRNA to the lung.

3.4.1 *Evaluation of nanocarriers for pulmonary delivery*

The aerodynamic particle size directly influences the regional deposition of the inhaled products in the lungs and respiratory tract and is thus a key parameter.[123] Its measurement is thus very critical during the product development cycle and for quality control. Although various sizing techniques are available, cascade impaction is the method of choice and specified for regulatory approval.[124]

The most important advantage offered by Anderson cascade impaction which is based on the principle of inertia, is that they allow for the determination of an assay for the FPF and other size fractions.[125] They also allow direct measurement if aerodynamic diameter which is a parameter most closely correlated with the particle behavior during inhalation (Figure 3.5).[126] Cascade impactors provide the required greatest degree of resolution for inhalation products in the particle size range of 0.5 to 5 microns. The Andersen cascade impactor gives five stages with cut-off diameters in the required range, over most operating conditions.[127]

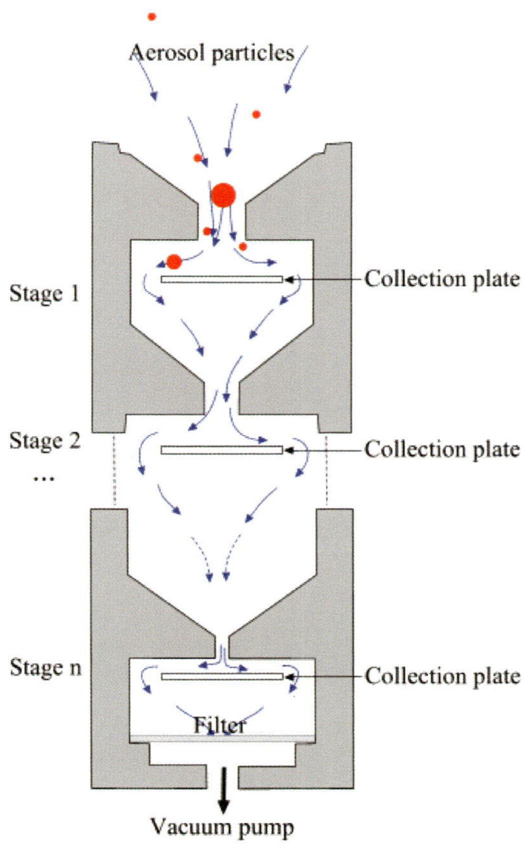

Figure 3.5 Schematic of flow in a cascade impactor. Reproduced from Ref. 128.

3.4.1.1 *Fine particle fraction*

The delivery of therapeutic aerosols to the lung by nebulization is a common method of treatment for various respiratory illnesses. However, the amount of drug delivered to the lung by different nebulizer models can vary dramatically. It is important to distinguish the proportion of mass from the inhaler that is contained in the "fine" particles that manage to penetrate beyond the oropharynx from the total mass emitted by the inhaler.[129,130] The fraction of a dose that because of its size, will deposit in the lung is often referred to as the fine particle or "respirable" fraction (FPF).[131] The fine particle fraction upper limit is defined in various ways but it mostly lies between the range of 4 – 6 microns, the optimal size for central airway deposition.

3.4.1.2 *Emitted dose*

Air flow characteristics through a passive dry powder inhaler (DPI) affect the amount of drug delivered from the device, which is an important consideration in patients where inhalation profiles are highly variable.[132] The Delivered or Emitted Dose is the total amount of drug emitted from the inhaler device and hence available to the user. Its uniformity is a Critical Quality Attribute (CQA) in determining the safety, quality and efficacy of an orally inhaled and drug product.[133]

3.4.1.3 *Mass median aerodynamic diameter (MMAD)*

The aerodynamic size of the particle is a very important parameter for all the inhalation based products. This directly influences the deposition of drugs in the lungs and the respiratory tract.[134] The MMAD is the diameter at which 50% of the particles by mass are larger and 50% are smaller. Therapeutic/diagnostic aerosols have an MMAD ranging from 0.5–5 μm depending on the intended site of deposition.[135] Aerosols of a particular MMAD may be generated but their size and site of impaction are influenced by the density, temperature and humidity of the carrier gas, toxicity of the aerosol itself, the ventilatory pattern and airway geometry. Thus the measurement of MMAD is critical during the product development cycle and for quality control. Cascade impaction is the method of choice for measuring the MMAD and is the method specified for regulatory approval.[125]

3.4.2 *Preclinical models and evaluation*

In the design and study of new regimens of cancer treatment, preclinical tumor models are considered as fundamental components. However, the models that have been used in the past and being used today have considerable shortcomings.[136] Various murine models have been developed over the years to study human cancer.[137] The models have been used to investigate various factors involved in malignant transformation, invasion and metastasis.

They have also been used to examine response to therapy. Some of the following commonly used important tumor models have been described in brief.

3.4.2.1 *Xenograft models*

The human tumor xenograft is one of the most widely used models.[137] The development of human tumor xenograft models was a big step in moving towards more clinically relevant tumor models.[138,139] In this model, human tumor cells are transplanted, either under the skin or into the organ type in which the tumor originated or into immunocompromised mice that do not reject human cells. There are various advantages of using human xenografts, the most important being that the malignant cells are human derived and the models produced are quite reproducible.[140] Also a wide variety of tumor cell lines are available to transplant into animals. However certain disadvantages are also associated with these models including high cost, the majority of animals being rodents and immunodeficient hosts.

3.4.2.2 *Orthotopic models*

Orthotopic transplantation of tumors in mice has been around for many years.[141] In this type of tumor models, the tumor cells are implanted into the organ of origin (orthotopically) which allows the tumor to develop within the site comparable to human tumoral development which metastasizes within a few weeks.[142] The advantage with this type of tumor model is that the tumor growth has some variety and is in the tissue of origin and the hosts are readily available. The main disadvantage is that the surgeries to introduce tumor cells are often very complex and the models are more costly.

3.4.2.3 *Transgenic tumor models*

The transgenic tumor models have been used for a number of years and are growing in sophistication.[140,143,144] In these type of models, the tumor arises in genetically engineered hosts. The main advantage with using transgenic tumor model is that the tumor arises naturally in the host and can be frequently followed over a long course of time. However there are certain disadvantages associated with using this models such as the high cost of maintenance, tumors arise at variable stages in animal-to-animal and nodule-to-nodule and the tumors are difficult to follow.[145,146]

There are various ways of administering the aerosolized nanoparticles to the animal *viz.* nose only exposure, intratracheal instillation, intratracheal nebulization, etc.[147–150]. The most conventional technique, nose only exposure involves exposing the animals to the aerosol atmosphere.[151] The drawback associated with this technique is that most of the particles administered will be deposited along the nasal passage and/or oropharyngeal activity. The majority of the particles will not reach the intended site of action due to being swallowed instead of entering the lungs.[152]

Another approach is to deliver the test solution *via* intratracheal instillation, in which it is directly administered into the tracheobronchial regions and the lungs.[153] The distribution is not uniform, with the insoluble suspension failing to reach the alveoli, localizing only in the medium-sized bronchi.[154] Intratracheal nebulization technique has been developed recently to improve the distribution of inhaled particles within the lungs. The technique delivers a small finite amount of aerosol to the bronchoalveolar regions of the lungs. The intratracheal nebulization techniques helped deliver small amounts of fluid into the tracheobronchial regions and alveoli without causing barotrauma.[152]

3.4.3 *In vivo characterization*

In vivo drug delivery targeting a specific region is a primary requirement of the effective therapy. Various routes such as intravenous or an airway route of administration have been explored in recent past for the delivery of gene to the respiratory tissues.[155] Non-localized drug delivery has low success rate and poor clinical outcome because of a sub-therapeutic concentration of therapeutic agent reaching the desired site of action. Also, most of the chemotherapeutic drugs administered by conventional route (i.e. oral or parenteral) also exert the cytotoxic effect on normal healthy cells inhibiting their growth and thereby resulting in toxic adverse effects.[156] Thus various pharmacokinetic parameters have to be taken into account in order to successfully deliver the gene to the pulmonary tract for the treatment of lung cancer and to avoid the possible adverse effects. *In vivo* characterization using different available methods are thus helpful in predicting the pulmonary deposition of nucleic acids in the lungs. *In vivo* characterization will also help in predicting systemic distribution of DNA administered by intravenous administration.

In this study the collision nebulizer connected to four port, nose-only exposure is used for the inhalation delivery of the developed nanoformulation labeled with Cy5.5 (amine-reactive dye used for molecular imaging) into nude mice with lung cancer.[157] The organ distribution of the nanoformulation can be studied *via* an imaging system after a certain time period. This study is helpful in demonstrating the selective organ (lung) distribution of the nano formulation. It also helps in determining the distribution of the nanoformulation in other organs.

In vivo studies with NLC demonstrated that the total amount of non-targeted NLC-TAX-siRNA retained in the lungs was around 83% after inhalation, was 3.5 times higher compared to the intravenously injected NLC-TAX-siRNA.[157] Similarly in another study, data demonstrated that the amount of LHRH-PEG-siRNA-DOX-MSN retained in the lungs (73%) was 14.6 times higher compared to i.v. injection of same formulation.

3.5 Examples of siRNA and miRNA based delivery systems

3.5.1 *Nanocarrier based inhalation delivery of siRNA*

3.5.1.1 *Folate–chitosan-graft-polyethylenimine based system*

H.-L. Jiang *et al.* developed a Folate–chitosan-graft-polyethylenimine (FC-g-PEI) copolymer by imine reaction between periodate-oxidized folate–chitosan (FC) and low molecular weight polyethylenimine (PEI).[159] The resulting copolymer exhibited higher gene transfer ability, cancer cell specificity and reduced toxicity compared to PEI25K. The copolymer was used to deliver the Akt1 shRNA and it exhibited good shRNA binding ability and was able to successfully protect the shRNA from degradation by DNAse1 compared to the naked shRNA. FC-g-PEI successfully delivered a reporter gene through a folate receptor-mediated mechanism *in vitro*. The potential of FC-g-PEI as an aerosol Akt1 shRNA delivery carrier in the lung cancer gene therapy was also investigated and it was found that FC-g-PEI significantly delivered Akt1 shRNA into the lungs when transfected with Akt1 shRNA. The tumor size decreased significantly along with suppressed tumor progression. The results clearly demonstrate that FC-g-PEI functions efficiently as a shRNA delivery carrier *in vivo.*

3.5.1.2 *Mesoporous silica nanoparticles*

Taratula *et al.* developed a modified mesoporous silica nanoparticles to deliver an anticancer drug (doxorubicin or cisplatin) and siRNA targeting to BCL2 or MRP1 mRNA.[158] The modified nanoparticles also consisted of a targeting moiety LHRH peptide for targeted delivery purpose. The modified silica nanoparticles was successfully able to deliver the drug payload into the cell cytoplasm. The two siRNA were able to suppress the BCL2 and MRP1 mRNA along with the anticancer activity of the delivered drugs resulting in cell death. The instillation containing a Collision nebulizer connected to four-port, nose only exposure chamber was used for inhalation delivery of the DDS into nude mice bearing lung cancer. The data showed that inhalation delivery substantially enhanced the lung exposure to nanoparticles compared to the i.v. administration. The total amount of MSN retained in the lung was approximately 15 times higher than i.v. administration. The distribution of the MSN to other organs (liver, kidney and spleen) was much lesser compared to i.v. administration indicating more localized distribution.

3.5.1.3 *Liposome based drug delivery system*

Taratula *et al.* used small "neutral" polyethylne glycol (PEG)ylated liposomes to load DOX and uncharged P-ethoxy ASO while positively charged DOTAP liposomes were used to complexate siRNA.[160] The study showed that the liposomes (small empty and large loaded) were successfully able to deliver their payload to the cellular cytoplasm and even nuclei. Liposomes of different sizes (100 nm) were used and were successfully able to provide efficient delivery of, siRNA into the cell cytoplasm. The data showed that the intratracheal administration of liposomes lead to sustained higher concentrations in lungs for up to three days. The liposome concentration in the tissue and their payload in lungs were significantly increased for prolonged period of time following intratracheal administration. The concentration of liposomes in non-target tissues was also significantly reduced.

3.5.2 *Nanocarrier based delivery of DNA*

3.5.2.1 *Polyethyleneimine based drug delivery*

A study by Gautam *et al.* demonstrated that the aerosol delivery of PEI-p53 complexes inhibits lung metastases. The inhibition of lung metastases was part in due to increase in the TSP-1 and down-regulation of VEGF.[161] The lung tissue and serum levels of the antiangiogenic factor TSP-1 was significantly increased in the normal and tumor cells following aerosol delivery of the PEI-p53 complexes. The delivery of the PEI-p53 complexes to the mice also lead to down-regulation of the angiogenic factor VEGF in the lung tissue and serum of mice.[161] The data demonstrated that aerosol delivery of PEI-P53 lead to inhibition of tumor lung metastases through suppression of angiogenesis. The aerosol delivery of PEI-P53 complexes to the mice lead to a significant increase in the antiangiogenic factor TSP-1 in the lung tissue and serum in both normal and tumor bearing mice. On transfection of PEI-p53 complexes in the lung tissue, significant down-regulation of the angiogenic factor VEGF was observed. There was a significant down-regulation of VEGF in serum of mice. *In situ* staining data revealed that p53 induced apoptosis of the peripheral cells of the tumor foci while not affecting much of the cells within the tumor foci. This might have been due to the fact that PEI-P53 DNA aerosol complexes could only transfect peripheral cells which and could not diffuse to the cells within the tumor foci.

Wilms tumor 1 (WT) 1 has been found to be over expressed in many types of neoplasias and thus can be considered to be an excellent potential target to treat cancer.[162,163] Zamora-Avila *et al.* showed that WT1 is expressed in B16F10 murine melanoma cell line and that *in vitro* RNAi silencing of WT1 expression results in the, induction of apoptosis, caspase-3 activation, inhibition of cell proliferation and poly-(ADP-ribose) polymerase cleavage *in vitro.*[164]

In a recent study by the same group, the recombinant plasmids (WT1-1 and WT1-2) were complexed with PEI resulting in formation of

PEI-WT-1-1 and PEI-WT1-2. The *in vivo* aerosol delivery of the two complexes using nose exposure system resulted in reduction of metastatic burden of tumor in lings of B16F10 mice. The number and size of tumor was significantly reduced following administration of PEI–WT1-1 RNAi or PEI–WT1-2 RNAi complexes. The WT1 RNA levels were also significantly reduced. They also observed a reduction in the number and size of blood vessels in tumor samples from treated mice and also the presence of cells in apoptosis induced by WT1 RNAi. WT1-1 RNAi treatment significantly increased the mean survival time of tumor-bearing mice in comparison with the mice treated with WT1-2 RNAi or the untreated mice.

3.5.2.2 *Spermine based drug delivery carrier*

In another attempt to deliver the DNA to the lung tumor, Jiang *et al.* developed a glycerol propoxylate triacrylate (GPT) and spermine (SPE) based drug delivery carrie.[165] The water soluble GPT-SPE polymer was synthesized using Michael addition reaction, which showed excellent DNA binding ability. The DNA binding ability determines the amount of DNA a polymer can bind, the higher the DNA binding ability the more a polymer can bind and deliver DNA. At weigh ratio 5, the binding ability of GPT-SPE was much stronger than naked DNA. The spherical shaped GPT-SPE/DNA complexes were successfully able to protect the DNA from nuclease degradation compared to naked DNA. The GPT-SPE/DNA complexes also induced high intracellular uptake and the transfection study showed that the transfection efficiency of the GPT-SPE/PGL3 complexes in A549 cells were significantly high compared to the naked PGL3. The toxicity study revealed that there was significantly higher cell viability of GPT-SPE treated cells compared with PEI-25K indicating that the polymer was itself nontoxic. *In vivo* study on K-rasLA1 mice using aerosol administered GPT-SPE/Akt1 shRNA complexes showed there was a significant anticancer effect without any toxicity. This indicates that the polymer GPT-SPE could be a good carrier for delivering nucleic acids to the lungs *via* aerosol delivery.

3.6 Imaging techniques for nanoparticles

3.6.1 *Computed tomography (CT)*

Patients with suspected lung cancer undergo computed tomography (CT) scans which is accepted as the standard procedure in the assessment of this disease.[72] This method utilizes X-rays to help visualize the body area in a series of slices. One of the main advantages of this procedure is its high spatial resolution found within each produced image. Contrasting agents are needed during the assessment. Iodine-based compounds are the most popular CT contrast agents. These contrast agents assist with blocking X-rays which allow for better contrast and enhancement of the area of interest.[72] One drawback is both malignant and benign lung nodules often show similar radiographic features during a CT scan.[166] With no distinction between both tumor types, the malignancy, if any, cannot always be properly identified and treated. In the National Lung Cancer Screening Trial (NLCST), CT screening of high-risk patients reduced lung cancer-specific mortality.[167] The United States Preventative Services Task Force (USPSTF) recommends routine CT lung cancer screening for high risk patients. The USPSTF classifies high risk patients as adults aged 55 to 80 years who have a 30 pack per year smoking history and currently smoke or have quit within the past 15 years. This screening process should be discontinued once a person has not smoked for 15 years or develops a health problem that substantially limits life expectancy or the ability or willingness to have curative lung surgery. However, this screening process yields small pulmonary nodules (~1 cm) which are not well-characterized by currently available imaging techniques (CT, MRI or PET).[168] The NLCST has proven that majority of these nodules are not clinically significant. Information already provided by CT scans may then extend into nodule characterization not only for characterizing known tumors but also for differentiating detected malignant tumors from benign nodules.[169]

Dual-energy (DE) CT imaging is an advanced technique that has recently risen to the forefront of CT technology[170] (Figure 3.6). The ability of contrast agents to adsorb X-rays relies heavily on their atomic weight and the energy of the incident X-ray. Materials having two distinct atomic weights can then be differentiated from one another based on their unique

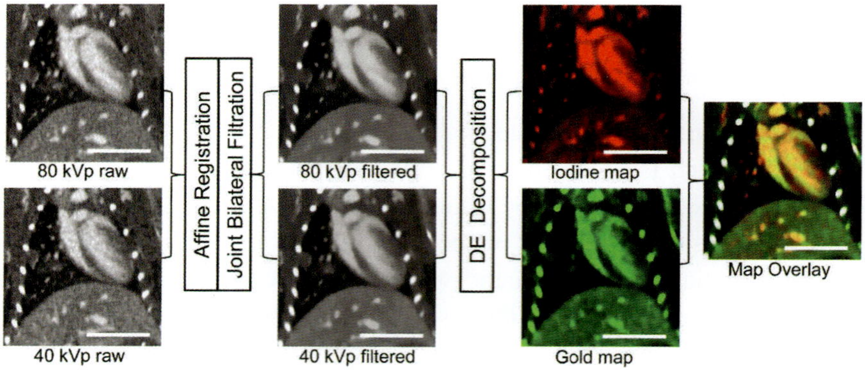

Figure 3.6 Dual-energy decomposition process diagram. Raw 40 kVp and 80 kVp data-sets were acquired and then they underwent affine registration and joint bilateral filtration to produce filtered data sets. Dual-energy decomposition was performed on these filtered images, which resulted in two independent images (maps) representing the iodine and gold concentration in each voxel. After obtaining the two maps, the images were overlaid and the bones were segmented out (colored white) to form the final concentration map overlay. These images were taken from CT scan 3, in which both iodine and gold were present in the blood stream. The scale bar represents 1 cm in all images. The 40 and 80 kVp images are windowed from –300 to 1200 HU, the iodine maps are windowed from 0.25 to 15 mg/mL, and gold maps are windowed from 0.25 to 6 mg/mL.[169]

attenuation coefficients at two different X-ray energies.[170] DE-CT allows for the quantification and visualization of multiple contrast agents with a single scan.[169] Significant progress has been made in clinical cancer imaging using this newly adopted DE-CT technique. Iodine enhancement in clinical DE-CT scans of lung tumors show a correlation with the SUV_{max} of ^{18}FDG- PET.[171] This correlation suggests that CT scans may potentially extend beyond anatomical imaging to functional and molecular imaging as well. DE-CT techniques show great promise for use in clinic cancer imaging, however, it has not been adopted largely by the preclinical domain due to several challenges associated with temporal resolution, spatial resolution and noise levels in micro-CT.[169] Aston *et al.* have been able to address some of these challenges. Their lab group demonstrated that DE micro-CT using nanoparticles as contrast agents can play a vital role in preclinical imaging for lung,[172] cardiac,[173] and cancer applications.[174,175]

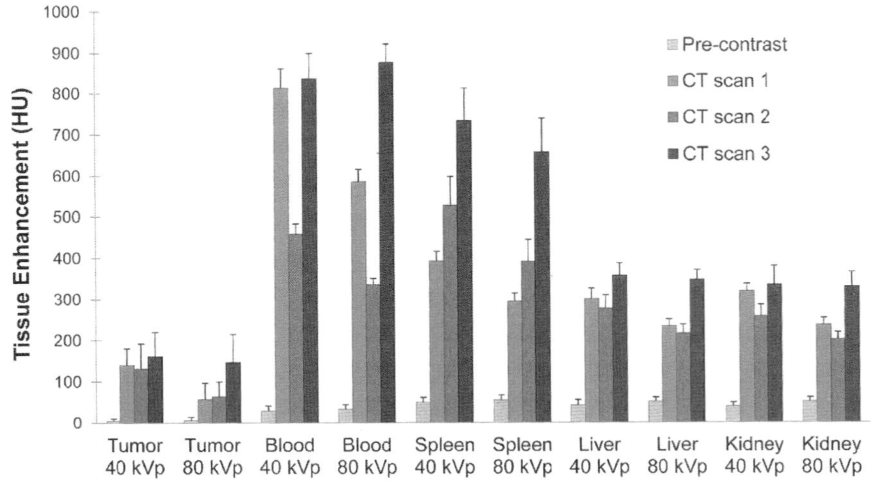

Figure 3.7 Average tissue enhancement. Tissue enhancement (average of five mice) is shown at 40 and 80 kVp for tumor, blood, spleen, liver, and kidneys in each of three CT scans. Pre-contrast values are the average tissue enhancement of the mice from pre-experiment monitoring scans.[169]

Conventional contrast agents are rapidly cleared from the bloodstream which does not allow for effective imaging of the tumor. For this reason contrast agent utilizing nanotechnology are essential to preclinical CT studies (Figure 3.7).[169] Preclinical animal models have high cardiac output, as compared to humans, allowing them to renally clear low molecular weight clinical contrast agents from their blood within seconds (*versus* 2–3 hours blood half-life in humans[176]). CT contrast agents utilizing nanotechnology tend to have a longer half-life (> 2 hours) and has been shown to accumulate in tumors themselves due to the enhanced permeation and retention (EPR) effect.[177] The range of use for these agents in preclinical CT imaging has been well established.[174,178–184,169]

3.6.2 *Magnetic resonance imaging (MRI)*

High spatial resolution and contrast are two of the main benefits for using magnetic resonance imaging (MRI) as a diagnostic tool within the clinic setting.[74] MRI takes advantage of the magnetism property of protons that align when passed through a very large magnetic field. The water molecules found within our body tissue are the major source of these protons. The protons spin can also be flipped through "resonance frequency" which is a radiofrequency generated at a particular frequency. Protons will flip back to their original state once the electromagnetic field has been turned off which also generates a radiofrequency signal. This process is called "relaxation." The relaxation process is measured by the receiver coils which is then turned into a visual image using computer algorithm.[185] Contrasting agents, T1 or T2, are used to alter these different relaxation time. T1 MRI contrasting agents help to amplify the positive signal on T1-weighted images. One particular example of a T1 contrasting agent is gadolinium chelators. On the other hand, T2 contrasting agents help to decrease the signal intensity on T2-weighted images. Superparamagnetic iron oxide nanoparticle (SPION) based agents may be used as a T2 contrasting agent.[74] SPION based agents contain a superparamagnetic iron core which is why they are utilized as T2 MRI contrast agents. They provide a high degree of sensitivity and detection when used for MRI with both the iron core and polymer components of SPION remaining biodegradable and biocompatible for clinical application.[186]

Biodistribution, stability and metabolism are influenced by a SPION's surface charge, size and surface chemistry.[187] With proper physiochemical properties, such as surface coatings that are biocompatible and biodegradable, a SPION agent can avoid both an immune response and serum protein absorption. Aggregation and the colloidal nature of these nanoparticles are controlled by its surface charge. Thus, neutral charges are more preferred as compared to a charged surface. These charges may also affect the way plasma protein binding during *in vivo* distribution which in-turn affects the nanoparticle biodistribution and clearance from the body.[74,188]

Magnetite (Fe_3O_4) or maghemite (Fe_2O_3) make up the typical SPION agent. The nanoparticles also maintain their aqueous solubility with appropriate surface coatings. These agents can be synthesized through various methods; co-precipitation, thermal decomposition and microemulsion (Figure 3.8 and 3.9).[74] Newly discovered methods include; microfluidic synthesis, microwave assisted synthesis and the hydrolysis

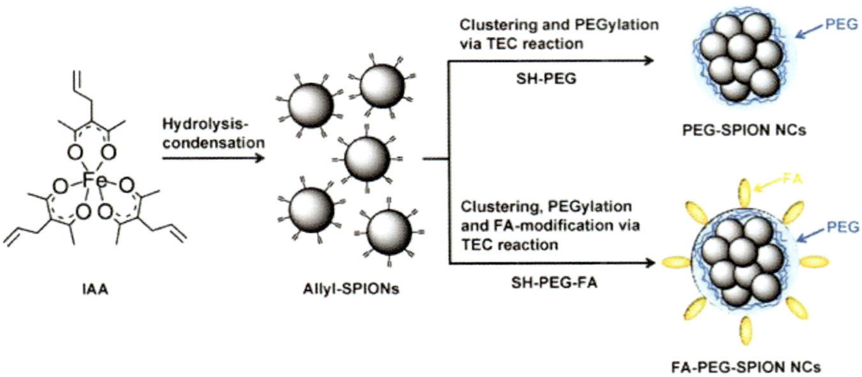

Figure 3.8 Synthesis scheme for PEG-SPION NCs and FA-PEG-SPION NCs.[197]

method.[189–191] The most popular method for the synthesis of SPION is thermal decomposition which yields nanoparticles with high crystallinity, high magnetization, and a narrow size distribution. The product nanoparticles are also hydrophobic which provide the basis for appropriate coating applications when used for biomedical operation.[192] A hydrophobic polymer coating will allow for high loading of hydrophobic drugs on the surface of the SPION, also if the drug has functional groups it can also be bound covalently linked to the SPIONs.[193] Another most commonly accepted process for SPION synthesis is the co-precipitation method. A concentrated base is added to a divalent or trivalent ferrous salt solution. This method yields nanoparticles in the size range from 2–15 nm with a narrow polydispersity index.[194,195] These particles will aggregate and grow within the solution thereby decreasing the surface free energy.[196] The last technique mentioned, the micro-emulsion technique has one main advantage of controlling the SPION size by acting as a nano-reactor. A water-in-oil or oil-in-water emulsions are used to formulate the SPION using this technique. The final result yields SPIONs with a controlled size distribution between 2 nm and 12 nm.[74,194]

Iron oxide nanoparticles must first undergo toxicity studies to better understand their mechanisms by which they can cause toxicity before they can be applied in a clinical setting. These nanoparticles will be converted to an elemental iron species which then merges to our body reserves or

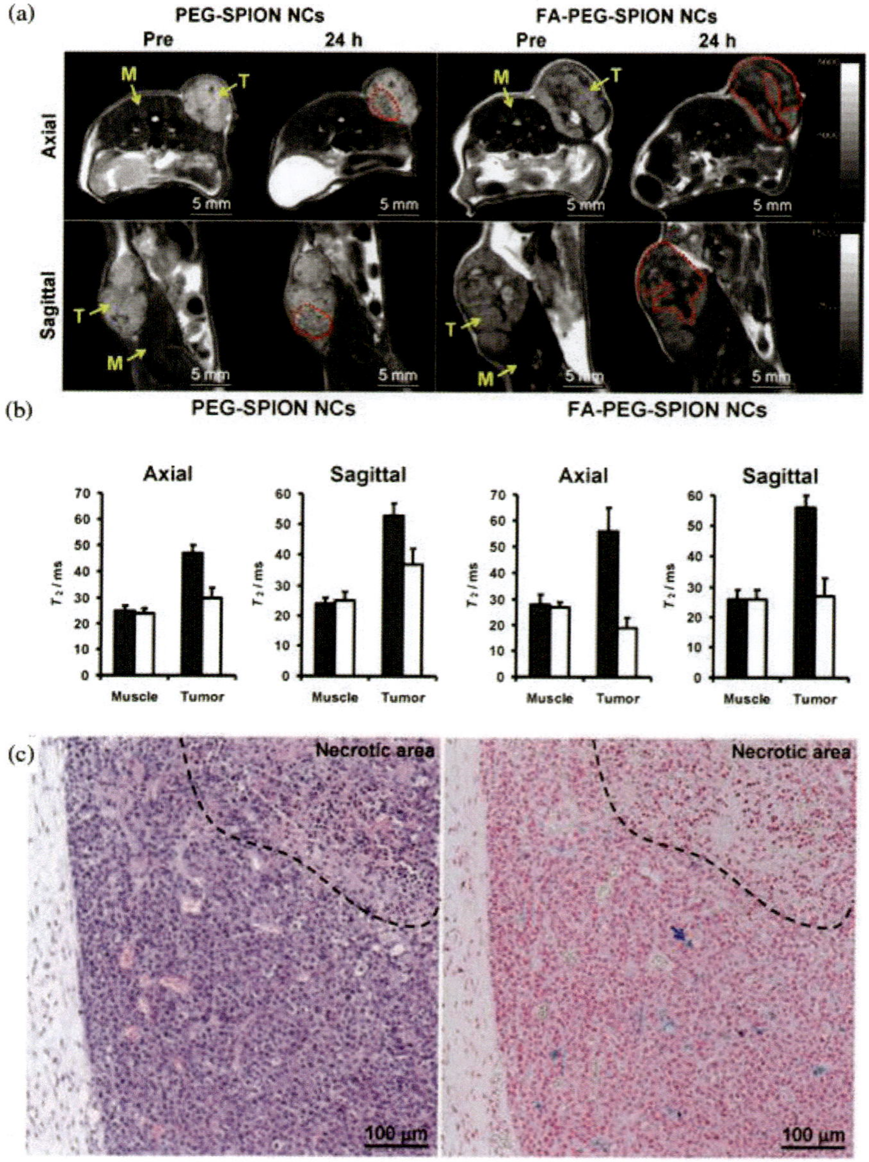

Figure 3.9 *In vivo* T2 axial and sagittal maps of mice before and 24 hours after intravenous injection of PEG-SPION NCs and FA-PEG-SPION NCs: M and T indicate muscle and tumor, respectively (a). Changes in T2 values in the tumors and muscles of mice before and 24h after intravenous injection of PEG-SPION NCs and FA-PEG-SPION NCs

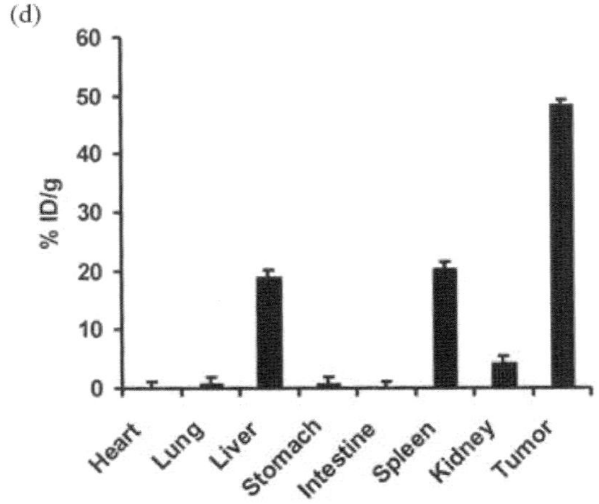

Figure 3.9 (*Continued*) (b). Histological observation of tumors 24 hours after intravenous injection of FA-PEG-SPION NCs visualized using hematoxylin and eosin staining (left) and prussian blue nuclear fast red staining (right): iron is stained blue as indicated by the arrow (c). Biodistributions of FA-PEG-SPION NCs at 24 hours postinjection (d). The iron amounts in tissue samples were assessed by nitroso-PSAP (n = 3).[197]

utilized for haemoglobin formation.[197] The biosafety of iron oxide nanoparticle consisting of Fe_3O_4 or $MNFe_3O_4$ have been established up to 200 μg/mL concentration.[197] Concentrations above the established biosafety level have been shown to affect normal cellular function, through formation of excess ROS taking place, leading to apoptosis and cell death.[74,189]

3.6.3 *Nuclear imaging of positron emission tomography (PET), single-photon emission computed tomography (SPECT)*

PET and SPECT are in the category of nuclear imaging. These techniques involve detecting radioisotopes which emit one or two gamma rays or positrons.[74] Both techniques are excellent imaging processes due to their specificity, sensitivity, and fast detection time.[198] However, both techniques lack the ability to display good spatial resolution.[74] PET uses isotopes such as ^{68}Ga, ^{76}Br, ^{94m}Tc, ^{11}C, ^{13}N, ^{15}O, ^{18}F, and ^{64}Cu. SPECT uses gamma-emitting heavy radioisotopes such as ^{123}I, ^{99m}Tc, and ^{133}Xe.[198]

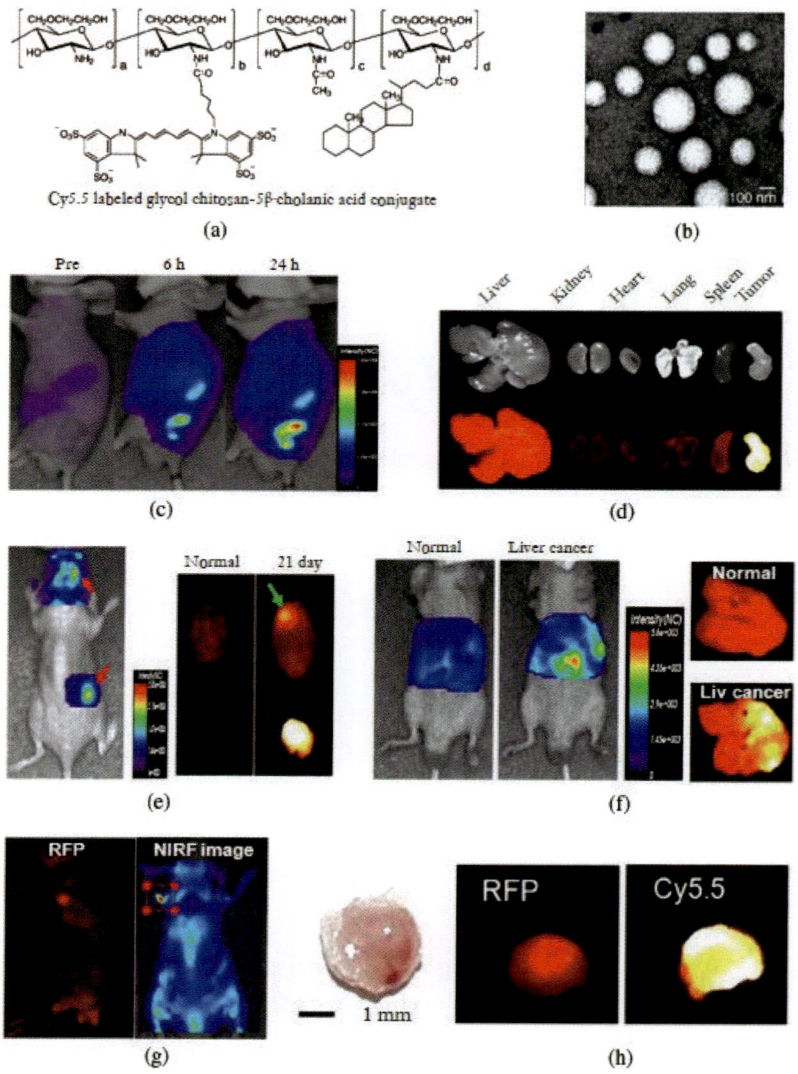

Figure 3.10 Self-assembled chitosan-based nanoprobes for cancer diagnosis. Chemical structure of Cy5.5-labeled glycol chitosan-5β cholanic acid conjugates (Cy5.5-GC) (a). TEM image of Cy5.5-GC nanoprobes (b). *In vivo* fluorescence imaging of tumor targeting characteristics of Cy5.5-GC nanoprobes in nude mice bearing SCC7 tumors (c). *Ex vivo* fluorescence imaging of organ distribution of Cy5.5-GC nanoprobes after one day post-injection (d). Brain tumor imaging of Cy5.5-GC nanoprobes (e). Liver tumor imaging of Cy5.5-GC nanoprobes (f). Secondary metastatic tumor imaging of Cy5.5-GC nanoprobes (g and h). *In vivo* RFP and NIRF imaging of secondary metastatic tumor with Cy5.5-GC nanoprobes (g). Photo, *ex vivo* RFP and NIRF images of secondary metastatic tumor (h).[202]

3.6.4 *Optical imaging techniques*

Optical imaging techniques (fluorescence, NIRF, bioluminescence and reflective imaging) rely on different physical parameters of light to interact with tissue. These techniques rely on fluorescence, absorption, reflectance, or bioluminescence as a source of contrast during imaging[74] (Figure 3.10). Majority of optical imaging consists of NIRF (near-infrared fluorescence), reflectance imaging, and bioluminescence imaging. Although optical imaging has only recently emerged, it is rapidly being incorporated in various clinical and research areas. The major advantages optimal imaging offers over those previously mentioned are; contrasting agents are non-toxic, inexpensive to conduct, versatile, and sensitive.[199–201] An inorganic or organic contrasting agent is used during optical imaging that fluoresces at various excitation wavelengths. Variations in endogenous fluorescence of neoplastic tissues were first used for diagnosis of cancer during early optical imaging procedures. Being able to distinguish the signal component from background fluorescence made this type of diagnosis extremely difficult to analyze. The development of exogenous contrast agents helped to eliminate this challenge.[200] Optical contrast agents composed of nanoparticles, such as "quantum dots," were synthesized giving superior imaging properties as compared to their organic counterparts. There are major drawbacks to using optical imaging techniques such as inability to quantify the image, and the auto-fluorescence of normal tissue against the fluorescence of the contrast agent casing severe impairment of the image quality.[74,86]

3.7 Conclusion and future directions

Over the years, extensive research has been performed to investigate an effective gene carrier system that can successfully deliver the gene at the site of action within tumor cells. However, very few systems have been able to make it to clinical trials due to the lack of safe and effective gene delivery vectors. Nanoparticulate systems have a great potential to encapsulate and deliver genes to the appropriate site of action i.e. tumor cells. The nanoparticles delivery *via* inhalation therapy holds great promise for treating pulmonary diseases including lung cancer, with reduced drug dosage and dosing frequency thereby minimizing associated systemic side-effects. Pulmonary delivery of the nanoparticulate system is a non-invasive approach that can be tailored for local sustained delivery of therapeutic agent(s) to treat lung cancer. The nanoparticles system can be formulated in various dosage forms such as pMDI, dry powder inhalation and nebulization. The main challenge facing the delivery of the nanoparticulate system to lungs is to understand the fate of particles and their interactions with biological systems. The nanoparticle delivery to healthy normal tissues results in side-effects, while rapid clearance might reduce the efficacy of therapeutic agents. The nanoparticle research for pulmonary applications should inspire the development of inhalable nanoparticle based system including surface modification to target-specific sites. With further knowledge of particle fate and toxicology after drug delivery nanoparticles materials as well as particle properties, can be tailored and optimized for effective delivery. Several non-invasive imaging techniques such as computed tomography, magnetic resonance imaging, nuclear imaging of positron emission tomography, single-photon emission computed tomography, and optical or fluorescence imaging either alone or in combination can be used for diagnosis of lung tumors. The advanced nanoparticle based imaging of lung tumor will help detect tumors early and initiate an effective treatment. A nanoparticle based theranostics approach enables diagnosis and treatment of lung cancer. A rationale based selection of gene delivery systems and designing specific targeting to tumor cells will improve effectiveness in the treatment of lung cancer thereby minimize the adverse side effects. The systematic evaluation of nanoparticulate systems under *in vitro* and *in vivo* systems will help to achieve the effective delivery gene at lung tumors while overcoming the associated limitations and side effects. The advantages of the nanoparticles are maximized in the cutting edge technologies such as multimodal and multifunctional probes. Future work on multimodal and multifunctional imaging probes must focus on correcting their current flaws by introducing new biocompatible polymers Finally, multimodal and multifunctional techniques may enhance clinical theranostics in the near future.

Acknowledgment

The Chougule group would like to acknowledge the 2013 George F. Straub Trust and Robert C. Perry Fund of the Hawai'i Community Foundation,

Honolulu, HI 96813, USA, for research support on lung cancer. Dr. Gandhi acknowledges the Daniel K. Inouye College of Pharmacy, University of Hawaii at Hilo, Hilo, HI 96720, USA, for providing graduate assistantship. The authors acknowledge the support of the National Institute of General Medical Science of the National Institutes of Health under award number SC3GM109873 and the Department of Pharmaceutics and Drug Delivery, School of Pharmacy, University of Mississippi, University, MS, USA for providing start-up financial support to Dr. Chougule's lab.

Declaration of Interest

The authors report no conflicts of interest.

References

1. Zarogoulidis P, Chatzaki E, Porpodis K, Domvri K, Hohenforst-Schmidt W, Goldberg EP, *et al.*, Inhaled chemotherapy in lung cancer: future concept of nanomedicine. *Int J Nanomed,* 2012; 7: 1551–1572.
2. Yang P. Epidemiology of lung cancer prognosis: quantity and quality of life. *Meth Mol Biol,* 2009; 471: 469–486.
3. Xu CX, Jere D, Jin H, Chang SH, Chung YS, Shin JY, *et al.,* Poly(ester amine)-mediated, aerosol-delivered Akt1 small interfering RNA suppresses lung tumorigenesis. *Am J Resp Crit Care,* 2008; 178: 60–73.
4. Hong SH, Kim JE, Kim YK, Minai-Tehrani A, Shin JY, Kang B, *et al.,* Suppression of lung cancer progression by biocompatible glycerol triacrylate-spermine-mediated delivery of shAkt1. *Int J Nanomed,* 2012; 7: 2293–2306.
5. Zarogouldis P, Karamanos NK, Porpodis K, Domvri K, Huang H, Hohenforst-Schimdt W, *et al.,* Vectors for inhaled gene therapy in lung cancer. Application for nano oncology and safety of bionanotechnology. *Int J Mol Sci,* 2012; 13: 10828–10862.
6. Blo M, Micklem DR, Lorens JB. Enhanced gene expression from retroviral vectors. *BMC Biotech,* 2008; 8: 19.
7. Romano G. Current development of nonviral-mediated gene transfer. *Drug News Perspect,* 2007; 20: 227–231.
8. Parker AL, Newman C, Briggs S, Seymour L, Sheridan PJ. Nonviral gene delivery: techniques and implications for molecular medicine. *Exp Rev Mol Med,* 2003; 5: 1–15.

9. Merlin JL, Dolivet G, Dubessy C, Festor E, Parache RM, Verneuil L, *et al.*, Improvement of nonviral p53 gene transfer in human carcinoma cells using glucosylated polyethylenimine derivatives. *Cancer Gene Ther*, 2001; 8: 203–210.
10. World Health Organization — Cancer. World Health Organization 2014.
11. Lung cancer 101
12. General Information About Small Cell Lung Cancer. National Cancer Institute 2014.
13. Molina JR, Yang P, Cassivi SD, Schild SE, Adjei AA. Non-small cell lung cancer: epidemiology, risk factors, treatment, and survivorship. *Mayo Clinic Proc,* 2008; 83: 584–594.
14. Youlden DR, Cramb SM, Baade PD. The International Epidemiology of Lung Cancer: geographical distribution and secular trends. *J Oncol*, 2008; 3: 819–831.
15. Stage IV NSCLC Treatment. 2014.
16. Adjei AA, Marks RS, Bonner JA. Current guidelines for the management of small cell lung cancer. *Mayo Clinic Proc,* 1999; 74: 809–816.
17. Cho K, Wang X, Nie S, Chen ZG, Shin DM. Therapeutic nanoparticles for drug delivery in cancer. *Clin Cancer Res*, 2008; 14: 1310–1316.
18. Yu KN, Minai-Tehrani A, Chang SH, Hwang SK, Hong SH, Kim JE, *et al.*, Aerosol delivery of small hairpin osteopontin blocks pulmonary metastasis of breast cancer in mice. *PloS One*, 2010; 5: e15623.
19. Torchilin VP. Drug targeting. *Eur J Pharm Sci,* 2000; 11, Suppl 2: S81–S91.
20. Luo Y, Zhai X, Ma C, Sun P, Fu Z, Liu W, *et al.*, An inhalable beta(2)-adrenoceptor ligand-directed guanidinylated chitosan carrier for targeted delivery of siRNA to lung. *J Control Release*, 2012; 162: 28–36.
21. Roy I, Vij N. Nanodelivery in airway diseases: challenges and therapeutic applications. *Nanomed-Nanotechnol,* 2010; 6: 237–244.
22. Gill DR, Davies LA, Pringle IA, Hyde SC. The development of gene therapy for diseases of the lung. *Cell Mol Life Sci*, 2004; 61: 355–368.
23. Sanders N, Rudolph C, Braeckmans K, De Smedt SC, Demeester J. Extracellular barriers in respiratory gene therapy. *Adv Drug Deliv Rev,* 2009; 61: 115–127.
24. Gunther M, Lipka J, Malek A, Gutsch D, Kreyling W, Aigner A. Polyethylenimines for RNAi-mediated gene targeting in vivo and siRNA delivery to the lung. *Eur J Pharm Biopharm,* 2011; 77: 438–449.

25. Conti DS, Bharatwaj B, Brewer D, da Rocha SR. Propellant-based inhalers for the non-invasive delivery of genes *via* oral inhalation. *J Control Release,* 2012; 157: 406–417.
26. Agu RU, Ugwoke MI, Armand M, Kinget R, Verbeke N. The lung as a route for systemic delivery of therapeutic proteins and peptides. *Respir Res,* 2001; 2: 198–209.
27. Yang W, Peters JI, Williams RO, 3rd. Inhaled nanoparticle — A recent review. *Int J Pharm,* 2008; 356: 239–247.
28. Lam JK, Liang W, Chan HK. Pulmonary delivery of therapeutic siRNA. *Adv Drug Deliv Rev,* 2012; 64: 1–15.
29. Eliyahu H, Barenholz Y, Domb A. Polymers for DNA delivery. *Molecules,* 2005; 10: 34–64.
30. Tiera MJ, Winnik FO, Fernandes JC. Synthetic and natural polycations for gene therapy: state of the art and new perspectives. *Curr Gene Ther,* 2006; 6: 59–71.
31. Ferrari S, Geddes DM, Alton EW. Barriers to and new approaches for gene therapy and gene delivery in cystic fibrosis. *Adv Drug Deliv Rev,* 2002; 54: 1373–1393.
32. Kim HW, Park IK, Cho CS, Lee KH, Beck GR, Jr., Colburn NH, *et al.,* Aerosol delivery of glucosylated polyethylenimine/phosphatase and tensin homologue deleted on chromosome 10 complex suppresses Akt downstream pathways in the lung of K-ras null mice. *Cancer Res,* 2004; 64: 7971–7976.
33. Hwang SK, Lim HT, Minai-Tehrani A, Lee ES, Park J, Park SB, *et al.,* Repeated aerosol delivery of carboxyl-terminal modulator protein suppresses tumor in the lungs of K-rasLA1 mice. *Am J Resp Crit Care Med,* 2009; 179: 1131–1140.
34. Hwang SK, Kwon JT, Park SJ, Chang SH, Lee ES, Chung YS, *et al.,* Lentivirus-mediated carboxyl-terminal modulator protein gene transfection via aerosol in lungs of K-ras null mice. *Gene Ther*, 2007; 14: 1721–1730.
35. Hess DR. Nebulizers: principles and performance. *Resp Care*, 2000; 45: 609–622.
36. Taylor KMG, McCallion ONM. Ultrasonic nebulisers for pulmonary drug delivery. *Int J Pharm*, 1997; 153: 93–104.
37. Harvey CJ, O'Doherty MJ, Page CJ, Thomas SH, Nunan TO, Treacher DF. Comparison of jet and ultrasonic nebulizer pulmonary aerosol deposition during mechanical ventilation. *Eur Resp J*, 1997; 10: 905–909.

38. Matah S. Aerosol therapy through nebulizer, 2008.
39. Le Brun PPH, de Boer AH, Frijlink HW, Heijerman HGM. A review of the technical aspects of drug nebulization. *Pharm World Sci*, 2000; 22: 75–81.
40. Terzano C. Pressurized metered dose inhalers and add-on devices. *Pulm Pharmacol Ther*, 2001; 14: 351–366.
41. Kim CS, Trujillo D, Sackner MA. Size aspects of metered-dose inhaler aerosols. *Am Rev Resp Dis*, 1985; 132: 137–142.
42. Shoyele SA, Slowey A. Prospects of formulating proteins/peptides as aerosols for pulmonary drug delivery. *Int J Pharm*, 2006; 314: 1–8.
43. Management of respiratory disease. 2014.
44. Prime D, Atkins PJ, Slater A, Sumby B. Review of dry powder inhalers. *Adv Drug Deliv Rev*, 1997; 26: 51–58.
45. Telko MJ, Hickey AJ. Dry powder inhaler formulation. *Respir Care*, 2005; 50: 1209–1227.
46. Alvarez-Salas LM. Nucleic acids as therapeutic agents. *Curr Topics Med Chem*, 2008; 8: 1379–1404.
47. Gavrilov K, Saltzman WM. Therapeutic siRNA: principles, challenges, and strategies. *Yale J Biol Med*, 2012; 85: 187–200.
48. Khatri N, Misra A. Development of siRNA lipoplexes for intracellular delivery in lung cancer cells. *J Pharm Bioall Sci*, 2012; 4: S1–S3.
49. Daka A, Peer D. RNAi-based nanomedicines for targeted personalized therapy. *Adv Drug Deliv Rev*, 2012; 64: 1508–1521.
50. Howard KA, Rahbek UL, Liu X, Damgaard CK, Glud SZ, Andersen MO, *et al.*, RNA interference *in vitro* and *in vivo* using a novel chitosan/siRNA nanoparticle system. *Mol Ther*, 2006; 14: 476–484.
51. das Neves J, Sarmento B, Amiji MM, Bahia MF. Development and validation of a rapid reversed-phase HPLC method for the determination of the non-nucleoside reverse transcriptase inhibitor dapivirine from polymeric nanoparticles. *J Pharm Biomed Anal*, 2010; 167–172.
52. Davis TD, Morris KR, Huang H, Peck GE, Stowell JG, Eisenhauer BJ, *et al.*, *In situ* monitoring of wet granulation using online X-ray powder diffraction. *Pharm Res*, 2003; 20: 1851–1857.
53. Bhavsar MD, Amiji MM. Gastrointestinal distribution and *in vivo* gene transfection studies with nanoparticles-in-microsphere oral system (NiMOS). *J Control Release*, 2007; 339–348.

54. Kimura T, Jiang S, Nishizawa M, Yoshigai E, Hashimoto I, Nishikawa M, *et al.,* Stabilization of human interferon-alpha1 mRNA by its antisense RNA. *Cell Mol Life Sci*, 2013; 70: 1451–167.
55. Chan DP, Deleavey GF, Owen SC, Damha MJ, Shoichet MS. Click conjugated polymeric immuno-nanoparticles for targeted siRNA and antisense oligonucleotide delivery. *Biomaterials*, 2013; 34: 8408–8415.
56. Jang YL, Yun UJ, Lee MS, Kim MG, Son S, Lee K, *et al.,* Cell-penetrating peptide mimicking polymer-based combined delivery of paclitaxel and siRNA for enhanced tumor growth suppression. *Int J Pharm,* 2012; 434: 488–493.
57. Ho JJ, Robb GB, Tai SC, Turgeon PJ, Mawji IA, Man HS, *et al.,* Active stabilization of human endothelial nitric oxide synthase mRNA by hnRNP E1 protects against antisense RNA and microRNAs. *Mol Cell Biol*, 2013; 33: 2029–2046.
58. Yoshikawa M, Iki T, Tsutsui Y, Miyashita K, Poethig RS, Habu Y, *et al.,* 3' fragment of miR173-programmed RISC-cleaved RNA is protected from degradation in a complex with RISC and SGS3. *Proc Natl Acad Sci USA*, 2013; 110: 4117–4122.
59. Reischl D, Zimmer A. Drug delivery of siRNA therapeutics: potentials and limits of nanosystems. *Nanomed-Nanotechnol*, 2009; 5: 8–20.
60. Sall A, Liu Z, Zhang HM, Yuan J, Lim T, Su Y, *et al.,* MicroRNAs-based therapeutic strategy for virally induced diseases. *Curr Drug Discov Technol*, 2008; 5: 49–58.
61. Winter J, Jung S, Keller S, Gregory RI, Diederichs S. Many roads to maturity: microRNA biogenesis pathways and their regulation. *Nat Cell Biol*, 2009; 11: 228–234.
62. Davidson BL, McCray PB, Jr. Current prospects for RNA interference-based therapies. *Nat Rev Genet*, 2011; 12: 329–340.
63. Cai Y, Yu X, Hu S, Yu J. A brief review on the mechanisms of miRNA regulation. *Genom Proteom Bioinform*, 2009; 7: 147–154.
64. Almog J, Cohen Y, Azoury M, Hahn TR. Genipin — A novel fingerprint reagent with colorimetric and fluorogenic activity. *J Foren Sci*, 2004; 49: 255–257.
65. Amati L, Pepe M, Passeri ME, Mastronardi ML, Jirillo E, Covelli V. Toll-like receptor signaling mechanisms involved in dendritic cell activation:

potential therapeutic control of T cell polarization. *Curr Pharm Design*, 2006; 12: 4247–4254.
66. Zhang B, Pan X, Cobb GP, Anderson TA. microRNAs as oncogenes and tumor suppressors. *Dev Biol*, 2007; 302: 1–12.
67. Allen TM, Hansen C. Pharmacokinetics of stealth versus conventional liposomes — effect of dose. *Biochim Biophys Acta*, 1991; 1068: 133–141.
68. Alminana N, Polo D, Alsina MA, Reig F. Optimization study of doxorubicin liposomal preparations coated with laminin fragments. *Prep Biochem Biotechnol*, 2002; 32: 1–16.
69. Ganesh S, Iyer AK, Weiler J, Morrissey DV, Amiji MM. Combination of siRNA-directed gene silencing with cisplatin reverses drug resistance in human non-small cell lung cancer. *Mol Ther Nucl Acids*, 2013; 2: e110.
70. Li S-D, Chen Y-C, Hackett MJ, Huang L. Tumor-targeted delivery of siRNA by self-assembled nanoparticles. *Mol Ther*, 2007; 16: 163–169.
71. Lee J-M, Yoon T-J, Cho Y-S. Recent developments in nanoparticle-based siRNA delivery for cancer therapy. *BioMed Res Int*, 2013; 2013: 10.
72. Grodzinski P, Silver M, Molnar LK. Nanotechnology for cancer diagnostics: promises and challenges. *Exp Rev Mol Diag*, 2006; 6: 307–318.
73. Schneider DF, Chen H. New developments in the diagnosis and treatment of thyroid cancer. *CA: Cancer J Clin*, 2013; 63: 373–394.
74. Thomas R, Park IK, Jeong YY. Magnetic iron oxide nanoparticles for multimodal imaging and therapy of cancer. *Int J Mol Sci*, 2013; 14: 15910–15930.
75. Na HB, Song IC, Hyeon T. Inorganic nanoparticles for MRI contrast agents. *Adv Mat*, 2009; 21: 2133–2148.
76. Willmann JK, van Bruggen N, Dinkelborg LM, Gambhir SS. Molecular imaging in drug development. *Nat Rev Drug Discov*, 2008; 7: 591–607.
77. Kirui DK, Khalidov I, Wang Y, Batt CA. Targeted near-IR hybrid magnetic nanoparticles for *in vivo* cancer therapy and imaging. *Nanomed-Nanotechnol*, 2013; 9: 702–711.
78. Misri R, Meier D, Yung AC, Kozlowski P, Hafeli UO. Development and evaluation of a dual-modality (MRI/SPECT) molecular imaging bioprobe. *Nanomed-Nanotechnol*, 2012; 8: 1007–1016.
79. Torres Martin de Rosales R, Tavaré R, Glaria A, Varma G, Protti A, Blower PJ. 99mTc-Bisphosphonate-iron oxide nanoparticle conjugates for dual-modality biomedical imaging. *Bioconj Chem*, 2011; 22: 455–465.

80. Jennings LE, Long NJ. 'Two is better than one'--probes for dual-modality molecular imaging. *Chem Comm*, 2009: 3511–3524.
81. Yoo D, Lee J-H, Shin T-H, Cheon J. Theranostic magnetic nanoparticles. *Acc Chem Res*, 2011; 44: 863–874.
82. Rosen JE, Chan L, Shieh D-B, Gu FX. Iron oxide nanoparticles for targeted cancer imaging and diagnostics. *Nanomed-Nanotechnol*, 2012; 8: 275–290.
83. Gu FX, Karnik R, Wang AZ, Alexis F, Levy-Nissenbaum E, Hong S, *et al.*, Targeted nanoparticles for cancer therapy. *Nano Today*, 2007; 2: 14–21.
84. Yu MK, Park J, Jon S. Targeting strategies for multifunctional nanoparticles in cancer imaging and therapy. *Theranostics*, 2012; 2: 3–44.
85. Nurunnabi M, Khatun Z, Moon WC, Lee G, Lee YK. Heparin based nanoparticles for cancer targeting and noninvasive imaging. *Quant Imaging Med Surg*, 2012; 2: 219–226.
86. Cai W, Chen X. Nanoplatforms for targeted molecular imaging in living subjects. *Small*, 2007; 3: 1840–1854.
87. Kreuter J. Nanoparticles — a historical perspective. *Int J Pharm*, 2007; 331: 1–10.
88. Panyam J, Labhasetwar V. Biodegradable nanoparticles for drug and gene delivery to cells and tissue. *Adv Drug Deliv Rev*, 2003; 55: 329–347.
89. Lobatto ME, Fuster V, Fayad ZA, Mulder WJ. Perspectives and opportunities for nanomedicine in the management of atherosclerosis. *Nat Rev Drug Dis*, 2011; 10: 835–852.
90. Lim EK, Jang E, Lee K, Haam S, Huh YM. Delivery of cancer therapeutics using nanotechnology. *Pharmaceutics*, 2013; 5: 294–317.
91. Mitragotri S, Lahann J. Physical approaches to biomaterial design. *Nat Mater*, 2009; 8: 15–23.
92. Tekade RK, Chougule MB. Formulation development and evaluation of hybrid nanocarrier for cancer therapy: Taguchi orthogonal array based design. *BioMed Res Int*, 2013; 2013: 712678.
93. Elzoghby AO, Samy WM, Elgindy NA. Albumin-based nanoparticles as potential controlled release drug delivery systems. *J Control Release*, 2012; 157: 168–182.
94. Martínez A, Iglesias I, Lozano R, Teijón JM, Blanco MD. Synthesis and characterization of thiolated alginate-albumin nanoparticles stabilized by disulfide bonds. Evaluation as drug delivery systems. *Carbohyd Polym*, 2011; 83: 1311–1321.

95. Youngren SR, Tekade RK, Gustilo B, Hoffmann PR, Chougule MB. STAT6 siRNA matrix-loaded gelatin nanocarriers: formulation, characterization, and *ex vivo* proof of concept using adenocarcinoma cells. *BioMed Res Int*, 2013; 2013: 858946.
96. Coester CJ, Langer K, van Briesen H, Kreuter J. Gelatin nanoparticles by two step desolvation — a new preparation method, surface modifications and cell uptake. *J Microencapsulation*, 2000; 17: 187–193.
97. Lee SJ, Huh MS, Lee SY, Min S, Lee S, Koo H, *et al.*, Tumor-homing poly-siRNA/glycol chitosan self-cross-linked nanoparticles for systemic siRNA delivery in cancer treatment. *Angewandte Chemie Int Ed*, 2012; 51: 7203–7207.
98. Stigliano C, Aryal S, de Tullio MD, Nicchia GP, Pascazio G, Svelto M, *et al.*, siRNA-chitosan complexes in poly(lactic-co-glycolic acid) nanoparticles for the silencing of aquaporin-1 in cancer cells, *Mol Pharm*, 2013; 10: 3186–3194.
99. Dang JM, Leong KW. Natural polymers for gene delivery and tissue engineering. *Adv Drug Deliv Rev,* 2006; 58: 487–499.
100. Ratner BD, Bryant SJ. Biomaterials: where we have been and where we are going. *Ann Rev Biomed Engin*, 2004 ; 6: 41–75.
101. Danhier F, Ansorena E, Silva JM, Coco R, Le Breton A, Préat V. PLGA-based nanoparticles: An overview of biomedical applications. *J Control Release*, 2012; 161: 505–522.
102. Pantazis P, Dimas K, Wyche JH, Anant S, Houchen CW, Panyam J, *et al.*, Preparation of siRNA-encapsulated PLGA nanoparticles for sustained release of siRNA and evaluation of encapsulation efficiency. *Meth Mol Biol*, 2012; 906: 311–309.
103. Cun D, Jensen DK, Maltesen MJ, Bunker M, Whiteside P, Scurr D, *et al.*, High loading efficiency and sustained release of siRNA encapsulated in PLGA nanoparticles: quality by design optimization and characterization. *Eur J Pharmaceut Biopharmaceut*, 2011; 77: 26–35.
104. Filippov SK, Chytil P, Konarev PV, Dyakonova M, Papadakis C, Zhigunov A, *et al.,* Macromolecular HPMA-based nanoparticles with cholesterol for solid-tumor targeting: detailed study of the inner structure of a highly efficient drug delivery system. *Biomacromolecules*, 2012; 13: 2594–2604.
105. Vangara KK, Liu JL, Palakurthi S. Hyaluronic acid-decorated PLGA-PEG nanoparticles for targeted delivery of SN-38 to ovarian cancer. *Anticancer Res*, 2013; 33: 2425–2434.

106. Morelli C, Maris P, Sisci D, Perrotta E, Brunelli E, Perrotta I, *et al.*, PEG-templated mesoporous silica nanoparticles exclusively target cancer cells. *Nanoscale*, 2011; 3: 3198–3207.
107. Wang X, Uto T, Akagi T, Akashi M, Baba M. Poly(gamma-glutamic acid) nanoparticles as an efficient antigen delivery and adjuvant system: potential for an AIDS vaccine. *J Med Virol*, 2008 ; 80: 11–19.
108. Li C. Poly(l-glutamic acid) — anticancer drug conjugates. *Adv Drug Deliv Rev*, 2002; 54: 695–713.
109. Martins S, Sarmento B, Ferreira DC, Souto EB. Lipid-based colloidal carriers for peptide and protein delivery — liposomes *versus* lipid nanoparticles. *Int J Nanomed*, 2007; 2: 595–607.
110. Misra A, Jinturkar K, Patel D, Lalani J, Chougule M. Recent advances in liposomal dry powder formulations: preparation and evaluation. *Expert Opin Drug Deliv*, 2009 ; 6: 71–89.
111. Chougule M, Padhi B, Misra A. Nano-liposomal dry powder inhaler of tacrolimus: preparation, characterization, and pulmonary pharmacokinetics. *Int J Nanomed*, 2007; 2: 675–688.
112. Chougule M, Padhi B, Misra A. Development of spray dried liposomal dry powder inhaler of Dapsone. *AAPS PharmSciTech*, 2008; 9: 47–53.
113. Wu J, Liu Q, Lee RJ. A folate receptor-targeted liposomal formulation for paclitaxel. *Int J Pharm*, 2006; 316: 148–153.
114. Wissing SA, Muller RH. The influence of solid lipid nanoparticles on skin hydration and viscoelasticity — *in vivo* study. *Eur J Pharm Biopharm*, 2003; 56: 67–72.
115. Zhang XG, Miao J, Dai YQ, Du YZ, Yuan H, Hu FQ. Reversal activity of nanostructured lipid carriers loading cytotoxic drug in multi-drug resistant cancer cells. *Int J Pharm*, 2008; 361: 239–244.
116. Bianco A, Kostarelos K, Partidos CD, Prato M. Biomedical applications of functionalised carbon nanotubes. *Chem Comm*, 2005: 571–577.
117. Wong BS, Yoong SL, Jagusiak A, Panczyk T, Ho HK, Ang WH, *et al.*, Carbon nanotubes for delivery of small molecule drugs. *Adv Drug Deliv Rev*, 2013; 65: 1964–2015.
118. Chougule MB, Padhi BK, Misra A. Nano-liposomal dry powder inhaler of amiloride hydrochloride. *J Nanosci Nanotechnol*, 2006; 6: 3001–3009.
119. Hall JB, Dobrovolskaia MA, Patri AK, McNeil SE. Characterization of nanoparticles for therapeutics. *Nanomedicine*, 2007; 2: 789–803.

120. Liu J-f, Yu S-j, Yin Y-g, Chao J-b. Methods for separation, identification, characterization and quantification of silver nanoparticles. *TrAC*, 2012; 33: 95–106.
121. Lopez-Serrano A, Olivas RM, Landaluze JS, Camara C. Nanoparticles: a global vision. Characterization, separation, and quantification methods. Potential environmental and health impact. *Anal Meth*, 2014; 6: 38–56.
122. Chu B, Liu T. Characterization of nanoparticles by scattering techniques. *J Nanopart Res*, 2000; 2: 29–41.
123. Berry J, Kline LC, Sherwood JK, Chaudhry S, Obenauer-Kutner L, Hart JL, *et al.*, Influence of the size of micronized active pharmaceutical ingredient on the aerodynamic particle size and stability of a metered dose inhaler. *Drug Dev Ind Pharm*, 2004; 30: 705–714.
124. Bonam M, Christopher D, Cipolla D, Donovan B, Goodwin D, Holmes S, *et al.,* Minimizing variability of cascade impaction measurements in inhalers and nebulizers. *AAPS PharmSciTech*, 2008; 9: 404–413.
125. Tougas TP, Christopher D, Mitchell JP, Strickland H, Wyka B, Van Oort M, *et al.,* Improved quality control metrics for cascade impaction measurements of orally inhaled drug products (OIPs). *AAPS PharmSciTech*, 2009; 10: 1276–1285.
126. Mitchell JP, Nagel MW. Cascade impactors for the size characterization of aerosols from medical inhalers: their uses and limitations. *J Aerosol Med*, 2003; 16: 341–377.
127. Mitchell JP, Nagel MW, Doyle CC, Ali RS, Avvakoumova VI, Christopher JD, *et al.,* Relative precision of inhaler aerodynamic particle size distribution (APSD) metrics by full resolution and abbreviated andersen cascade impactors (ACIs): part 1. *AAPS PharmSciTech*, 2010; 11: 843–851.
128. Chapter 15: Monitoring Air Pollution, 2010.
129. Finlay WH, Stapleton KW, Zuberbuhler P. Fine particle fraction as a measure of mass depositing in the lung during inhalation of nearly isotonic nebulized aerosols. *J Aerosol Sci, 1997*; 28: 1301–1309.
130. Claus S, Weiler C, Schiewe J, Friess W. Optimization of the fine particle fraction of a lyophilized lysozyme formulation for dry powder inhalation. *Pharm Res*, 2013; 30: 1698–1713.
131. Nagel JPMaMW. Particle size analysis of aerosols from medicinal inhalers† KONA powder and particle, 2004: 34.

132. Tarsin WY, Pearson SB, Assi KH, Chrystyn H. Emitted dose estimates from Seretide® Diskus® and Symbicort® Turbuhaler® following inhalation by severe asthmatics. *Int J Pharm*, 2006; 316: 131–137.
133. Abdelrahim ME. Emitted dose and lung deposition of inhaled terbutaline from Turbuhaler at different conditions. *Respir Med*, 2010; 104: 682–689.
134. Patel G, Chougule M, Singh M, Misra A. Chapter 9 — Nanoliposomal dry powder formulations. *Methods Enzymol*, 2009; 464: 167–191.
135. Labiris NR, Dolovich MB. Pulmonary drug delivery. Part I: physiological factors affecting therapeutic effectiveness of aerosolized medications. *Br J Clin Pharmacol*, 2003; 56: 588–599.
136. Francia G, Cruz-Munoz W, Man S, Xu P, Kerbel RS. Mouse models of advanced spontaneous metastasis for experimental therapeutics. *Nat Rev Cancer*, 2011; 11: 135–141.
137. Richmond A, Su Y. Mouse xenograft models *vs* GEM models for human cancer therapeutics. *DMM*, 2008; 1: 78–82.
138. Sausville EA, Newell DR. Preclinical models in cancer drug discovery and development. *Eur J Cancer*, 2004; 40: 783–784.
139. Graham C, Tucker C, Creech J, Favours E, Billups CA, Liu T, *et al.*, Evaluation of the antitumor efficacy, pharmacokinetics, and pharmacodynamics of the histone deacetylase inhibitor depsipeptide in childhood cancer models *in vivo*. *Clin Cancer Res*, 2006; 12: 223–234.
140. Teicher BA. Tumor models for efficacy determination. *Mol Cancer Ther*, 2006; 5: 2435–2443.
141. Tan MH, Holyoke ED, Goldrosen MH. Murine colon adenocarcinoma: syngeneic orthotopic transplantation and subsequent hepatic metastases. *J Natl Cancer Inst*, 1977; 59:1 537–1544.
142. Bibby MC. Orthotopic models of cancer for preclinical drug evaluation: advantages and disadvantages. *Eur J Cancer*, 2004; 40: 852–857.
143. Dinulescu DM, Ince TA, Quade BJ, Shafer SA, Crowley D, Jacks T. Role of K-ras and Pten in the development of mouse models of endometriosis and endometrioid ovarian cancer. *Nat Med*, 2005; 11: 63–70.
144. Kim CF, Jackson EL, Woolfenden AE, Lawrence S, Babar I, Vogel S, *et al.*, Identification of bronchioalveolar stem cells in normal lung and lung cancer. *Cell*, 2005; 121: 823–835.

145. Bergers G, Song S, Meyer-Morse N, Bergsland E, Hanahan D. Benefits of targeting both pericytes and endothelial cells in the tumor vasculature with kinase inhibitors. *J Clin Invest*, 2003; 111: 1287–1295.
146. Huss WJ, Barrios RJ, Greenberg NM. SU5416 selectively impairs angiogenesis to induce prostate cancer-specific apoptosis. *Mol Cancer Ther*, 2003; 2: 611–616.
147. Rudolph C, Gleich B, Flemmer AW. Magnetic aerosol targeting of nanoparticles to cancer: nanomagnetosols. *Meth Mol Biol*, 2010; 624: 267–280.
148. Wong BA. Inhalation exposure systems: design, methods and operation. *Toxicol Pathol*, 2007; 35: 3–14.
149. Thomson EM, Williams A, Yauk CL, Vincent R. Impact of nose-only exposure system on pulmonary gene expression. *Inhal Toxicol*, 2009; 21 Suppl 1: 74–82.
150. Osier M, Oberdorster G. Intratracheal inhalation *vs* intratracheal instillation: differences in particle effects. *Fundam Appl Toxicol*, 1997; 40: 220–227.
151. Brain JD, Valberg PA. Deposition of aerosol in the respiratory tract. *Am Rev Respir Dis*, 1979; 120: 1325–1373.
152. Leong BK, Coombs JK, Sabaitis CP, Rop DA, Aaron CS. Quantitative morphometric analysis of pulmonary deposition of aerosol particles inhaled via intratracheal nebulization, intratracheal instillation or nose-only inhalation in rats. *J Appl Toxicol*, 1998; 18: 149–160.
153. Enna SJ, Schanker LS. Absorption of saccharides and urea from the rat lung. *Am J Physiol*, 1972; 222: 409–414.
154. Brain JD, Knudson DE, Sorokin SP, Davis MA. Pulmonary distribution of particles given by intratracheal instillation or by aerosol inhalation. *Environ Res*, 1976; 11: 13–33.
155. Koshkina NV, Agoulnik IY, Melton SL, Densmore CL, Knight V. Biodistribution and pharmacokinetics of aerosol and intravenously administered DNA-polyethyleneimine complexes: optimization of pulmonary delivery and retention. *Mol Ther*, 2003; 8: 249–254.
156. Patel AR, Chougule MB, I T, Patlolla R, Wang G, Singh M. Efficacy of aerosolized celecoxib encapsulated nanostructured lipid carrier in non-small cell lung cancer in combination with docetaxel. *Pharm Res*, 2013; 30: 1435–1446.
157. Taratula O, Kuzmov A, Shah M, Garbuzenko OB, Minko T. Nanostructured lipid carriers as multifunctional nanomedicine platform for pulmonary co-delivery of anticancer drugs and siRNA. *J Control Release*, 2013.

158. Taratula O, Garbuzenko OB, Chen AM, Minko T. Innovative strategy for treatment of lung cancer: targeted nanotechnology-based inhalation co-delivery of anticancer drugs and siRNA. *J Drug Target*, 2011; 19: 900–914.
159. Jiang HL, Xu CX, Kim YK, Arote R, Jere D, Lim HT, *et al.*, The suppression of lung tumorigenesis by aerosol-delivered folate-chitosan-graft-polyethylenimine/Akt1 shRNA complexes through the Akt signaling pathway. *Biomaterials*, 2009; 30: 5844–5852.
160. Garbuzenko OB, Saad M, Betigeri S, Zhang M, Vetcher AA, Soldatenkov VA, *et al.,* Intratracheal versus intravenous liposomal delivery of siRNA, antisense oligonucleotides and anticancer drug. *Pharm Res*, 2009; 26: 382–394.
161. Gautam A, Densmore CL, Melton S, Golunski E, Waldrep JC. Aerosol delivery of PEI-p53 complexes inhibits B16-F10 lung metastases through regulation of angiogenesis. *Cancer Gene Ther*, 2002; 9: 28–36.
162. Wagner N, Panelos J, Massi D, Wagner KD. The Wilms' tumor suppressor WT1 is associated with melanoma proliferation. *Eur J Physiol*, 2008; 455: 839–847.
163. Oji Y, Ogawa H, Tamaki H, Oka Y, Tsuboi A, Kim EH, *et al.,* Expression of the Wilms' tumor gene WT1 in solid tumors and its involvement in tumor cell growth. *Jpn J Cancer Res*, 1999; 90: 194–204.
164. Zamora-Avila DE, Franco-Molina MA, Trejo-Avila LM, Rodriguez-Padilla C, Resendez-Perez D, Zapata-Benavides P. RNAi silencing of the WT1 gene inhibits cell proliferation and induces apoptosis in the B16F10 murine melanoma cell line. *Melanoma Res*, 2007; 17: 341–348.
165. Jiang HL, Hong SH, Kim YK, Islam MA, Kim HJ, Choi YJ, *et al.,* Aerosol delivery of spermine-based poly(amino ester)/Akt1 shRNA complexes for lung cancer gene therapy. *Int J Pharm*, 2011; 420: 256–265.
166. Markowitz SB, Miller A, Miller J, Manowitz A, Kieding S, Sider L, *et al.,* Ability of low-dose helical CT to distinguish between benign and malignant noncalcified lung nodules. *Chest*, 2007; 131: 1028–1034.
167. Kramer BS, Berg CD, Aberle DR, Prorok PC. Lung cancer screening with low-dose helical CT: results from the National Lung Screening Trial (NLST). *J Med Screen*, 2011; 18: 109–111.
168. Hodnett PA, Ko JP. Evaluation and management of indeterminate pulmonary nodules. *Radiol Clin North Am*, 2012; 50: 895–914.
169. Ashton JR, Clark DP, Moding EJ, Ghaghada K, Kirsch DG, West JL, *et al.,* Dual-energy micro-CT functional imaging of primary lung cancer in mice

using gold and iodine nanoparticle contrast agents: a validation study. *PloS One*, 2014; 9: e88129.

170. Fornaro J, Leschka S, Hibbeln D, Butler A, Anderson N, Pache G, *et al.*, Dual- and multi-energy CT: approach to functional imaging. *Insights Imaging*, 2011; 2: 149–159.
171. Schmid-Bindert G, Henzler T, Chu TQ, Meyer M, Nance JW, Jr., Schoepf UJ, *et al.*, Functional imaging of lung cancer using dual energy CT: how does iodine related attenuation correlate with standardized uptake value of 18FDG-PET-CT? *Eur Radiol*, 2012; 22: 93–103.
172. Badea CT, Guo X, Clark D, Johnston SM, Marshall CD, Piantadosi CA. Dual-energy micro-CT of the rodent lung. *Am J Physiol Lung Cell Mol Physiol*, 2012; 302: L1088-L1097.
173. Ashton JR, Befera N, Clark D, Qi Y, Mao L, Rockman HA, *et al.*, Anatomical and functional imaging of myocardial infarction in mice using micro-CT and eXIA 160 contrast agent. *Contrast Media Mol Imaging*, 2014; 9: 161–168.
174. Clark DP, Ghaghada K, Moding EJ, Kirsch DG, Badea CT. *In vivo* characterization of tumor vasculature using iodine and gold nanoparticles and dual energy micro-CT. *Phys Med Biol*, 2013; 58: 1683–1704.
175. Moding EJ, Clark DP, Qi Y, Li Y, Ma Y, Ghaghada K, *et al.*, Dual-energy micro-computed tomography imaging of radiation-induced vascular changes in primary mouse sarcomas. *Int J Radiat Oncol Biol Phys*, 2013; 85: 1353–1359.
176. Bourin M, Jolliet P, Ballereau F. An overview of the clinical pharmacokinetics of x-ray contrast media. *Clin Pharmacokinet*, 1997; 32: 180–193.
177. Maeda H, Nakamura H, Fang J. The EPR effect for macromolecular drug delivery to solid tumors: Improvement of tumor uptake, lowering of systemic toxicity, and distinct tumor imaging *in vivo. Adv Drug Deliv Rev*, 2013; 65: 71–79.
178. Au JT, Craig G, Longo V, Zanzonico P, Mason M, Fong Y, *et al.*, Gold nanoparticles provide bright long-lasting vascular contrast for CT imaging. *Am J Roentgenol*, 2013; 200: 1347–1351.
179. Danila D, Johnson E, Kee P. CT imaging of myocardial scars with collagen-targeting gold nanoparticles. *Nanomed-Nanotechnol*, 2013; 9: 1067–1076.

180. Ghann WE, Aras O, Fleiter T, Daniel M-C. Syntheses and characterization of lisinopril-coated gold nanoparticles as highly stable targeted CT contrast agents in cardiovascular diseases. *Langmuir*, 2012; 28: 10398–10408.
181. Badea CT, Johnston SM, Qi Y, Ghaghada K, Johnson GA. Dual-energy micro-CT imaging for differentiation of iodine- and gold-based nanoparticles. 2011. p. 79611X-X-10.
182. Hainfeld JF, O'Connor MJ, Dilmanian FA, Slatkin DN, Adams DJ, Smilowitz HM. Micro-CT enables microlocalisation and quantification of Her2-targeted gold nanoparticles within tumour regions. *Bri J Radiol*, 2011; 84: 526–533.
183. Cormode DP, Roessl E, Thran A, Skajaa T, Gordon RE, Schlomka JP, *et al.*, Atherosclerotic plaque composition: analysis with multicolor CT and targeted gold nanoparticles. *Radiology*, 2010; 256: 774–782.
184. Zheng J, Allen C, Serra S, Vines D, Charron M, Jaffray DA. Liposome contrast agent for CT-based detection and localization of neoplastic and inflammatory lesions in rabbits: validation with FDG-PET and histology. *Contrast Media Mol Imaging*, 2010; 5: 147–154.
185. Hendee WR, Morgan CJ. Magnetic resonance imaging. Part I — Physical principles. *West J Med*, 1984; 141: 491–500.
186. Yoo MK, Kim IY, Kim EM, Jeong HJ, Lee CM, Jeong YY, *et al.*, Superparamagnetic iron oxide nanoparticles coated with galactose-carrying polymer for hepatocyte targeting. *J Biomed Biotechnol*, 2007; 2007: 94740.
187. Wang YX. Superparamagnetic iron oxide based MRI contrast agents: Current status of clinical application. *Quant Imaging Med Surg*, 2011; 1: 35–40.
188. Liu G, Gao J, Ai H, Chen X. Applications and potential toxicity of magnetic iron oxide nanoparticles. *Small*, 2013; 9: 1533–1545.
189. Zeng L, Ren W, Zheng J, Cui P, Wu A. Ultrasmall water-soluble metal-iron oxide nanoparticles as T1-weighted contrast agents for magnetic resonance imaging. *Phys Chem Chem Phys*, 2012; 14: 2631–2636.
190. Kumar K, Nightingale AM, Krishnadasan SH, Kamaly N, Wylenzinska-Arridge M, Zeissler K, *et al.*, Direct synthesis of dextran-coated superparamagnetic iron oxide nanoparticles in a capillary-based droplet reactor. *J Mater Chem*, 2012; 22: 4704–4708.

191. Osborne EA, Atkins TM, Gilbert DA, Kauzlarich SM, Liu K, Louie AY. Rapid microwave-assisted synthesis of dextran-coated iron oxide nanoparticles for magnetic resonance imaging. *Nanotechnology*, 2012; 23: 215602.
192. Belaid S, Laurent S, Vermeech M, Vander Elst L, Perez-Morga D, Muller RN. A new approach to follow the formation of iron oxide nanoparticles synthesized by thermal decomposition. *Nanotechnology*, 2013; 24: 055705.
193. Zhang MQ, Kievit FM. Surface engineering of iron oxide nanoparticles for targeted cancer therapy. *Abstr Pap Am Chem Soc*, 2012; 243: 1.
194. Thorek DL, Chen AK, Czupryna J, Tsourkas A. Superparamagnetic iron oxide nanoparticle probes for molecular imaging. *Ann Biomed Eng*, 2006; 34: 23–38.
195. Ling D, Hyeon T. Chemical design of biocompatible iron oxide nanoparticles for medical applications. *Small*, 2013; 9: 1450–1466.
196. Gupta AK, Gupta M. Synthesis and surface engineering of iron oxide nanoparticles for biomedical applications. *Biomaterials*, 2005; 26: 3995–4021.
197. Hayashi K, Nakamura M, Sakamoto W, Yogo T, Miki H, Ozaki S, *et al.*, Superparamagnetic nanoparticle clusters for cancer theranostics combining magnetic resonance imaging and hyperthermia treatment. *Theranostics*, 2013; 3: 366–376.
198. Jun YW, Lee JH, Cheon J. Chemical design of nanoparticle probes for high-performance magnetic resonance imaging. *Angew Chemie Int Ed*, 2008; 47: 5122–5135.
199. Cassidy PJ, Radda GK. Molecular imaging perspectives. *J R Soc Interface*, 2005; 2: 133–144.
200. Ntziachristos V, Bremer C, Weissleder R. Fluorescence imaging with near-infrared light: new technological advances that enable *in vivo* molecular imaging. *Eur Radiol*, 2003; 13: 195–208.
201. Pierce MC, Javier DJ, Richards-Kortum R. Optical contrast agents and imaging systems for detection and diagnosis of cancer. *Int J Cancer*, 2008; 123: 1979–1990.
202. Park K. Polysaccharide-based near-infrared fluorescence nanoprobes for cancer diagnosis. *Quant Imaging Med Surg*, 2012; 2: 106–113.

Chapter 4

QUANTITATIVE CONTRAST ENHANCED ULTRASOUND IMAGING IN CANCER THERAPY

Shashank R. Sirsi and Mark A. Borden*[†]

**Eric Johnson School of Engineering & Computer Science,*
University of Texas at Dallas,
800W Campbell Road EC39, Richardson, Texas 75080, USA
shashank.sirsi@utdallas.edu

[†]*Department of Mechanical Engineering*
University of Colorado
1111 Engineering Drive
Boulder, CO 80309-0427, USA
mark.borden@colorado.edu

Abstract

Quantitative contrast enhanced ultrasound (CEUS) imaging has the potential to advance methods of cancer diagnosis and treatment by providing clinicians more suitable benchmarks for monitoring tumor perfusion and response to cancer therapies. In this chapter, we describe the basic methods of how quantitative CEUS imaging is performed in tumors with the use of ultrasound contrast agents (UCAs) and discuss the challenges that still need to be overcome to fully realize the potential of CEUS imaging in a cancer therapy.

4.1 Introduction

Ultrasound (US) offers a unique modality for *in vivo* imaging that has the potential to revolutionize cancer diagnosis and therapy. Traditional US diagnostic modalities use high frequency sound waves (clinical scanners operate typically between 1–10 MHz) to visualize anatomical structures and monitor disease progression *in vivo*. In cancer diagnostics, US imaging can be especially useful for monitoring tumor growth, and tumor response to drug therapy, safely and non-invasively. Overall, US is a low-cost and safe imaging modality that has numerous advantages for monitoring disease progression in a clinical setting, such as the ability to perform repetitive imaging procedures without depositing harmful ionizing radiation. In clinical practice, conventional US imaging provides mainly qualitative data, and interpretation of such data is operator dependent. For this reason, US is not currently accepted as a viable method of assessing tumor response in cancer therapies, as indicated by current Response Evaluation Criteria in Solid Tumours (RECIST) protocols used in anti-cancer drug trials.[1] US imaging is often used as an adjunct to other imaging modalities (CT and MRI) and can provide clinicians with useful quantitative parameters, such as tumor lesion size (B-mode imaging) and blood flow rates (Doppler Imaging). While US imaging is not currently the preferred imaging modality in quantitative assessment of cancer therapy, there is an ever growing demand for developing standardized clinical protocols with US imaging. This demand is especially apparent in monitoring therapies of pediatric patients where concerns about repetitive CT scanning exist.[2] US imaging for evaluating tumors is currently widely used in pediatric oncology practice, for example to identify and diagnose lesions, but more developed and standardized methods of assessment are needed for further clinical applications, namely monitoring tumor response to drug therapy.[3]

While the growth in demand for US imaging techniques over the past 40 years may be sufficient reason to pursue the further development of quantitative US imaging techniques, the full potential of US imaging remains yet to be fully realized. A major drawback of conventional US imaging (and other imaging modalities as well) is that more detailed information about tumor biology, such as the microvasculature, is difficult to resolve due to the inability to differentiate fine tissue structures within the tumor. Furthermore, measurements of blood flow with Doppler imaging are restricted to large vessels with fast moving blood, due to the limited ability of blood components to scatter US energy compared to tissue structures.[4] The versatility of US imaging can be significantly improved by the introduction of UCAs into the bloodstream that can enhance the contrast between vascular and non-vascular compartments within the tumor. The perfusion of UCAs through the tumor allows clinicians to perform more detailed and accurate measurements of the tumor microvasculature through CEUS imaging that complements conventional B-mode and Doppler US imaging techniques. UCAs can also be designed to target and bind vascular epitopes. Thus, CEUS can be used to identify molecular patterns present in the tumor vasculature well below the US image resolution (typically 0.1–1 mm). Overall, CEUS imaging can significantly improve sensitivity for monitoring tumor blood flow and evaluating molecular expression in the tumor vasculature, allowing more quantitative approaches to be developed for cancer imaging. CEUS has the potential to push the field of cancer imaging and therapy beyond conventional RECIST techniques for monitoring tumor response and fundamentally change the way drug therapies are evaluated in a clinical setting. In this chapter, we aim to introduce the basic concepts of CEUS and how such quantitative analysis can be performed using UCAs. We will discuss current methodologies of CEUS in tumor imaging, limitations to CEUS in evaluating tumor response, and discuss novel work that is being performed to overcome these limitations.

4.2 Basics of contrast enhanced ultrasound imaging

4.2.1 *The ultrasound contrast agent*

UCAs, commonly referred to as "microbubbles," are small gas-filled spheres (typically 1–10 μm in diameter) stabilized by thin lipid, protein or polymer shells.[5] (Figure 4.1). Microbubbles are fabricated by a variety of techniques, including sonication, shaking and microfluidics.[6–8] Microbubbles are small enough to circulate effectively in the bloodstream, but are too large to extravasate beyond the vascular space. Due to their inherent size, they are

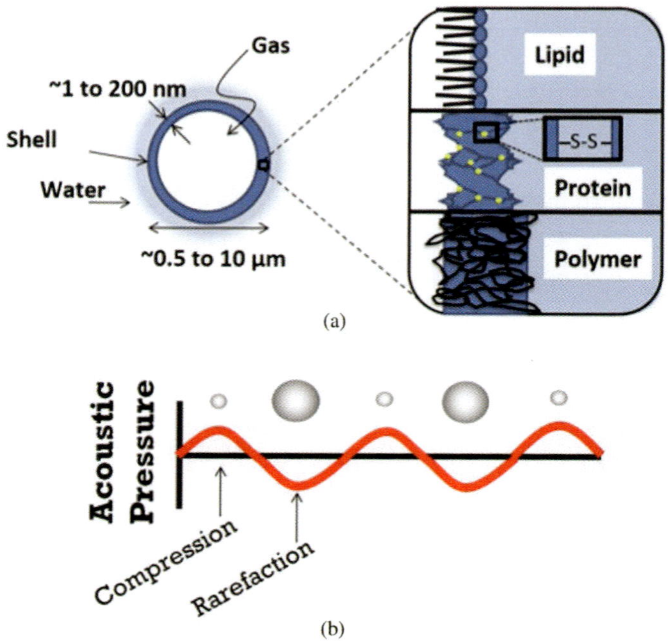

Figure 4.1 (a) Schematic of microbubble composition depicting gas core and shell material. Reproduced with permission from Ref. 5 (b) Microbubble response in an ultrasound field showing volumetric expansion and compression in response to the oscillating pressure of the sound wave.

excellent intravascular probes. Microbubbles are effective US contrast agents because they scatter US energy significantly more than the surrounding blood and tissue in the tumor. This ability to scatter US energy relies on the mechanical response of the microbubble in an ultrasound field. The gas core of the microbubble is highly compressible causing the microbubble to undergo rapid volumetric changes in response to the oscillating pressure of the sound wave.[9,10] The mechanical response of the microbubble creates an acoustic backscatter, which is detectable by the clinical US scanner. The US signals received from the microbubbles are much greater than those scattered by surrounding blood and tissue, which translates to an increase in pixel intensity (brightening) in an US B-mode image. The increase in overall pixel intensity is proportional to the concentration of microbubbles in the dilute regime (10^{-15} to 10^{-11} mM), which allows quantitative measurements to be performed with relatively simple techniques. Furthermore, many clinical US scanners are able to isolate the signal from the UCAs from tissue using linear or non-linear imaging techniques, further improving the ability to resolve the tissue vasculature. Currently, microbubble UCAs are widely used in Europe and Asia, but were not approved by the FDA in non-cardiac imaging applications until recently.[11,12] However, the potential of microbubbles in cancer treatment is being explored in preclinical studies and will likely revolutionize this field of medicine.

4.3 CEUS imaging in cancer

UCAs are fundamental in CEUS imaging to provide quantitative measures of blood perfusion in tumors and quantitative evaluation of vascular epitope expression. CEUS can provide real-time imaging at the bedside for qualitative tracking and quantitative measurement of perfusion-related biomarkers.[13] In preclinincal studies, CEUS is currently used to assess malignancy of liver, bowel, kidney, pancreas, breast and prostate lesions by measuring perfusion-related parameters, such as uptake and washout rates of UCAs.[4,11] CEUS also has strong potential to monitor the efficacy of other cancer therapies including radiotherapy and focused ultrasound mediated tumor ablation by immediately monitoring changes in tumor response.[14] CEUS is can also be used to identify the response of tumor models to anti-angiogenic therapy,[15–19] and has been used in clinical pilot studies to predict improvement in progression-free and overall survival of patients with renal cell carcinoma treated with commercial tyrosine-kinase inhibitors, such as sorafenib (Nexavar).[20] and imatinib (Glivec).[21]

One of the major dilemmas in cancer therapy is that drug therapies which target specific biological processes in tumors can show high variability in efficacy for unknown reasons. This variability suggests that rapid non-invasive methods of benchmarking tumor responses are needed in a clinical setting to identify if a particular treatment is viable. CEUS imaging has a strong potential to provide such assessment of benchmarking tumor responses to cancer therapy by providing quantitative measures of tumor parameters such as blood volumes, blood flow rates and molecular expression of vascular epitopes. Here, we describe the basic techniques of how UCAs can be used to quantitatively evaluate such parameters and how these parameters can be used to monitor tumor progression and efficacy of therapy.

4.4 Methods of quantitative CEUS imaging

4.4.1 *Perfusion imaging in tumors*

After UCAs are administered systemically as a bolus, they begin to rapidly mix and circulate with blood. By monitoring a particular region in the body with US, such as a solid tumor, the in-flow and washout of the agents can be observed by the change in video signal intensity. Microbubbles increase the pixel intensities proportionally to the concentration of microbubbles in the tumor. By continuously monitoring a region of interest (ROI) in the tumor, this change in overall intensity can be observed by averaging or summing the pixel intensities within the ROI in each sequential US image. The data can then be plotted to generate a time-intensity curve, so long as the microbubble concentration in blood is below ~10^{-11} mM (i.e., the linear regime between acoustic backscattered power and microbubble concentration). An example of a time-intensity curve within a tumor is shown in Figure 4.2b. The ROI be drawn around the tumor using the B-mode image (in this case, defined by the hypoechoic region outlined in white; Figure 4.2a) while the signal from UCAs is isolated using linear or non-linear imaging techniques, either available on the scanner or using post-image acquisition software. The initial bolus of injected microbubbles is often used to evaluate perfusion and blood flow rates in tumors (Figure 4.3). However, this technique may cause operator dependent variability depending on the rate of injection.

When possible, destruction-replenishment (DR) imaging should be used to obtain quantitative measures. In DR imaging, the operator waits until the microbubbles have reached a steady-state level following the bolus injection (or during constant infusion) and a low-frequency US pulse is used to destroy the microbubbles in the US field of view. Immediately following the destruction pulse, the microbubble re-perfusion into the ROI is evaluated (Figure 4.4). The time-intensity curve can be fit to a first-order exponential model:

$$I(t) = A(1 - e^{-B^*t}) \quad (1)$$

where I is the video signal intensity, A is the maximum signal enhancement in the ROI, and B is a rate constant describing the influx of contrast agent into the ROI. Since video intensity of the contrast agent is

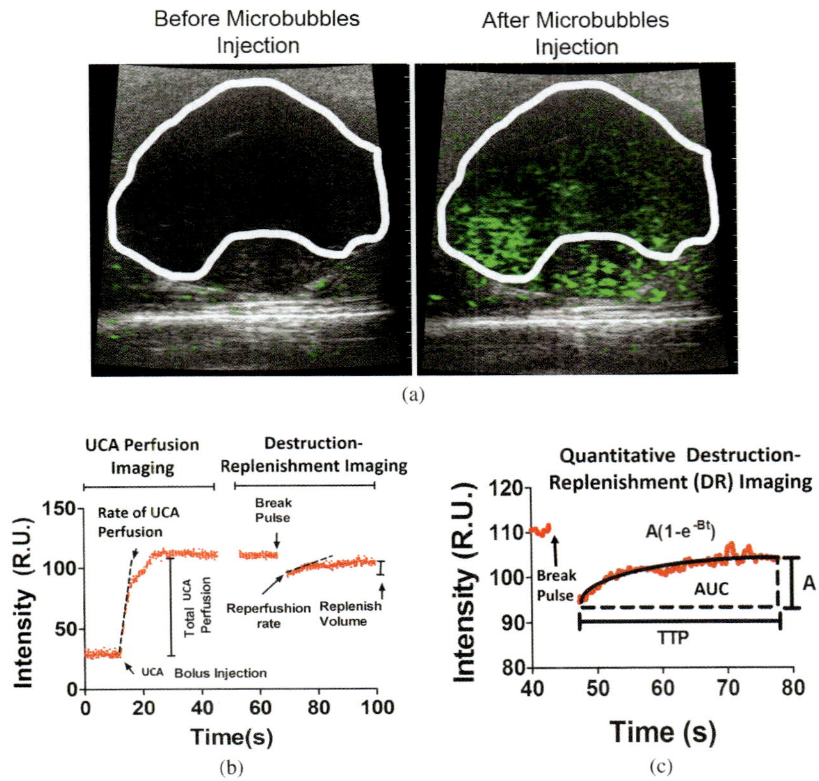

Figure 4.2 (a) Ultrasound Images of SKNEP-1 tumors implanted intrarenally in mice (outlined in white) before and after microbubble injection (contrast enhancement shown in green). Reproduced with permission from Sirsi *et al.* [40]. (b) The contrast within the tumor can be evaluated by averaging the pixel intensities within the ROI to generate a time-intensity curve from the initial bolus (left) or from destruction-replenishment imaging (right) contrast quantification. (c) Quantitative parameters of tumor perfusion can be extracted including the relative blood volume (A), perfusion rate (B), area under the curve (AUC), and time-to-peak (TTP).

representative of blood flow, the parameter A is most closely related to the relative blood volume (relative units, R.U.), whereas the B parameter can be used to describe the blood flow rate (sec^{-1}). Using this simple model (Eq. 1), additional quantitative metrics can be determined, such as area-under-the-curve (AUC) and time-to-peak (TTP). Using these parameters, quantitative comparisons can be made between different tumors or on the

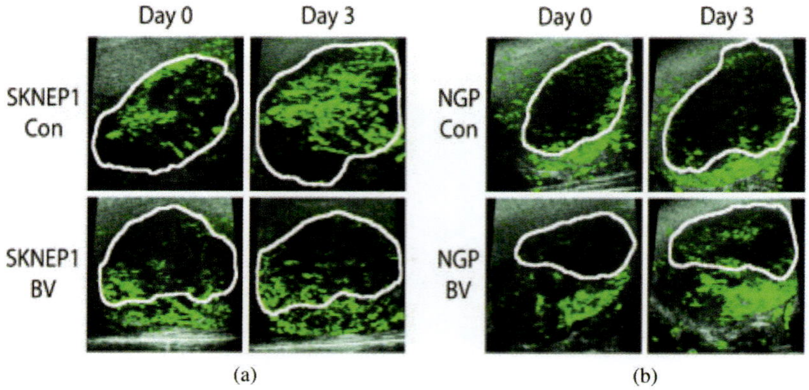

Figure 4.3 CEUS images of relative tumor blood volume measured by microbubble signal enhancement. Representative tumors with total microbubble signal enhancement overlays from the (a) SK-NEP-1 and (b) NGP groups are shown at day 0 and day 3. Hypoechoic tumor regions are outlined in white and microbubble signal enhancement is colored green. In contrast to the SK-NEP-1 Con, NGP Con and NGP BV groups, the BV-treated SK-NEP-1 tumors showed no increase in size or microbubble signal enhancement. The field of view for each of the ultrasound images was 17 × 17 mm^2. Reproduced with permission from Sirsi *et al.*[40]

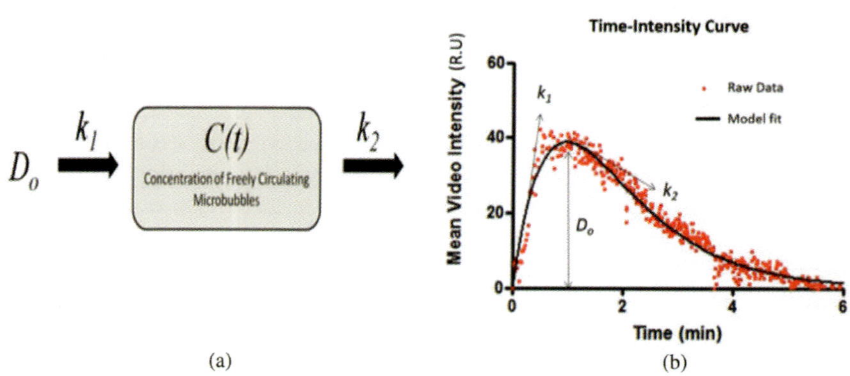

Figure 4.4 (a) Single compartment model used to fit time-intensity curve from UCA bolus injection. (b) the raw data from microbubble signal enhancement fit to single compartment model.

same tumor over varying time-periods to evaluate malignancy or monitor changes in tumor perfusion in response to therapy. Currently, there is no clinical standard of comparisons using parametric imaging. However, several preclinical studies have been performed in various disease models to evaluate the power of different quantitative parameters. Eisenbrey *et al.* performed quantitative CEUS imaging using the subharmonic signal from injected contrast agents to characterize breast lesions in cancer diagnosis.[22] In this study, a parametric map was generated to compare different parameters from the time intensity curve, namely perfusion level (A; defined as cumulative maximum intensity (CMI) in this study), perfusion rate (B; defined as estimated perfusion (EP) in this study, which is a modified definition derived from the slope of the contrast agent arrival), AUC and TTP. The results of this study showed that the EP parameter was able to accurately diagnose carcinomas from cysts and fibroadenomas whereas the A, AUC and TTP were not. Similar quantitative CEUS analysis has also been used to successfully diagnose pancreatic masses[23] and differentiate chronic pseudotumoral pancreatitis and pancreatic cancer using quantitative low mechanical index (MI) CEUS imaging.[24] Quantitative CEUS has also been shown to effectively monitor tumor disease progression in response to therapy.[25] Hoyt *et al.* demonstrated that CEUS destruction-replenishment imaging could be used to predict whether breast cancer was responsive to bevazicumab (BV) therapy within three days of treatment with a 10% accuracy in murine models[18] using A, AUC and TTP parameters. Overall, these studies demonstrate the potential of CEUS perfusion imaging for cancer diagnosis, monitoring tumor vascular growth, and assessing relative changes in perfusion in response to therapy.

4.5 Methods of quantitative CEUS imaging

4.5.1 *Microbubble pharmacodynamics*

the relative blood volume of the tumor and blood flow rates are particularly valuable metrics for evaluating tumor response to therapy, especially in therapies that utilize vascular targeting agents meant to induce regression of blood vessels within tumors. However, these quantitative measures focus solely on the in-flow of microbubble contrast agents into the tumor. Previous reports have demonstrated that both early-phase contrast agent in-flow and late-phase contrast agent washout can provide valuable information towards identifying tumors phenotypes and determining malignancy.[26–28] In these studies, early-phase contrast agent perfusion patterns could be used to identify lesion types, whereas late-phase washout of contrast agents was able to accurately differentiate between benign and malignant liver masses. Ideally, methods of quantifying both in-flow and out-flow of contrast agents from tumors could be beneficial in a variety of cancers and should be explored. The pharmacodynamics of UCAs can be examined by measuring the full time-intensity curve and measuring the in-flow and out-flow rates following systemic delivery. We have previously presented such a method of evaluating the full pharmacodynamics of UCAs in circulation by fitting the time-intensity curves to a simple single-compartment model.[29,30] (Figure 4.4):

$$C(t)=\frac{D_o k_1}{k_2-k_1}\left(e^{-k_1 * t}-e^{-k_2 * t}\right) \quad (2)$$

where C is the contrast enhancement from the intravascular compartment, t is time, D_o is the "dose" of contrast (given in units of relative video intensity), k_1 and k_2 are pseudo-first-order rate constants describing the influx and efflux of contrast from the circulation. The proposed model was shown to accurately fit full-length time-intensity curves when evaluating microbubble perfusion in the kidney.[29] and could be applied towards monitoring wash-in and wash-out rates in tumor lesions as well.

4.5.2 *Quantitative molecular imaging in tumors*

Perfusion monitoring is the most commonly used technique in CEUS imaging. However, the use of UCAs for molecular imaging significantly enhances their versatility in tumor imaging (Figure 4.5). In order to target microbubbles to specific vascular epitopes, the surface of microbubbles must be modified to include specific targeting ligands using simple covalent or non-covalent conjugation techniques.[31] As an example, microbubbles can be surface modified to include cyclic RGD peptides that preferentially bind $\alpha_V\beta_3$ integrins.[32] $\alpha_V\beta_3$ integrins are vascular epitopes overexpressed in angiogenic tissue, providing an excellent vascular target for monitoring blood vessel growth. Microbubbles can be targeted to this epitope by surface conjugation of RGD peptides. When administered systemically, RGD-coated microbubbles preferentially accumulate in a manner proportional to $\alpha_V\beta_3$ integrin expression.

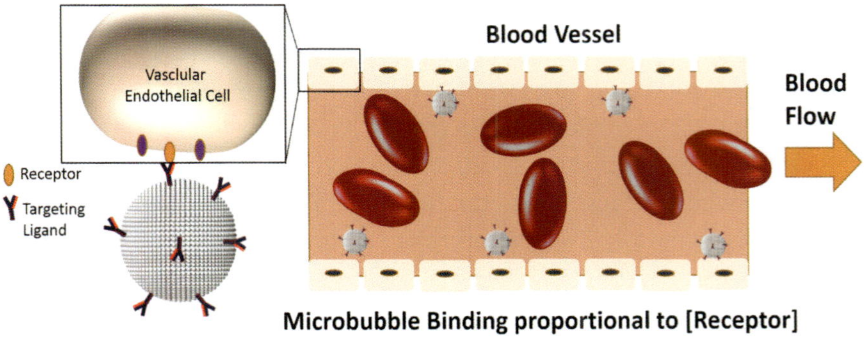

Figure 4.5 Schematic depicting a targeted microbubble binding to a vascular epitope in a blood vessel. The signal enhancement seen in an US image is proportional to the concentration of the targeted receptor, allowing relative quantification of molecular expression to be achieved.

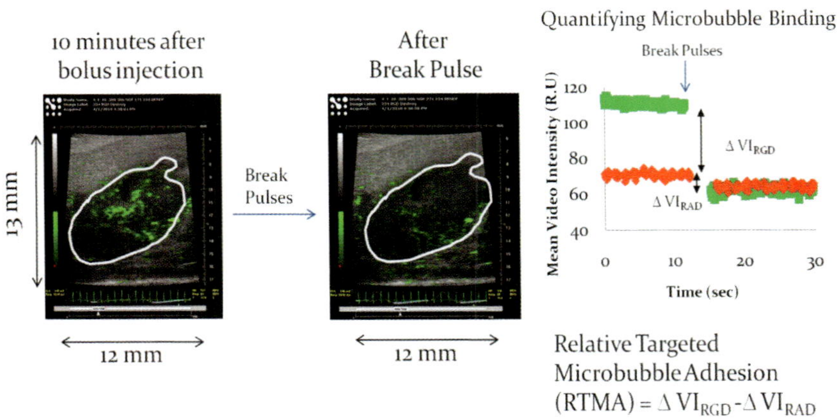

Figure 4.6 Method of determining relative molecular expression of $\alpha_v\beta_3$ integrin using targeted microbubbles with RGD loaded on the surface.

The accumulation of microbubbles can be detected by ultrasound imaging using a "dwell-destroy" technique.[33] (Figure 4.6). In this technique, a specific dwell period following the bolus injection is selected, determined by the amount of time it takes for microbubbles to be removed from circulation (~5–10 minutes). Only the targeted microbubbles that are bound to $\alpha_v\beta_3$ integrins remain in the imaging plane. Following the dwell period, a series of low-frequency higher-pressure pulses are applied to destroy the remaining microbubbles. US images are captured immediately before and after the microbubble fragmentation pulse. The decrease in video signal intensity following microbubble destruction is used to quantify relative targeted microbubble adhesion (*rtma*). Non-targeted isotype microbubbles should be used in separate injections to determine levels of non-specific microbubble adhesion (in this example RAD coated microbubbles are used as a control). The overall *rtma* can be calculated as follows:

$$rtma = \left(\overline{VI}_{RGD_{before}} - \overline{VI}_{RGD_{after}}\right) - \left(\overline{VI}_{RAD\_before} - \overline{VI}_{RAD\_after}\right) \quad (2)$$

where $\overline{VI}_{RGD\_before}$ and $\overline{VI}_{RGD\_after}$ are the average video intensities before and after the break pulse from the RGD-coated microbubbles, and $\overline{VI}_{RAD\_before}$ and $\overline{VI}_{RAD\_after}$ are the average video intensities before and after the break pulse from the RAD-coated microbubbles.

Variability in tumor perfusion and levels of epitope expression may influence the microbubble concentrations and dwell times using this technique. Recent studies have shown that the diagnostic sensitivity in molecular imaging can be significantly improved using novel size-selected microbubbles,[34] as well as the application of US radiation forces (low amplitude US pulses near the microbubble resonance frequency) that can facilitate interactions between the microbubble and vascular endothelium.[35] Methods of eliminating long dwell-times are also being developed that will ultimately reduce variability in molecular imaging procedures and allow this technique to become more clinically viable.[36]

The procedure for determining relative microbubble binding to vascular epitopes provides a quantitative parameter for evaluating relative levels of expression for a wide variety of diseases. Currently, microbubbles have been targeted to numerous epitopes such as VEGFR2, VECAM-1, I-CAM and PECAM, as well as for various diseases including atherosclerosis and cancer.[37] In cancer imaging, targeted microbubbles (VEGR2) have demonstrated highly accurate (>95%) diagnostic performance in preclinical screening of malignant vs. benign breast cancer lesions,[38] indicating that molecular imaging using UCAs may be promising for early breast cancer detection and monitoring breast cancer progression *in vivo*.

4.6 Challenges in quantitative CEUS imaging in cancer

While the CEUS techniques that have been described in this chapter have the potential to provide valuable clinical data for identifying tumor lesions and assessing tumor response to therapies, there are several challenges in performing quantitative CEUS that must be considered. A full discussion of variability in CEUS imaging due to ultrasound parameters and physical properties of the microbubbles are beyond the scope of this chapter, but are discussed in detail by Tang *et al.*[39] Briefly, sources of variability may include variation in scanner settings (frequency, gain, power, etc.), attenuation of the US signal caused by different tumor depths, contrast agent type, and complex interactions of the microbubbles in the US field. For these reasons, CEUS is often referred to as a "pseudo-quantitative" imaging technique. Quantitative parameters obtained from CEUS are therefore more suited towards monitoring relative changes in the tumor microvasculature rather than physical metrics of perfusion or molecular expression. Currently, there is no clinical standard for CEUS tumor imaging, making direct comparisons between patients difficult as well. Further work in modeling the microbubble response in the US field, and standardized protocols for imaging are needed to make CEUS a truly quantitative imaging modality.

4.6.1 *Tumor variability*

Quantitative CEUS imaging can provide valuable metrics for evaluating tumor perfusion and longitudinal monitoring of tumor response to therapy. However, there are several obstacles that must be considered when applying such metrics in a clinical setting. One of the issues with conventional 2D imaging is that only a single subjectively chosen imaging plane within a tumor is being evaluated. Furthermore, conventional 2D imaging may be problematic when evaluating tumors longitudinally as the tumor itself may change, thus identifying the same 2D tumor cross-section can be difficult in repetitive imaging procedures. Studies have shown that 3D imaging techniques can provide superior diagnostic results over the entire tumor volume in terms of quantifying vascularity through Doppler imaging[33] and response of tumors to therapy using contrast-enhanced molecular imaging.[10,34] While, 3D-CEUS is a promising approach for addressing heterogeneity in tumors, this technique remains underutilized

Table 4.1 Advantages and disadvantages of CEUS over other quantitative imaging modalities.

Imaging Modality	Advantages	Disadvantages
Ultrasound	Inexpensive Portable Point-of-care Real-time imaging Safe	Hand held probe; not whole body Operator-dependent
MRI	Whole body scan Detailed diagnostic pictures	Expensive imaging device Risks with contrast agents (NSF) Risks with anesthesia Problems with medical implants Time consuming
PET/CT	Whole body scan Multi-target imaging	Requires expensive isotopes Harmful radiation Time consuming

due to hardware restrictions and increased image acquisition/processing times. Further work in this area is needed to make 3D CEUS more amenable for clinical use (see Table 4.1).

4.7 Conclusion

CEUS has the potential to significantly improve cancer therapies by providing quantitative benchmarks for assessing changes in tumor vasculature longitudinally in a safe and cost-effective manner. Clinical use of quantitative vascular perfusion or molecular imaging data could provide significant benefits for patients by enabling earlier and more effective clinical decision-making. Given the high cost of biologically-targeted therapies, such early assessment of drug effectiveness could reduce the economic strains of cancer treatment for patients and families.

References

1. Van Persijn van Meerten EL, Gelderblom H, Bloem JL. RECIST revised: implications for the radiologist. A review article on the modified RECIST guideline. *Eur Radiol*, 2010; 20(6): 1456–1467.

2. McHugh K, Kao S. Response evaluation criteria in solid tumours (RECIST): problems and need for modifications in paediatric oncology? *Br J Radiol*, 2003; 76(907): 433–436.
3. Eisenhauer EA, Therasse P, Bogaerts J, Schwartz LH, Sargent D, Ford R, *et al.*, New response evaluation criteria in solid tumours: revised RECIST guideline (version 1.1). *Eur J Cancer Oxf Engl.* (1990), 2009; 45(2): 228–247.
4. Wilson SR, Burns PN. Microbubble-enhanced US in body imaging: what role? *Radiology*, 2010; 257(1): 24–39.
5. Sirsi S, Borden M. Microbubble compositions, properties and biomedical applications. *Bubble Sci Eng Technol*, 2009; 1(1–2): 3–17.
6. Feshitan JA, Chen CC, Kwan JJ, Borden MA. Microbubble size isolation by differential centrifugation. *J Colloid Interface Sci*, 2009; 329(2): 316–324.
7. Stride E, Edirisinghe M. Novel microbubble preparation technologies. *Soft Matter*, 2008; 4(12): 2350–2359.
8. Talu E, Lozano MM, Powell RL, Dayton PA, Longo ML. Long-Term Stability by lipid coating monodisperse microbubbles formed by a flow-focusing device. *Langmuir*, 2006; 22(23): 9487–9490.
9. Qin S, Caskey CF, Ferrara KW. Ultrasound contrast microbubbles in imaging and therapy: physical principles and engineering. *Phys Med Biol*, 2009; 54(6): R27–R57.
10. Postema M, Schmitz G. Bubble dynamics involved in ultrasonic imaging. *Expert Rev Mol Diagn*, 2006; 6(3): 493–502.
11. Wilson SR, Greenbaum LD, Goldberg BB. Contrast-enhanced ultrasound: what is the evidence and what are the obstacles? *AJR Am J Roentgenol*, 2009; 193(1): 55–60.
12. Quaia E. Microbubble ultrasound contrast agents: an update. *Eur Radiol*, 2007; 17(8): 1995–2008.
13. Williams R, Hudson JM, Lloyd BA, Sureshkumar AR, Lueck G, Milot L, *et al.*, Dynamic microbubble contrast-enhanced US to measure tumor response to targeted therapy: a proposed clinical protocol with results from renal cell carcinoma patients receiving antiangiogenic therapy. *Radiology*, 2011; 260(2): 581–590.
14. Cosgrove D. Ultrasound contrast agents: an overview. *Eur J Radiol*, 2006; 60(3): 324–330.
15. Cheung AMY, Brown AS, Cucevic V, Roy M, Needles A, Yang V, *et al.*, Detecting vascular changes in tumour xenografts using micro-ultrasound and

micro-ct following treatment with VEGFR-2 blocking antibodies. *Ultrasound Med Biol*, 2007; 33(8): 1259–1268.
16. Forsberg F, Ro RJ, Potoczek M, Liu JB, Merritt CRB, James KM, *et al.*, Assessment of angiogenesis: implications for ultrasound imaging. *Ultrasonics*, 2004; 42(1–9): 325–330.
17. Guibal A, Taillade L, Mulé S, Comperat E, Badachi Y, Golmard JL, *et al.*, Noninvasive contrast-enhanced US quantitative assessment of tumor microcirculation in a murine model: effect of discontinuing anti-VEGF therapy. *Radiology*, 2010; 254(2): 420–429.
18. Hoyt K, Warram JM, Umphrey H, Belt L, Lockhart ME, Robbin ML, *et al.*, Determination of breast cancer response to bevacizumab therapy using contrast-enhanced ultrasound and artificial neural networks. *J Ultrasound Med Off J Am Inst Ultrasound Med*, 2010; 29(4): 577–585.
19. Watson KD, Hu X, Lai C-Y, Lindfors HA, Hu-Lowe DD, Tuthill TA, *et al.*, Novel ultrasound and DCE-MRI analyses after antiangiogenic treatment with a selective VEGF receptor inhibitor. *Ultrasound Med Biol*, 2011; 37(6): 909–921.
20. Lamuraglia M, Escudier B, Chami L, Schwartz B, Leclère J, Roche A, *et al.*, To predict progression-free survival and overall survival in metastatic renal cancer treated with sorafenib: pilot study using dynamic contrast-enhanced Doppler ultrasound. *Eur J Cancer Oxf Engl 1990*, 2006; 42(15): 2472–2479.
21. Lassau N, Lamuraglia M, Chami L, Leclère J, Bonvalot S, Terrier P, *et al.*, Gastrointestinal stromal tumors treated with imatinib: monitoring response with contrast-enhanced sonography. *Am J Roentgenol*, 2006; 187(5): 1267–1273.
22. Eisenbrey JR, Dave JK, Merton DA, Palazzo JP, Hall AL, Forsberg F. Parametric imaging using subharmonic signals from ultrasound contrast agents in patients with breast lesions. *J Ultrasound Med*, 2011; 30(1): 85–92.
23. Seicean A, Badea R, Stan-Iuga R, Mocan T, Gulei I, Pascu O. Quantitative contrast-enhanced harmonic endoscopic ultrasonography for the discrimination of solid pancreatic masses. *Ultraschall in Med* 2010; 31(6): 571–576 DOI: 10.1055/s-0029-1245833.
24. Gheonea DI, Streba CT, Ciurea T, Săftoiu A. Quantitative low mechanical index contrast-enhanced endoscopic ultrasound for the differential diagnosis of chronic pseudotumoral pancreatitis and pancreatic cancer. *BMC Gastroenterol*, 2013; 13: 2.

25. Dickson PV, Hamner JB, Sims TL, Fraga CH, Ng CYC, Rajasekeran S, *et al.*, Bevacizumab-induced transient remodeling of the vasculature in neuroblastoma xenografts results in improved delivery and efficacy of systemically administered chemotherapy. Clin Cancer Res, 2007; 13(13): 3942–3950.
26. Bhayana D, Kim TK, Jang H-J, Burns PN, Wilson SR. Hypervascular liver masses on contrast-enhanced ultrasound: the importance of washout. *Am J Roentgenol*, 2010; 194(4): 977–983.
27. Wilson SR, Burns PN. An algorithm for the diagnosis of focal liver masses using microbubble contrast-enhanced pulse-inversion sonography. *Am J Roentgenol*, 2006; 186(5): 1401–1412.
28. Von Herbay A, Vogt C, Willers R, Häussinger D. Real-time imaging with the sonographic contrast agent SonoVue: differentiation between benign and malignant hepatic lesions. *J Ultrasound Med*, 2004; 23(12): 1557–1568.
29. Sirsi S, Feshitan J, Kwan J, Homma S, Borden M. Effect of microbubble size on fundamental mode high frequency ultrasound imaging in mice. *Ultrasound Med Biol*, 2010; 36(6): 935–948.
30. Sirsi SR, Hernandez SL, Zielinski L, Blomback H, Koubaa A, Synder M, *et al.*, Polyplex-microbubble hybrids for ultrasound-guided plasmid DNA delivery to solid tumors. *J Control Release*, 2012; 157(2): 224–234.
31. Bulte JWM, Modo MMJJ, editors. Nanoparticles in Biomedical Imaging: Emerging Technologies and Applications. New York: Springer; 2008. p. 524
32. Anderson CR, Hu X, Zhang H, Tlaxca J, Declèves A-E, Houghtaling R, *et al.*, Ultrasound molecular imaging of tumor angiogenesis with an integrin targeted microbubble contrast agent. *Invest Radiol*, 2011; 46(4): 215–224.
33. Dayton PA, Rychak JJ. Molecular ultrasound imaging using microbubble contrast agents. *Front Biosci J Virtual Libr*, 2007; 12: 5124–5142.
34. Streeter JE, Gessner R, Miles I, Dayton PA. Improving sensitivity in ultrasound molecular imaging by tailoring contrast agent size distribution: in vivo studies. *Mol Imaging*, 2010; 9(2): 87–95.
35. Borden MA, Streeter JE, Dayton PA, Sirsi SR. *In vivo* demonstration of cancer molecular imaging with ultrasound radiation force and buried-ligand microbubbles. *Mol Imaging*, 2013; 12: 1.
36. Pysz MA, Guracar I, Tian L, Willmann JK. Fast microbubble dwell-time based ultrasonic molecular imaging approach for quantification and monitoring of angiogenesis in cancer. *Quant Imaging Med Surg*, 2012; 2(2): 68–80.

37. Kiessling F, Fokong S, Koczera P, Lederle W, Lammers T. Ultrasound microbubbles for molecular diagnosis, therapy, and theranostics. *J Nucl Med*, 2012; 53(3): 345–348.
38. Bachawal SV, Jensen KC, Lutz AM, Gambhir SS, Tranquart F, Tian L, *et al.*, Earlier detection of breast cancer with ultrasound molecular imaging in a transgenic mouse model. *Cancer Res*, 2013; 73(6): 1689–1698.
39. Tang M-X, Mulvana H, Gauthier T, Lim AKP, Cosgrove DO, Eckersley RJ, *et al.*, Quantitative contrast-enhanced ultrasound imaging: a review of sources of variability. *Interface Focus*, 2011; 1(4): 520–539.
40. Shashank R. Sirsi *et al.* Contrast ultrasound imaging for identification of early responder tumor models to anti-angiogenic therapy. *Ultrasound Med Biol*, 2012, 38(6), 1019–1029.

Chapter 5

MULTIFUNCTIONAL DENDRITIC NANOPARTICLES AS A NANOMEDICINE PLATFORM

Hao-jui Hsu, Ryan M. Pearson* and Seungpyo Hong*,†*

**Division of Pharmaceutical Sciences*
School of Pharmacy, University of Wisconsin, Madison, WI 53705, USA
†Division of Integrated Science and Engineering,
Underwood International College, Yonsei University, Seoul, Korea 03706
†seungpyo.hong@wisc.edu

Abstract

Dendrimers are a class of well-defined, highly branched, multifunctional polymeric nanomaterials that have been largely studied as delivery platforms to target therapeutics or diagnostics to various sites of interest. In this chapter, we systematically describe the use of dendrimers for biomedical applications and how their physicochemical properties can be tuned to achieve desired outcomes in areas such as drug and gene delivery, anti-pathogen, and diagnostics. Furthermore, we introduce dendritic polymers and how their unique properties merge advantageous properties of dendrimers with other drug delivery platforms such as linear-block polymers and liposomes to improve biological responses. Several examples of the use of dendrimers and dendritic polymers will be detailed with the objective to reveal critical nanocarriers design features to achieve specific delivery goals. Although challenges still remain to be addressed to enable the clinical translation of this nanotechnology, it is clear that the future for dendritic nanomaterials is highly promising.

5.1 Introduction

Nanotechnology has impacted a wide range of biomedical applications.[1–3] A variety of nanoparticles have been developed to deliver therapeutic (e.g. small molecules and nucleic acids) and/or diagnostic agents (e.g. MRI contrast agents) to specific targets in the body with improved efficacies and efficiencies.[4–7] Polymers are promising candidates to be fabricated as nano-scale particles due to their relatively simple synthesis, biocompatibility, versatility, ability to self-assemble, and functionality to be stimuli responsive.[8]

Recently, a new design concept has emerged focusing on producing polymers with precisely controlled molecular weight distributions, as well as well-defined particle size and shape for biomedical applications. Dendrimers, a class of highly branched, deformable, tree-like macromolecules that display a high density of peripheral functional groups, fit in this category perfectly. The unique iterative synthetic procedure of dendrimers produces nano-scale macromolecules with close-to-monodisperse molecular weight distributions.[9] Moreover, owing to their large number of surface groups and high degree of molecular flexibility, strong binding interactions are promoted between surface conjugated targeting ligands and complimentary cell surface receptors,[10,11] which makes them an ideal platform to deliver various therapeutics or diagnostics.[12]

The unique physicochemical and biological properties of dendrimers have spawned the development of novel polymer classes that combine the dendritic wedge-shaped segments of dendrimers (or dendrons) with other polymeric structures.[13] For example, amphiphilic linear dendritic-block copolymers, which are a dendritic scaffold with multiple linear polymer chains, are capable of self-assembling into core-shell micelles that possess larger loading capacity compared to dendrimers while maintaining the multivalency.[14] Although these dendritic polymers are not as widely studied as dendrimers to date, they have shown their potential to greatly impact the nanomedicine arena.

In this chapter, we aim to describe the relationship between the molecular architecture of dendritic polymers and their biological properties, along with discussion on how those polymers are used in constructing advanced nanomedicine. The discussion will progress through the important design parameters of dendrimer-based nanoparticles, followed by their therapeutic and diagnostic applications. Following this, we will list a few types of novel dendritic polymers and discuss their design rationales for specific applications. This chapter will be concluded with a brief discussion about the major challenges that these dendritic nanoparticles are facing in clinical translation and a future prospective.

5.2 Dendrimers used in biomedical applications

Dendrimers are regularly hyperbranched, flexible, monodispersed macromolecules with sizes ranging from 2 to 10 nm, depending on their generation and surface modifications.[9,15,16] Because dendrimers possess a large number of surface groups, a multifunctional platform can be achieved *via* functionalization of a single dendrimer molecule with targeting, therapeutic, and diagnostic moieties.[12] In the following sections, we will describe the synthetic process, development history, and biological properties of various types of dendrimers used in the biomedical field, followed by an application-oriented discussion of the recent advances in dendrimer-based biomedical devices.

5.2.1 *Synthetic routes of dendrimers*

Dendrimers are synthesized primarily using one of two major synthetic routes (divergent and convergent) as depicted in Figure 5.1. The divergent method, developed by Tomalia *et al.*, constructs a poly(amidoamine) (PAMAM) dendrimer macromolecule outward from a poly-functional core using a reiterative sequence consisting of activation and condensation reactions.[17] Each successive growth step of the dendrimer results in an increased generation (G). As G increases, the number of peripheral groups and molecular weight are doubled. Although the divergent method is the most widely adopted, it is prone to produce defective products, especially for higher G dendrimers, due to increasing peripheral steric hindrance and difficulties in purification.[15] The convergent method, on the other hand, can minimize the production of defective products since each dendrimer is built from individually prepared dendrons and then assembled into a dendrimer through a multifunctional core. However, the convergent method requires additional synthetic steps and provides relatively lower

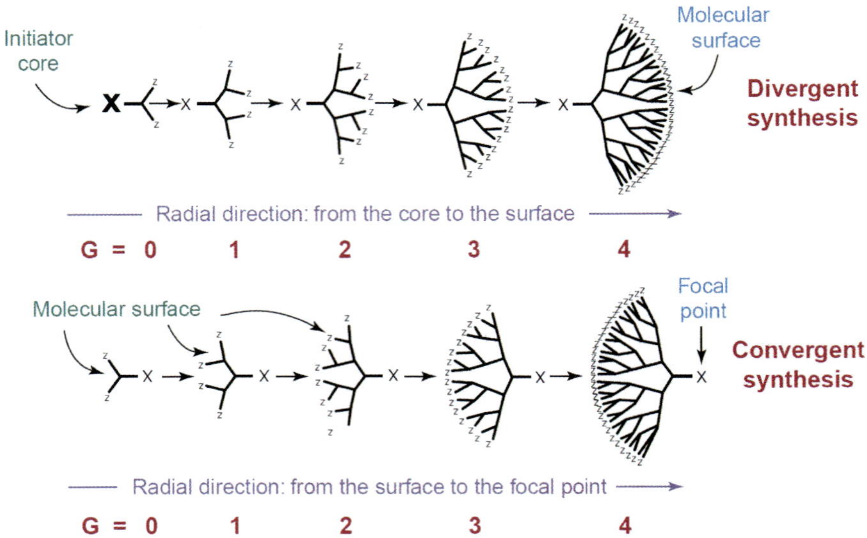

Figure 5.1 Major synthetic routes for dendrimers. The divergent synthesis route and the convergent synthesis route of dendrimer. Reprinted with permission from Elsevier.[15]

yields for higher generation dendrimers. In addition to these two methods, synthetic methods such as double-stage convergent growth,[18] orthogonal synthesis,[19] double exponential growth,[20] and orthogonal coupling,[21] have been developed to synthesize dendrimers in less steps, with less defects, and with increased yields.

5.2.2 *Types of dendrimers*

Dendrimers are comprised of three functional domains (core, branches and periphery), as depicted in Figure 5.2a. The central core dictates shape and directionality, the interior branches control size and biocompatibility, and the peripheral functional groups mainly determine their biological interactions.[22] Through manipulation of these domains, a variety of molecular structures can be made and tailored. In following subsections, we introduce PAMAM, poly(propylenimine) (PPI), poly(L-lysine) (PLL), and carbosilane (C–Si) dendrimers and their distinct features in biomedical applications (Figures 5.2a–5.2d).

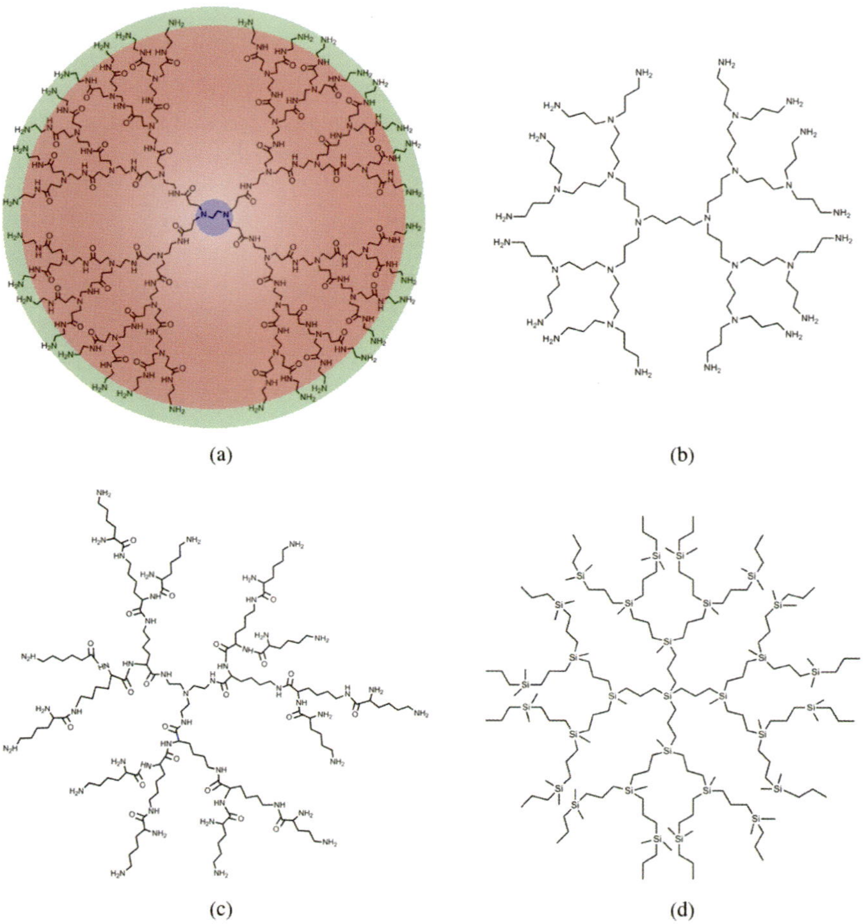

Figure 5.2 Chemical structures of commonly used dendrimers. (a) polyamidoamine (blue: core, red: branches, green: periphery), (b) poly(propyleneimine), (c) poly(L-lysine), and (d) carbosilane dendrimers. Generation 3 (G3) of each dendrimer is illustrated.

PAMAM dendrimers are synthesized using the divergent method from amine-functional cores (Figure 5.2a).[17] They are synthesized through successive alkylation and amidation reactions, which account for each generational growth step. Complete generations of PAMAM dendrimers carry positively charged surface groups (primary amine groups), while half-generations possess carboxyl termini and thereby present negatively charged surfaces.[23] The amine termini are ideal functional handles to enable facile conjugation with various bioactive moieties. In addition, because of the long development history and a broad range of applications, PAMAM dendrimers have become one of the most extensively investigated dendrimers.

PPI dendrimers are synthesized by the divergent method starting from 1,4-diaminobutane (DAB) and are also referred to as DAB-dendrimers (Figure 5.2b).[24,25] PPI dendrimers are synthesized by repeated double alkylation of the amines with acrylonitrile by "Michael addition" followed by reduction. Therefore, the surface modification of PPI dendrimers can be easily achieved using the similar methods described above for PAMAM dendrimers.

PLL dendrimers are prepared from BOC-protected lysines using an iterative sequence of peptide synthetic process (condensation) and BOC de-protection (activation) (Figure 5.2c).[26] PLL dendrimers were first introduced to the biomedical field as a platform for synthetic vaccines, where peptide antigens were conjugated to the surface and enhanced immunogenicity was observed.[27] Recently, a breakthrough has been made in preventing sexually transmitted disease using naphthalene disulfonic acid (DNAA)-conjugated PLL dendrimers.[28] Several clinical trials have been done and a condom coated with this DNAA-conjugated dendrimer has been approved in Japan. More details regarding the clinical applications of this DNAA-conjugated dendrimer can be found in the subsection of "dendrimers for therapeutic applications."

C–Si dendrimers were first synthesized using the divergent method in 1990s.[29–31] These dendrimers possesses many distinct features, especially high energy C–Si bonds that provide better kinetic and thermodynamic stability, and low polarity making these macromolecules more hydrophobic than polyamine dendrimers (Figure 5.2d). Therefore, hydrophilic functional groups must be attached to the surface in order to make C–Si dendrimers water soluble.[32]

5.2.3 *Biological properties of dendrimers*

When introducing dendrimers to a biological environment, a number of parameters need to be considered and optimized in order to achieve desired biological interactions. In this section, we discuss the general principles for designing and modulating dendrimers to obtain a favorable pharmacokinetic profile, tolerable cytotoxicity, and highly specific biological interactions.

5.2.4 *Pharmacokinetics and biodistribution*

The pharmacokinetics and biodistribution of dendrimers primarily depends on their generation and surface composition. In general, the kidney and reticuloendothelial system (RES) are major targets of dendrimer-based nanomedicines. While lower generation dendrimers result in more effective renal clearance due to lower molecular weight and smaller size, larger dendrimers induce more effective uptake by the RES that leads to increased RES clearance.[33,34] Kobayashi *et al.* examined the blood residence of a series of PAMAM dendrimers (G2–G9) and reported that the plasma circulation time was increased with an increase of the generation, and G7 PAMAM dendrimers reached at the maximum time.[35,36] G8 and G9 PAMAM dendrimers exhibited an increased uptake in liver and spleen in lieu of plasma circulation time.

Surface charge of dendrimers is another important parameter that affects their pharmacokinetic profiles. Amine-terminated dendrimers (e.g. PAMAM, PPI, and PLL dendrimers) carry positive surface charges under physiological conditions. Those cationic dendrimers were shown to strongly bind to vasculature of highly perfused organs due to non-specific electrostatic interactions, resulting in rapid clearance from the blood and high initial volume of distribution compared to anionic dendrimers.[37,38] Malik *et al.* used ^{125}I-labeled dendrimers to study the charge effects of dendrimers using Wistar rats.[37] This study revealed that the circulation time of anionic dendrimers is 10–20 fold longer than that of cationic dendrimers.

Importantly, the charge-based non-specific interactions between positively charged dendrimers and cell membranes are typically associated with toxicity, which is more pronounced with the generational growth of dendrimers likely due to the higher number of charged groups in high generation dendrimers. Although the exact mechanism that causes toxicity is still not clear, Hong *et al.* showed that the electrostatic interactions of cationic dendrimers with cell membranes induced the formation of nano-scale holes through a reversible membrane destabilization mechanism.[39–41] In addition, upon the removal of positive surface charges on dendrimers, a tolerable cytotoxicity profile could be obtained.[39] Kaminskas *et al.* studied the effect of replacing terminal amine groups with different chemical structures on the pharmacokinetics and disposition of PLL dendrimers. When the peripheral amine groups were modified by benzene sulfonate, benzene disulfonate, or succinate groups, PLL dendrimers showed significantly lower vasculature binding and enhanced metabolic stability.[42] Coating the dendrimer surface with polyethylene glycol (PEGylation) is another widely adopted method to mask the surface charges.[43] This highly flexible, hydrophilic polymer improves the pharmacokinetic profiles of dendrimers by endowing stealth properties that help avoid opsonization and reduce RES recognition.[44] Kojima *et al.* showed that PEGylated PAMAM dendrimers exhibited longer circulation time compared to non-PEGylated dendrimers.[45] The same phenomenon was shown on PEGylated PLL dendrimers as well.[46] The prolonged circulation time can be attributed to the high degree of surface coverage of PEG and the increased molecular weight. Taken together, all the reports described here support that the pharmacokinetic and toxicological profile of dendrimers can be controlled by their size and surface modifications.

5.2.5 *Multivalent interactions*

Multivalent interactions are characterized by the simultaneously binding of multiple ligands to multiple receptors, which have been found play a critical role in many physiological and pathological processes such as cell-to-cell binding and communication, antigen-antibody interactions, and adhesion of a virus particle to a host cell membrane.[47–49] Depending on the biological event, these multivalent interactions result in higher binding strength, amplified signals for cell-to cell communication, enhancements in binding specificity, or prevention of undesired interactions. This biological phenomenon has inspired the development of biomimetic materials that are configured to perform multivalency and thus magnify ligand-receptor interactions. Dendrimers are some of the most ideal platforms for exploiting the multivalent interactions, because they possess numerous terminal groups for conjugation with multiple ligands and their highly deformable structure are able to optimize the ligand orientations for maximum substrate binding.[50] The multivalent binding effect of dendrimers was first described in a study using mannoside-functionalized dendrimers for investigating the interactions between mannoside and concanavalin A or pea lectins.[51] The dendrimer conjugated with 16 mannoside moieties showed significantly stronger binding avidity toward concanavalin A and pea lectins compared to free mannoside.[52] To quantify the dendrimer-mediated multivalent binding effect, Hong *et al.* measured the binding avidities between G5 PAMAM with different numbers of surface-functionalized folic acid (FA) and folate-binding protein using surface plasmon resonance (SPR).[10] G5 PAMAM dendrimers were found to have enhanced binding avidities up to 170,000-fold higher that of free folic acid. However, the material properties were deteriorated when the number of conjugated FA increased (increased polydisperisty index (PDI) and poorer water solubility). Furthermore, fluorescence-activated cell sorting showed that when the number of FA per dendrimer was above seven, only marginal increases in cell binding were observed demonstrating that there is an optimized number of conjugated ligands which maximizes the multivalent binding effects.

PEGylation could be used to enhance multivalent binding, because PEG and other flexible linkers are able to facilitate a more desirable geometric orientation of ligand towards their receptors in a thermodynamically favorable manner.[53–56] Importantly, the length of those flexible linkers needs to be optimized to prevent undesirable back-folding of the conjugated ligands which masks the binding.[53] Overall, the design of dendritic multivalent ligands is different from case to case due to the large variations in dendritic polymers, ligands and receptors. Therefore, optimization is always needed when different ligands are applied.

5.2.6 *Dendrimers for therapeutic applications*

Surface decoration of dendrimers plays a crucial role in dictating the biological properties of nanomaterials. The surface properties of the dendrimers can be modified with different bioactive entities (targeting, therapeutic or imaging agents) for various purposes, such as targeted drug delivery, gene delivery, anti-microbial, and diagnosis. In this section, we will discuss the basic design strategies of dendrimer-based biomedicines in three major therapeutic applications, drug delivery, gene delivery, and anti-pathogen. Furthermore, examples of application will be provided in each category to demonstrate how to use dendrimers-based platforms for additional therapeutic benefits.

5.2.7 *Drug delivery*

Dendrimers have been widely investigated as a drug delivery platform for small molecules *via* multiple administration routes. Besides the commonly used parenteral route, dendrimer-based drug delivery systems can be administered through oral,[57] transdermal,[58–60] pulmonary,[61] and ocular routes.[62] Regardless of the routes of administration, a few design strategies, such as methods of drug incorporation and surface decoration, are crucial for all types of dendrimer-based carriers. There are two major methods that have been employed which allow dendrimers to "carry" small molecules. The first method is physical adsorption or encapsulation, in which the small molecule drugs interact with the dendrimers non-covalently and form a host-guest complex;[63,64] and the second, drugs are conjugated to the peripheral groups of dendrimers through chemical reactions.[65–67] Although the drug encapsulation is relatively simple, several drawbacks, such as low drug loading efficiency and uncontrolled drug release, limit the utility of this method. Chemical conjugation, on the other hand, is more preferable and has been used in most of the dendrimer-based delivery systems despite the fact that a single dendrimer molecule can only carry a limited amount of molecules in order to maintain structural regularity and solubility.[10,68]

Conjugation of drug molecules to the surface of dendrimers can negatively alter their observed biological properties. For example, Lim *et al.* studied a series of PEGylated G2 triazine dendrimers with 12 paclitaxel molecules using PC-3 prostate cancer xenografted SCID mice.[69] This study showed that the dendrimer-based carrier increased the tolerated dose of paclitaxel up to 200% compared to free paclitaxel. However, lower accumulation at the tumor site and shorter circulation times were observed. The commonly used PEGylation technique can also affect the efficacy of conjugated drugs. Zhu *et al.* compared the degree of PEGylation using doxorubicin (DOX)-PAMAM dendrimers conjugates.[70] While the increasing degree of PEGylation improved the circulation time and tumor accumulation of DOX-PAMAM conjugates *in vivo*, the cytotoxicity of DOX-PAMAM significant decreased due to lower cellular uptake by tumor cells. Since the stealth layer unavoidably reduces the interaction between nanoparticles and tumor cells, the incorporation of targeting ligands[71] and stimuli-responsive linkers[70] have been used to achieve the site-specific delivery.

5.2.8 *Gene delivery*

Since Haensler *et al.* first reported that PAMAM dendrimers could mediate the transfection of mammalian cells in culture,[72] various dendrimers were developed to deliver nucleic acids (NA) including antisense oligonucleotides,[73] miRNA,[74] siRNA and pDNA.[75] Just like other cationic polymers, cationic dendrimers can form complexes with negatively charged NAs based on electrostatic interactions.[76] Since the NA is highly condensed in the dendrimer-NA complex (dendriplex), the enzymatic degradation of the NA is prevented. The transfection efficiency of dendrimers is mainly determined by their generation and the N/P ratio (ratio of amines on the polymer to phosphates on the NA) of dendrimer-NA complex. Haensler *et al.* investigated the transfection efficiency of PAMAM dendrimers with generations ranging from G2 to G10 at various N/P ratios using CV-1 cells.[72] They reported that the optimal generations ranged from G7 to G10 and excess of dendrimers to NA enhanced the transfection efficiency. When they compared different types of cationic dendrimers, PAMAM-DNA complexes at N/P = 6 performed 1000-fold higher transfection efficiency than complexes made by PLL dendrimers, which indicated that the transfection efficiency also depends on the molecular structure of dendrimers. Interestingly, partial degradation of the dendrimer structure resulted in a 50-fold enhancement in the transfection efficiency compared to intact dendrimers, which is primarily attributed to a higher degree of flexibility of degraded dendrimer.[77] This finding led to the development of a commercialized dendrimer-based transfection reagent, SuperFect®, a partially degraded PAMAM dendrimer.

Surface modification and/or spatial modulation of the positive charges can also enhance the transfection efficiency. Huang *et al.* conjugated lactoferrin to PAMAM G5 through PEG spacers, and showed 2.25-fold higher brain uptake than PAMAM and PAMAM-PEG *in vivo*.[78]

PAMAM-arginine-DNA complexes showed 100-fold higher and 10-fold higher transfection efficiency than lipofectamine and PEI-DNA complexes.[79] Conjugation of glucocorticoids, such as dexamethasone and triamcinolone acetonide, to dendrimers has been shown to enhance the transport of dendrimer-NA complexes to the cell nucleus.[80,81] Those glucocorticoid-conjugated dendrimer-NA complexes bound to glucocorticoid receptors presented in the cytosol, inducing the dilation of the nucleus pore complex, and facilitating the translocation of dendrimer-NA complexes. Recently, a PAMAM-like dendrimer (quaternized PAMAM) with a fully acetylated or hydroxyl-terminated surface and the internal tertiary amine groups were developed.[82] The neutral surface and additional internal positive charges decreased the cytotoxicity drastically and did not sacrifice the efficiency of cellular uptake.

5.2.9 *Anti-pathogen*

The anti-pathogenic activities of dendrimer-based nanomaterials have been evaluated on a variety of pathogens. The mechanisms of the antibacterial and antifungal activities of dendrimers are primarily related to the formation of electrostatic interactions between the membrane of pathogens and dendrimers, followed by membrane disruption and cytoplasm leakage.[83] Since membrane disruption has also been observed in human cells, their selectivity towards those pathogens is still an issue. Nevertheless, dendrimers have been used as a carrier of anti-microbial drugs for improved solubility, bioavailability, and activity.[84,85]

Antiviral dendrimers, on the other hand, mainly prevent the binding of viruses to the host cell surface or inhibit replication of the viral genome. Due to the multivalent binding effect, the interactions between dendrimer-based nanoparticles and the viral surface are extremely stable, which effectively prevent the virus from interacting with host cells. Reuter *et al.* showed sialic acid-functionalized dendrimers enhanced the activity of sialic acid to inhibit the haemagglutination caused by influenza by 5×10^4-fold.[86] In a follow-up study, sialic acid-functionalized PAMAM dendrimers completely prevented mice from being infected by murine influenza pneumonitis. Furthermore, sialyllactose-functionalized C–Si dendrimers and polysulfated galactose-derived PPI dendrimers were shown to inhibit the infection of influenza and HIV-1, respectively.[87,88] Recently, a DNAA-based dendrimer, denoted as SPL7013, was shown to inhibit human immunodeficiency virus (HIV) and herpes simplex virus (HSV).[28,89] SPL7031 has been used as an active ingredient in a topical microbicide gel, VivaGel®, to prevent sexually transmitted diseases and bacterial vaginosis. VivaGel® is currently under phase II of clinical trials for prevention of recurrence of bacterial vaginosis. In addition, a VivaGel®-coated antiviral condom has been approved in Japan.

In addition to inhibiting viral binding, dendrimers have been shown to exhibit the ability to disturb the life cycle of viruses.[90] Unmodified PAMAM dendrimers have been reported to interact with transacting response elements RNA and thus affect the viral life cycle.[91] Viral protease and reverse transcriptase were shown to be inhibited by carboxylated fullerene-based dendrimers and sulfonated dendrimers, respectively.[92,93]

5.2.10 *Dendrimers for diagnostics*

While dendrimers are widely investigated as therapeutic agents, they are also employed to enhance the sensitivity of current imaging techniques *via* conjugating with contrast agents. The mutivalency, monodisperse characteristics, and relatively high molecular weight of dendrimers make them well suited for carrying T1 magnetic resonance (MR) imaging agents such as gadolinium (Gd).[94] Gd(III) complexes, such as Gd(III)DTPA (diethylenetriaminepentaacetic acid) and Gd(III)DOTA (1,4,7,10-tetraazacyclododecane-1,4,7,10-tetraacetic acid), have been proven to increase the T1 relaxation time and enhance the contrast in T1-weighted images. However, due to the low molecular weight, those agents possessed high renal clearance, and low molar relaxivity properties, which resulted in a loss of specificity and contrast efficiency.

It has been shown by many research groups that conjugating Gd(III) chelators onto dendrimers are able to achieve contrast efficiency enhancement which is attributed to high relaxivities and longer circulation time of the conjugates.[95–97] Interestingly, higher organ specificities can be achieved by optimizing the generation of those dendrimer-based MRI contrast agents. The diameters of dendrimers increase with the increasing generations, which result in differences in organ or tissue accumulation. For example, smaller contrast agents (3–6 nm in diameter) could be applied for functional renal imaging,[98] while larger ones (7–12 nm) are more suitable for blood pool imaging.[95] Backer *et al.* reported that FA and DOTA-Gd(III) conjugated G5 PAMAM dendrimers targeted to xenograft tumor cells *in vivo*.[99] Carboxyl-terminated FA-PEG-PAMAM dendrimers were reported to form complexes with Gd(III) ions for tumor targeting without using conventional chelators.[100] In addition to Gd(III), dendrimers have been used to improve the colloid stability of iron oxide nanoparticles for T2 MRI.[101]

5.3 Dendritic polymers used in biomedical applications

Dendritic polymers are unique materials that have received much attention recently for use in biomedical applications. Due to their ability to combine the advantageous properties of dendrimers such as low polydispersity, facile multifunctionalization, and multivalency with those of other inorganic or organic nanomaterials, novel materials that display significantly enhanced physicochemical and biological properties can be achieved.[13] Of particular interest are self-assembling dendritic polymers such as linear dendritic-block copolymers and dendrimersomes. Through the appropriate choice of the chemical composition and amphiphilicity, nanostructures such as micelles and vesicles that possess features amenable for applications in drug and gene delivery can be engineered. As a consequence of the self-assembly process, the nanocarriers are typically larger than 20 nm, which is favorable to enable passive targeting of the tumor sites through utilization of the enhanced permeability and retention (EPR) effect. Furthermore, delivery of hydrophobic and/or hydrophilic drug molecules without the need for chemical modification of the molecular structure of the drug offers an additional advantage.

5.3.1 *Dendritic-block copolymers*

Dendritic polymers that combine one or multiple linear polymer chains with a dendritic scaffold are termed linear dendritic block copolymers (LDBCs).[14] Given the increasing interest in the development of dendritic macromolecules in the early 1990s,[18,102,103] Frechet *et al.* developed the first LDBC based on a 3,5-dihydroxybenzyl alcohol backbone in 1992.[104] In this study, two generation 3 or 4 (G3/G4) dendrons were conjugated to the distal ends of a single PEG linker with various weights ranging from approximately 1 kDa to 20 kDa to study how conjugation efficiency was influenced by dendrons and how the molecular size was affected by dendronization of the PEG chain. Since that time, LDBCs have received considerable attention because of their modular capability to be multifunctionalized with imaging agents, targeting ligands, drugs, and other bioactive molecules for therapeutic applications.[9,14]

Dendrons, which are the highly branched segments that comprise dendrimers are synthesized using either divergent or convergent synthesis methods analogous to dendrimers.[9] The synthesis of LDBCs follow three possible pathways: (1) coupling; (2) chain-first, or (3) dendron-first (Figure 5.3).[14] In the coupling strategy, polymers and dendritic blocks are individually synthesized and conjugated in a single step. This method is used less frequently for the preparation of linear-block copolymers due to unfavorable coupling yields and difficulty to purify reaction mixtures. Since dendritic blocks are typically of modest molecular weight and display distinctly different physiochemical properties than linear-block copolymer counterparts, purification is typically more easy perform. In the chain-first strategy, a polymer with a reactive functional group is used to grow a dendron outwards by divergent synthesis. This method is

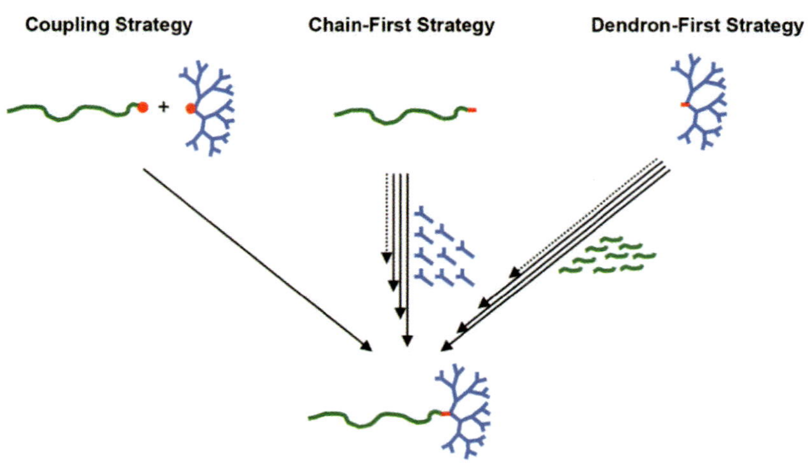

Figure 5.3 Different pathways to synthesize linear dendritic-block copolymers. Reprinted with permission from Elsevier.[14]

frequently used due to its robustness and versatility to adapt to a variety of dendrimer chemistries. However, it is difficult to produce dendron generations larger than G3 because during each generational growth step the number of reactions that are required to occur simultaneously doubles that ultimately decreases their yields. The dendron-first strategy utilizes the core functional group of the dendron to initiate the polymer chain growth. Here, dendrons are usually synthesized using the convergent synthesis strategy, where the dendron is built from the periphery to the core thus allowing for higher dendron generations to be synthesized increasing the number of potential structures available.

To obtain a more in-depth understanding of how the unique molecular architecture of LDBCs affects the self-assembly of copolymers into micelles, previous studies by our group provided a systematic comparison of the self-assembly process of PEGylated dendron-based copolymers (PDCs) over linear-block copolymer counterparts at similar hydrophilic-lipophilic balances (HLBs) (Figure 5.4a).[105] It was found that the pre-organized assembly of PEG chains on the surface of the dendron dramatically decreased the entropic cost for self-assembly and was responsible for PDCs displaying critical micelle concentrations 1–2 orders of magnitude ($\sim 10^{-8}$ M) lower than linear-block copolymers at similar HLBs. A similar trend showing that dendritic polymers have much lower CMCs than linear counterparts at similar HLBs was also observed recently with poly(lactide-co-glycolide) (PLGA)-b-dendron-b-PEG.[106] Additionally, molecular dynamics simulations were used to demonstrate that the surface of dendron-based micelles was almost completely covered by dense PEG layers whereas substantial portions of the hydrophobic core of linear-block copolymer micelles were uncovered by PEG.[105] Surface-modified dendron micelles that presented acetamide-, carboxyl-, or amine-functional groups were compared to G4 PAMAM dendrimers in terms of their cellular interactions.[107] Interestingly, the cellular interactions of the dendron micelles investigated were found to be negligible and surface-charge independent compared to dendrimers that are well known to display surface-charge dependent cellular interactions.[107] These results provide evidence that dendron-based micelles have greater thermodynamic stability than similarly comprised linear-block copolymer micelles, which may be translatable to drug delivery potential.

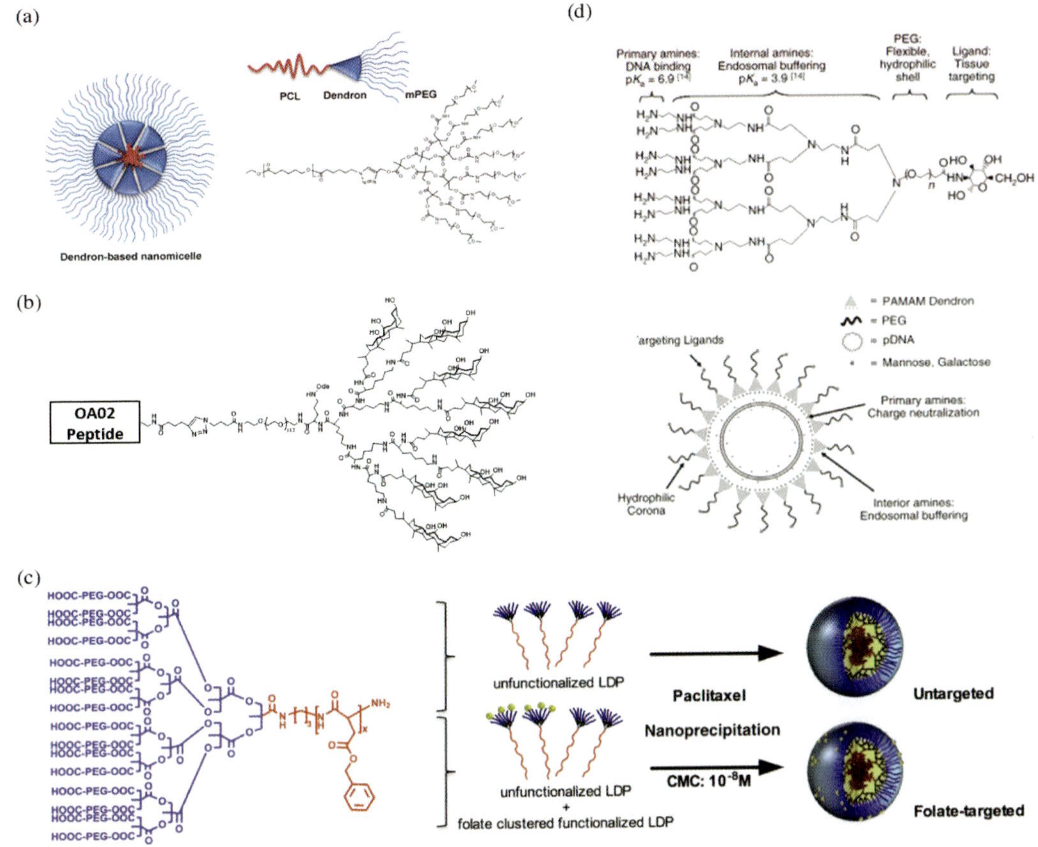

Figure 5.4 Examples of linear dendritic-block copolymers used for drug and gene delivery. (a) PEGylated dendron-based copolymers and their self-assembly into dendron-based nanomicelles.[105] (b) Telodendrimers for active targeting of ovarian carcinoma using OA02 peptide.[108] (c) Paclitaxel-encapsulated FA-targeted linear dendritic block copolymers for breast cancer treatment. Reprinted with permission from Elsevier.[114] (d) Mannose-PEG-PAMAM dendrons for delivery of plasmid DNA. Reprinted with permission from John Wiley and Sons.[117]

A number of groups have investigated LDBCs for the delivery of drugs and other genetic materials. For example, multi-functional LDBCs have been developed for intracellular delivery of antitumor drugs. Amphiphilic LDBCs comprised of semi-polyamidoamine-b-poly(D,L-lactic acid) (PALA) were self-assembled into micelles and used to deliver encapsulated docetaxel (DTX) to MCF-7 cells *in vitro* and a pharmacokinetic study using rats demonstrated that DTX-loaded PALA micelles enhanced the plasma concentration of DTX and prolonged drug clearance in comparison to Taxotere®, a clinically used formulation of DTX. [108] Other groups have demonstrated the use of LDBC-based micelles to deliver chemotherapeutics such as paclitaxel.[109] Linear PEG-block-dendritic cholic acid copolymers, or telodendrimers were investigated for their ability to use active targeting to enhance the delivery of paclitaxel (PTX) in a nanoparticle (NP) formulation.[110] The distal azide ($-N_3$) group on the PEG chain of telodendrimers was functionalized with a high-affinity peptide targeting ligand "OA02" (cdG-HoCit-GPQc-Ebes-K-alkyne) that binds selectively to alpha-3 integrin receptor using click chemistry for the targeted delivery of paclitaxel to ovarian cancer (Figure 5.4b). Decoration of the surface of dendritic NP lead to increased uptake efficiency in SKOV-3 and ES-2 ovarian cancer cells *in vitro*, greater tumor accumulation, superior anti-tumor efficacy, and improved toxicity profiles, when compared to non-targeted NPs and Taxol®, a clinically used formulation of PTX in SKOV-3 tumor xenografts. The hydrophobic chemotherapeutic camptothecin (CPT) was conjugated to dendritic polylysine to tune the hydrophobicity of the amphiphilic dendron-based copolymers (PEG-block-dendritic polylysine-CPT). Interestingly, by modulating the hydrophobic drug, various self-assembled morphologies could be achieved such as short nanorods, long nanorods, and nanospheres.[111] Endoxifen (EDX) was encapsulated into dendron micelles comprised of poly(ε-caprolactone)-b-G3 polyester dendron-b-PEG_8 and evaluated for their use as polymeric penetration enhancers.[112] Carboxylated dendron micelles were able to enhance the permeation of EDX across mouse skin up to 20-fold and in human skin up to 4-fold relative to ethanol. This study demonstrated that the use of dendron-based micelles for transdermal administration of chemotherapeutics for breast cancer prevention.

Besides utilizing polymers to direct the assembly of LDBCs and facilitate drug delivery, peptide-based LDBCs offer another level of biocompatibility. Hammond and colleagues have extensively investigated peptide-based LDBCs. An amphiphilic comb-dendritic block copolymer using poly(gamma-n-dodecyl-L-glutamate) (PDLG) as the hydrophobic block and a hydrophilic polyester dendron modified with PEG chains was investigated for its ability to self-assemble and encapsulate a model drug triclosan.[113] The PDLG hydrophobic backbone provided rigidity to the nanostructure resulting in a "tree-like" architecture. The critical micelle concentration was on the order of 10^{-8} M and approximately 30 wt% of triclosan could be encapsulated. Besides micelle structures being formed, the un-PEGylated version of the monomer formed polymeric vesicles.[114] Additionally, the cellular interactions of targeted LDBCs were investigated. Hydrophobic poly(benzyl-L-aspartate) (PBLA) was used to form the polypeptide linear block that was coupled to a generation 4 (G4; 16 end groups) polyester dendron to which 16 PEG (600 g/mol) chains were attached. FA was used as the targeting ligand and it was conjugated to the distal ends of the PEG chains to study the effect of ligand clustering on the cellular interactions of dendritic micelles.[115] They found that LDCs with ~3.2 FA/polymer mixed in a 60:40 ratio to untargeted polymer resulted in the highest level of cell uptake and highest binding avidity and greatest *in vivo* tumor localization. This study demonstrated that just as observed using dendrimers, the cluster arrangement of targeting ligands plays a significant role in determining the effectiveness of the nanocarrier. Additionally, similar polypeptide-based LDBCs were evaluated for their ability to localize at tumor xenograft sites and direct a four-fold enhancement in efficacy of paclitaxel with reduced toxicity (Figure 5.4c).[116]

In addition to uses in hydrophobic drug delivery, LDBCs have been used for the delivery of nucleic acid payloads. Wood *et al.* developed a library of linear-dendritic hybrid polymers that could form polyplexes with luciferase DNA using dendritic PAMAM-PEG-(galactose or mannose) of G0 to G6 (1 to 64 surface groups) (Figure 5.4d).[117] The polyplexes (G4–G6) displayed the ability to transfect P388D1 macrophages and HepG2 hepatocytes more effectively than commercially available polyethylenimine (PEI) under serum-containing culture conditions demonstrating their potential for use *in vivo*. Linear dendritic cationic vectors were developed for efficient DNA grasp and delivery.[118] Methoxy-PEG5K-dendritic polyglycerol-graft-tris(2-aminoethyl)amine (mPEG-DPG-g-TAEA) was evaluated for its pH buffering capacity, size, and zeta potential. It was found that each of those parameters could be tailored by varying the dendritic DPG block and the degree of TAEA substitution. Due to the molecular topology, the cationic TAEA was able to bind strongly to DNA with low amine content density. This resulted in significantly less cytotoxicity compared to PEI5K, while displaying similar transfection efficiency to that of lipofectamine 2000. Unique material properties of LDBCs are possible through the combination linear-block copolymers and dendrimers.

5.3.2 *Dendrimersomes*

Dendrimersomes are stable, monodisperse unilamellar vesicles self-assembled in water from amphilphilic Janus dendrimers.[119] Compared to liposomes and polymersomes prepared from phospholipids and linear-block copolymers, respectively, dendrimersomes display favorable material properties. For example, it is necessary for vesicles to exhibit stability in various media, low polydispersity, near impermeability, good mechanical properties, and they should be easily prepared.[120] Although liposomes have been frequently used in drug delivery applications, their preparation is more complicated due to the fact that various components such as phospholipids, PEGylated phospholipids, and cholesterol must be co-assembled. Polymersomes are generally considered to have greater mechanical stability than liposomes but the methods used for their preparation are difficult. The vesicles formed are usually polydisperse, and they lack biocompatibility in most cases. Additionally, to control the sizes and polydispersity of the formed vesicles, extrusion is necessary.

The Percec lab pioneered the development of dendrimersomes. Notably, the formation of dendrimersomes can be achieved using simple methods such as injection of ethanol or other water miscible solvents into water or buffer and display material properties that are superior to liposomes or polymersomes individually. Percec *et al.* synthesized 11 distinct libraries that contained a total of 107 amphiphilic Janus dendrimers and described their modular preparation methodology.[119] By tailoring the composition of the hydrophilic and hydrophobic branched segments, a variety of morphologies such as dendrimersomes, cubosomes, disks, tubular vesicles, and helical ribbons were observed. Although relatively less investigated than LDBCs, dendrimersomes combine the properties of liposomes and polymersomes to produce a more versatile vesicular delivery system.

5.4 Challenges with clinical translation

Dendrimer-based technologies have made significant progress in the biomedical field over the past two decades. However, there is only one dendrimer-based nanomedicine that has made it to clinical trials so far. Apparently, there are several limitations that impede their advancement of clinical translation. First, the highly uniform and well-defined dendritic architecture comes with the cost of a complex and multistep synthetic procedure. Second, while the covalent drug conjugation is preferred over the physical drug complexation, limited capacity and efficiency of drug loading becomes an issue. Moreover, excessive conjugation with drug payload or targeting moieties might result in undesirable change of the material properties, such as polydispersity and solubility. Those issues affect the scalability and reproducibility of dendrimer-based nanomedicines. In addition, just like other nanomedicines, the development of dendrimers faces a major challenge that the correlation between animal studies and clinical trials are poor and sometimes unpredictable. Those challenges seem inevitable for every emerging technique in the biomedical field. There are a few review articles discussing how to address the challenges that nanomedicines are facing which are recommended to read.[121,122] Undeniably, more efforts needed to be placed to convert the promising preliminary results into clinical products.

5.5 Future perspective

In spite of the challenges mentioned above, dendritic nanoparticles are still one of the most promising nanomedicine platforms. As discussed in this chapter, extensive *in vitro* and *in vivo* research has been carried out and demonstrated the advantages of the use of dendritic architectures in nanomedicine. The many drawbacks of dendrimers such as the high toxicity and low circulation time can be overcome by the structure engineering (e.g. PEGylation and terminal group modification), and the incorporation of functional moieties (e.g. targeting ligands). The scalability issue should be solved in the ever-growing development in dendrimer chemistry. Finally, the hybridization of dendritic and linear structures creates new platforms that possess additional benefits that are not seen in either system alone. While the intensive biological validation of dendritic nanoparticles is still ongoing, our knowledge in this field is accumulating at an unprecedented rate. With these efforts, a great advancement in clinical translation and a profound impact on human health is foreseeable for dendritic nanomedicines in the future.

References

1. Brannon-Peppas L, Blanchette JO, Nanoparticle and targeted systems for cancer therapy. *Adv Drug Deliv Rev*, 2012; 64: 206–212.
2. Mura S, Couvreur P, Nanotheranostics for personalized medicine. *Adv Drug Deliv Rev*, 2012; 64(13): 1394–1416.
3. Chauhan VP, Jain RK, Strategies for advancing cancer nanomedicine. *Nat Mater*, 2013; 12(11): 958–962.
4. Peer D, Karp JM, Hong S, Farokhzad OC, Margalit R, Langer R, Nanocarriers as an emerging platform for cancer therapy. *Nat Nanotechnol*, 2007; 2(12): 751–760.
5. Doane TL, Burda C, The unique role of nanoparticles in nanomedicine: imaging, drug delivery and therapy. *Chem Soc Rev*, 2012; 41(7): 2885–2911.
6. Lee DE, Koo H, Sun IC, Ryu JH, Kim K, Kwon IC, Multifunctional nanoparticles for multimodal imaging and theragnosis. *Chem Soc Rev*, 2012; 41(7): 2656–2672.
7. Pearson RM, Sunoqrot S, Hsu H-J, Bae JW, Hong S, Dendritic nanoparticles: the next generation of nanocarriers? *Ther Deliv*, 2012; 3(8): 941–959.
8. Ulery BD, Nair LS, Laurencin CT, Biomedical Applications of Biodegradable Polymers. *J Polym Sci B Polym Phys*, 2011; 49(12): 832–864.
9. Lee CC, MacKay JA, Fréchet JM, Szoka FC, Designing dendrimers for biological applications. *Nature biotechnol*, 2005; 23(12): 1517–1526.
10. Hong S, Leroueil PR, Majoros IJ, Orr BG, Baker Jr JR, Banaszak Holl MM, The binding avidity of a nanoparticle-based multivalent targeted drug delivery platform. *Chem Biol*, 2007; 14(1): 107–115.

11. Myung JH, Gajjar KA, Saric J, Eddington DT, Hong S, Dendrimer-Mediated Multivalent Binding for the Enhanced Capture of Tumor Cells. *Angew Chem Int Ed*, 2011; 50(49): 11769–11772.
12. Baars MW, Karlsson AJ, Sorokin V, de Waal BF, Meijer E, Supramolecular modification of the periphery of dendrimers resulting in rigidity and functionality. *Angew Chem Int Ed*, 2000; 39(23): 4262–4265.
13. Rosen BM, Wilson CJ, Wilson DA, Peterca M, Imam MR, Percec, Dendron-mediated self-assembly, disassembly, and self-organization of complex systems. *Chem Rev*, 109(11): 6275–6540.
14. Wurm F, Frey H, Linear–dendritic block copolymers: The state of the art and exciting perspectives. *Prog Polym Sci*, 2011; 36(1): 1–52.
15. Esfand R, Tomalia DA, Poly (amidoamine)(PAMAM) dendrimers: from biomimicry to drug delivery and biomedical applications. *Drug Discov Today*, 2001; 6(8): 427–436.
16. Svenson S, Tomalia DA, Dendrimers in biomedical applications — reflections on the field. *Adv Drug Deliv Rev,* 2012; 64: 102–115.
17. Tomalia D, Baker H, Dewald J, Hall M, Kallos G, Martin S, Roeck J, Ryder J, Smith P, A new class of polymers: starburst-dendritic macromolecules. *Polymer J,* 1985; 17(1): 117–132.
18. Wooley KL, Hawker CJ, Fréchet J, Hyperbranched macromolecules via a novel double-stage convergent growth approach. *J Am Chem Soc*, 1991; 113(11): 4252–4261.
19. Freeman AW, Fréchet JM, A rapid, orthogonal synthesis of poly (benzyl ester) dendrimers via an activated monomer approach. *Org Lett*, 1999; 1(4): 685–688.
20. Kawaguchi T, Walker KL, Wilkins CL, Moore JS, Double exponential dendrimer growth. *J Am Chem Soc*, 1995; 117(8): 2159–2165.
21. Zeng F, Zimmerman SC, Rapid synthesis of dendrimers by an orthogonal coupling strategy. *J Am Chem Soc*, 1996; 118(22): 5326–5327.
22. Duncan R, Izzo L, Dendrimer biocompatibility and toxicity. *Adv Drug Deliv Rev*, 2005; 57(15): 2215–2237.
23. Medina SH, El-Sayed ME, Dendrimers as carriers for delivery of chemotherapeutic agents. *Chem Rev*, 2009; 109(7): 3141–3157.
24. Buhleier E, Wehner W, Vögtle F, Cascade and nonskid-chain-like synthesis of molecular cavity topologies. *Synthesis*, 1978; 2: 155–158.

25. de Brabander-van den Berg E, Meijer E, Poly (propylene imine) dendrimers: Large-Scale synthesis by hetereogeneously catalyzed hydrogenations. *Angew Chem Int Ed*, 1993; 32(9): 1308–1311.
26. Denkewalter RG, Kolc J, Lukasavage WJ, Macromolecular highly branched homogeneous compound based on lysine units, 1981. US Patent No. 4,289,872.
27. Posnett DN, McGrath H, Tam JP, A novel method for producing anti-peptide antibodies. Production of site-specific antibodies to the T cell antigen receptor beta-chain. *Biol Chem*, 1988; 263(4): 1719–1725.
28. Rupp R, Rosenthal SL, Stanberry LR, VivaGel™(SPL7013 Gel): A candidate dendrimer–microbicide for the prevention of HIV and HSV infection. *Int J Nanomed*, 2007; 2(4): 561.
29. van der Made AW, van Leeuwen PWNM, Silane dendrimers. *J Chem Soc Chem Comm*, 1992; 19: 1400–1401.
30. Muzafarov A, Gorbatsevich O, Rebrov E, ignat'eva G, Chenskaya T, Myakushev V, Bulkin A, Papkov V, Organosilicon dendrimers: volume-growing polyallylcarbosilanes. *Polymer Sci*, 1993; 35(11): 1575–1580.
31. Zhou LL, Roovers J, Synthesis of novel carbosilane dendritic macromolecules. *Macromolecules*, 1993; 26(5): 963–968.
32. Roovers J, Ding J, Carbosilane dendrimers. *Silicon-Containing Dendritic Polymers*, Springer: 2009; 31–74.
33. Kaminskas, LM, Boyd BJ and Porter CJ, Dendrimer pharmacokinetics: the effect of size, structure and surface characteristics on ADME properties. *Nanomedicine*, 2011; 6(6): 1063–1084.
34. Wijagkanalan W, Kawakami S, Hashida M, Designing dendrimers for drug delivery and imaging: Pharmacokinetic Considerations. *Pharm Res*, 2011; 28(7): 1500–1519.
35. Kobayashi H, Kawamoto S, Saga T, Sato N, Hiraga A, Konishi J, Togashi K, Brechbiel MW, Micro-MR angiography of normal and intratumoral vessels in mice using dedicated intravascular MR contrast agents with high generation of polyamidoamine dendrimer core: Reference to pharmacokinetic properties of dendrimer-based MR contrast agents. *J Magn Reson Imaging*, 2001; 14(6): 705–713.
36. Kobayashi H, Sato N, Hiraga A, Saga T, Nakamoto Y, Ueda H, Konishi J, Togashi K, Brechbiel MW, 3D-micro-MR angiography of mice using macromolecular MR contrast agents with polyamidoamine dendrimer core with reference to their pharmacokinetic properties. *Magn Reson Med*, 2001; 45(3): 454–460.

37. Malik N, Wiwattanapatapee R, Klopsch R, Lorenz K, Frey H,Weener J, Meijer E, Paulus W, Duncan R, Dendrimers: Relationship between structure and biocompatibility *in vitro*, and preliminary studies on the biodistribution of 125I-labelled polyamidoamine dendrimers *in vivo*. *J Control Release*, 2000; 65(1): 133–148.
38. Boyd BJ, Kaminskas LM, Karellas P, Krippner G, Lessene R, Porter CJ, Cationic poly-L-lysine dendrimers: pharmacokinetics, biodistribution, and evidence for metabolism and bioresorption after intravenous administration to rats. *Mol Pharm*, 2006; 3(5): 614–627.
39. Hong S, Bielinska AU, Mecke A, Keszler B, Beals JL, Shi X, Balogh L, Orr BG, Baker JR, Banaszak Holl MM, Interaction of poly (amidoamine) dendrimers with supported lipid bilayers and cells: hole formation and the relation to transport. *Bioconjug Chem*, 2004; 15(4): 774–782.
40. Hong S, Leroueil PR, Janus EK, Peters JL, Kober MM, Islam MT, Orr BG, Baker JR, Banaszak Holl MM, Interaction of polycationic polymers with supported lipid bilayers and cells: nanoscale hole formation and enhanced membrane permeability. *Bioconjug Chem*, 2006; 17(3): 728–734.
41. Hong S, Rattan R, Majoros IJ, Mullen DG, Peters JL, Shi X, Bielinska AU, Blanco L, Orr BG, Baker Jr JR, The role of ganglioside GM1 in cellular internalization mechanisms of poly (amidoamine) dendrimers. *Bioconjug Chem* , 2009; 20(8): 1503–1513.
42. Kaminskas L M, Boyd BJ, Karellas P, Henderson SA, Giannis MP, Krippner GY, Porter CJ, Impact of surface derivatization of poly-L-lysine dendrimers with anionic arylsulfonate or succinate groups on intravenous pharmacokinetics and disposition. *Mol Pharm*, 2007; 4(6): 949–961.
43. Jain K, Kesharwani P, Gupta U, Jain N, Dendrimer toxicity: Let's meet the challenge. *Int J Pharm*, 2010; 394(1): 122–142.
44. Owens III DE, Peppas NA, Opsonization, biodistribution, and pharmacokinetics of polymeric nanoparticles. *Int J Pharm*, 2006; 307(1): 93–102.
45. Kojima C, Regino C, Umeda Y, Kobayashi H, Kono K, Influence of dendrimer generation and polyethylene glycol length on the biodistribution of PEGylated dendrimers. *Int J Pharm*, 2010; 383(1–2): 293–296.
46. Kaminskas, LM, Boyd BJ, Karellas P, Krippner GY, Lessene R, Kelly B, Porter CJ, The impact of molecular weight and PEG chain length on the systemic pharmacokinetics of PEGylated poly l-lysine dendrimers. *Mol Pharm*, 2008; 5(3): 449–463.

47. Mammen M, Choi S-K, Whitesides GM, Polyvalent interactions in biological systems: implications for design and use of multivalent ligands and inhibitors. *Angew Chem Int Ed* 1998; 37(20): 2754–2794.
48. Kiessling LL, Gestwicki JE, Strong LE, Synthetic multivalent ligands in the exploration of cell-surface interactions. *Curr Opin Chem Biol,* 2000; 4(6): 696–703.
49. Kiessling LL, Gestwicki JE, Strong LE, Synthetic multivalent ligands as probes of signal transduction. *Angew Chem Int Ed*, 45(15): 2348–2368.
50. Mecke A, Lee I, Holl MB, Orr B, Deformability of poly (amidoamine) dendrimers. *Euro Phys J E Soft Matter*, 2004; 14(1): 7–16.
51. Pagé D, Aravind S, Roy R, Synthesis and lectin binding properties of dendritic mannopyranoside. *Chem Commun*, 1996; (16): 1913–1914.
52. Ashton PR, Hounsell EF, Jayaraman N, Nilsen TM, Spencer N, Stoddart JF, Young M, Synthesis and biological evaluation of α-D-mannopyranoside-containing dendrimers. *J Org Chem*, 1998; 63(10): 3429–3437.
53. Rele SM, Cui W, Wang L, Hou S, Barr-Zarse G, Tatton D, Gnanou Y, Esko JD, Chaikof EL, Dendrimer-like PEO glycopolymers exhibit anti-inflammatory properties. *J Am Chem Soc*, 2005; 127(29): 10132–10133.
54. Shewmake TA, Solis FJ, Gillies RJ, Caplan MR, Effects of linker length and flexibility on multivalent targeting. *Biomacromolecules*, 2008; 9(11): 3057–3064.
55. Kim Y, Lee JH, Ryu J, Kim DJ, Multivalent Multifunctional Ligands to-Amyloid. *Curr Pharm Des*, 2009; 15(6): 637–658.
56. Kane RS, Thermodynamics of multivalent interactions: influence of the linker. *Langmuir*, 2010; 26(11): 8636–8640.
57. D'Emanuele A, Jevprasesphant R, Penny J, Attwood D, The use of a dendrimer-propranolol prodrug to bypass efflux transporters and enhance oral bioavailability. *J Control Release*, 2004; 95(3): 447–453.
58. Chauhan AS, Sridevi S, Chalasani KB, Jain AK, Jain SK, Jain N, Diwan PV, Dendrimer-mediated transdermal delivery: enhanced bioavailability of indomethacin. *J Control Release*, 2003; 90(3): 335–343.
59. Yiyun C, Na M, Tongwen X, Rongqiang F, Xueyuan W, Xiaomin W, Longping W, Transdermal delivery of nonsteroidal anti-inflammatory drugs mediated by polyamidoamine (PAMAM) dendrimers. *J Pharm Sci*, 2007; 96(3): 595–602.

60. Yang Y, Sunoqrot S, Stowell C, Ji J, Lee C-W, Kim JW, Khan SA, Hong S, Effect of size, surface charge, and hydrophobicity of poly (amidoamine) dendrimers on their skin penetration. *Biomacromolecules* 13(7): 2154–2162.
61. Bai S, Thomas C, Ahsan F, Dendrimers as a carrier for pulmonary delivery of enoxaparin, a low-molecular weight heparin. *J Pharm Sci*, 2007; 96(8): 2090–2106.
62. Vandamme TF, Brobeck L, Poly (amidoamine) dendrimers as ophthalmic vehicles for ocular delivery of pilocarpine nitrate and tropicamide. *J Control Release*, 2005; 102(1): 23–38.
63. Kojima C, Kono K, Maruyama K and Takagishi T, Synthesis of polyamidoamine dendrimers having poly (ethylene glycol) grafts and their ability to encapsulate anticancer drugs. *Bioconjug Chem*, 2000; 11(6): 910–917.
64. Beezer A, King A, Martin I, Mitchel J, Twyman L Wain C, Dendrimers as potential drug carriers; encapsulation of acidic hydrophobes within water soluble PAMAM derivatives. *Tetrahedron*, 2003; 59(22): 3873–3880.
65. Khandare J, Kolhe P, Pillai O, Kannan S, Lieh-Lai M,Kannan RM, Synthesis, cellular transport, and activity of polyamidoamine dendrimer-methylprednisolone conjugates. *Bioconjug Chem*, 2005;16(2): 330–337.
66. Majoros IJ, Myc A, Thomas T, Mehta CB, Baker JR, PAMAM dendrimer-based multifunctional conjugate for cancer therapy: synthesis, characterization, and functionality. *Biomacromolecules*, 2006; 7(2): 572–579.
67. Lai P-S, Lou P-J, Peng C-L, Pai C-L, Yen W-N, Huang M-Y, Young T-H, Shieh M-J, Doxorubicin delivery by polyamidoamine dendrimer conjugation and photochemical internalization for cancer therapy. *J Control Release*, 2007; 122(1): 39–46.
68. Tekade RK, Kumar PV, Jain NK, Dendrimers in Oncology: An Expanding Horizon. *Chem Rev*, 2008; 109(1): 49–87.
69. Lim J, Chouai A, Lo S-T, Liu W, Sun X, Simanek EE, Design, synthesis, characterization, and biological evaluation of triazine dendrimers bearing paclitaxel using ester and ester/disulfide linkages. *Bioconjug Chem*, 2009; 20(11): 2154–2161.
70. Zhu S, Hong M, Zhang L, Tang G, Jiang Y, Pei Y, PEGylated PAMAM dendrimer-doxorubicin conjugates: *in vitro* evaluation and *in vivo* tumor accumulation. *Pharm Res*, 2010; 27(1): 161–174.

71. Zhang Y, Thomas TP, Desai A, Zong H, Leroueil PR, Majoros IJ, Baker Jr JR, Targeted dendrimeric anticancer prodrug: a methotrexate-folic acid-poly (amidoamine) conjugate and a novel, rapid,one pot synthetic approach. *Bioconjug Chem*, 2010; 21(3): 489–495.
72. Haensler J, Szoka Jr FC, Polyamidoamine cascade polymers mediate efficient transfection of cells in culture. *Bioconjug Chem*, 1993; 4(5): 372–379.
73. Yoo H, Juliano R, Enhanced delivery of antisense oligonucleotides with fluorophore-conjugated PAMAM dendrimers. *Nucleic Acids Res*, 2000; 28(21): 4225–4231.
74. Ren Y, Kang C-S, Yuan X-B, Zhou X, Xu P, Han L, Wang GX, Jia Z, Zhong Y, Yu S, Co-delivery of as-miR-21 and 5-FU by poly (amidoamine) dendrimer attenuates human glioma cell growth *in vitro*. *J Biomater Sci Polym Ed*, 2010; 21(3): 303–314.
75. Zhou J, Wu J, Hafdi N, Behr J-P, Erbacher P, Peng L, PAMAM dendrimers for efficient siRNA delivery and potent gene silencing. *Chem Commun*, 2006; (22): 2362–2364.
76. Chen W, Turro NJ, Tomalia DA, Using ethidium bromide to probe the interactions between DNA and dendrimers. *Langmuir*, 2000; 16(1): 15–19.
77. Tang MX, Redemann CT, Szoka FC, *In vitro* gene delivery by degraded polyamidoamine dendrimers. *Bioconj Chem*, 1996; 7(6): 703–714.
78. Huang, R-Q., Qu Y-H, Ke W-L, Zhu J-H, Pei Y-Y, Jiang C, Efficient gene delivery targeted to the brain using a transferrin-conjugated polyethyleneglycol-modified polyamidoamine dendrimer. *FASEB J*, 2007; 21(4): 1117–1125.
79. Choi JS, Nam K, Park J-Y, Kim J-B, Lee J-K, Park J-S, Enhanced transfection efficiency of PAMAM dendrimer by surface modification with L-arginine. *J Control Release*, 2004; 99(3): 445–456.
80. Choi JS, Ko KS, Park JS, Kim Y-H, Kim SW, Lee M, Dexamethasone conjugated poly (amidoamine) dendrimer as a gene carrier for efficient nuclear translocation. *Int J Pharma*, 2006; 320(1): 171–178.
81. Ma K, Hu M-X, Qi Y, Zou J-H, Qiu L-Y, Jin Y, Ying X-Y and Sun H-Y, PAMAM–Triamcinolone acetonide conjugate as a nucleus-targeting gene carrier for enhanced transfer activity. *Biomaterials*, 2009; 30(30): 6109–6118.
82. Patil ML, Zhang M, Betigeri S, Taratula O, He H, Minko T, Surface-modified and internally cationic polyamidoamine dendrimers for efficient siRNA delivery. *Bioconj chem*, 2008; 19(7): 1396–1403.

83. Polcyn P, Jurczak M, Rajnisz A, Solecka J, Urbanczyk-Lipkowska Z, Design of antimicrobially active small amphiphilic peptide dendrimers. *Molecules*, 2009; 14(10): 3881–3905.
84. Cheng Y, Qu H, Ma M, Xu Z, Xu P, Fang Y, Xu T, Polyamidoamine (PAMAM) dendrimers as biocompatible carriers of quinolone antimicrobials: an *in vitro* study. *Euro J Medi Chem*, 2007; 42(7): 1032–1038.
85. Mishra MK, Kotta K, Hali M, Wykes S, Gerard HC, Hudson AP, Whittum-Hudson JA, Kannan RM, PAMAM dendrimer-azithromycin conjugate nanodevices for the treatment of *Chlamydia trachomatis* infections. *Nanomed: Nanotech Biol Med,* 2011; 7(6): 935–944.
86. Reuter JD, Myc A, Hayes MM, Gan Z, Roy R, Qin D, Yin R, Piehler LT,. Esfand R, Tomalia DA, Inhibition of viral adhesion and infection by sialic-acid-conjugated dendritic polymers. *Bioconjug Chem*, 1999; 10(2): 271–278.
87. Kensinger RD, Catalone BJ, Krebs FC, Wigdahl B, Schengrund C-L, Novel polysulfated galactose-derivatized dendrimers as binding antagonists of human immunodeficiency virus type 1 infection. *Antimicrob Agents Chemother*, 2004; 48(5): 1614–1623.
87. Landers JJ, Cao Z, Lee I, Piehler LT, Myc PP, Myc A, Hamouda T, Galecki AT, Baker JR, Prevention of influenza pneumonitis by sialic acid–conjugated dendritic polymers. *J Infect Dis*, 2002; 186(9): 1222–1230.
88. Matsuoka K, Terunuma D, Kuzuhara H, Suzuki Y, Natori Y, Nishikawa K, Sugar chain-containing carbosilane dendrimer compounds, process for producing the same verotoxin neutralizers and antiviral agents, 2004. US Patent 20,040,040,554.
89. Telwatte S, Moore K, Johnson A, Tyssen D, Sterjovski J, Aldunate M, Gorry PR, Ramsland PA, Lewis GR, Paull JR, Virucidal activity of the dendrimer microbicide SPL7013 against HIV-1. *Antiviral Res*, 2011; 90(3): 195–199.
90. Matthews BR, Holan G, Antiviral dendrimers, 2001. US Patent 6,190,650
91. Zhao H, Li J, Xi F, Jiang L, Polyamidoamine dendrimers inhibit binding of Tat peptide to TAR RNA. *FEBS Lett*, 2004; 563(1): 241–245.
92. Witvrouw M, Fikkert V, Pluymers W, Matthews B, Mardel K, Schols D, Raff J, Debyser Z, De Clercq E, Holan G, Polyanionic (ie, polysulfonate) dendrimers can inhibit the replication of human immunodeficiency virus by interfering with both virus adsorption and later steps (reverse transcriptase/integrase) in the virus replicative cycle. *Mol Pharm*, 2000; 58(5): 1100–1108.

93. Schinazi RF, Brettreich M, Hirsch A, Water-soluble dendrimeric fullerene as anti-HIV therapeutic, 2003. US Patent 20,030,036,562.
94. Langereis S, Dirksen A, Hackeng TM, Van Genderen MH, Meijer E, Dendrimers and magnetic resonance imaging. *New J Chem*, 2007; 31(7): 1152–1160.
95. Kobayashi H, Brechbiel MW, Nano-sized MRI contrast agents with dendrimer cores. *Adv Drug Deliv Rev*, 2005; 57(15): 2271–2286.
96. Venditto VJ, Regino CAS, Brechbiel MW, PAMAM dendrimer based macromolecules as improved contrast agents. *Mol Pharm*, 2005; 2(4): 302–311.
97. Villaraza LAJ, Bumb A, Brechbiel MW, Macromolecules, dendrimers, and nanomaterials in magnetic resonance imaging: the interplay between size, function, and pharmacokinetics. *Chem Rev*, 2010; 110(5): 2921–2959.
98. Kobayashi H, Jo SK, Kawamoto S, Yasuda H, Hu X, Knopp MV, Brechbiel MW, Choyke PL, Star RA, Polyamine dendrimer-based MRI contrast agents for functional kidney imaging to diagnose acute renal failure. *J Magn Reson Imaging*, 2004; 20(3): 512–518.
99. Swanson SD, Kukowska-Latallo JF, Patri AK, Chen C, Ge S, Cao Z, Kotlyar A, East AT, Baker JR, Targeted gadolinium-loaded dendrimer nanoparticles for tumor-specific magnetic resonance contrast enhancement. *Int J Nanomedicine,* 2008; 3(2): 201–210.
100. Zhang WL, Li N, Huang J, Yu JH, Wang DX, Li YP, Liu SY, Gadolinium-conjugated FA-PEG-PAMAM-COOH nanoparticles as potential tumor-targeted circulation-prolonged macromolecular MRI contrast agents. *J Appl Polym Sci*, 2010; 118(3): 1805–1814.
101. Strable E, Bulte JW, Moskowitz B, Vivekanandan K, Allen M, Douglas T, Synthesis and characterization of soluble iron oxide-dendrimer composites. *Chem Mater*, 2001; 13(6): 2201–2209.
102. Hawker CJ, Frechet JMJ, A New Convergent Approach to monodisperse dendritic macromolecules. *J Chem Soc Chem Commun*, 1990; (15): 1010–1013.
103. Hawker CJ, Frechet JMJ, Preparation of polymers with controlled molecular architecture. A new convergent approach to dendritic macromolecules. *J Am Chem Soc*, 1990; 112(21): 7638–7647.
104. Gitsov I, Wooley KL, Fréchet JMJ, Novel Polyether Copolymers Consisting of Linear and Dendritic Blocks. *Angew Chem Int Ed*, 1992; 31(9): 1200–1202.

105. Bae JW, Pearson RM,Patra N,Sunoqrot S, Vukovic L, Kral P, Hong S, Dendron-mediated self-assembly of highly PEGylated block copolymers: a modular nanocarrier platform. *Chem Commun*, 2011; 47(37): 10302–10304.
106. Santos JL, Herrera-Alonso M, Kinetically Arrested Assemblies of Architecturally Distinct Block Copolymers. *Macromolecules* 2013; 47(1): 137–145.
107. Pearson RM, Patra N, Hsu H-J, Uddin S, Kral P, Hong S, Positively Charged Dendron Micelles Display Negligible Cellular Interactions. *ACS Macro Lett*, 2013; 2: 77–81.
108. Qiao H, Li J, Wang Y, Ping Q, Wang G, Gu X, Synthesis and characterization of multi-functional linear-dendritic block copolymer for intracellular delivery of antitumor drugs. *Int J Pharm*, 2013; 452(1–2): 363–373.
109. Zhou ZY, D'Emanuele A, Attwood D, Solubility enhancement of paclitaxel using a linear-dendritic block copolymer. *Int J Pharm*, 2013; 452(1–2): 173–179.
110. Xiao K, Li YP, Lee JS, Gonik AM, Dong T, Fung G, Sanchez E, Xing L, Cheng HR, Luo JT, Lam KS, "OA02" Peptide facilitates the precise targeting of paclitaxel-loaded micellar nanoparticles to ovarian cancer *in vivo*. *Cancer Res*, 2012; 72(8): 2100–2110.
111. Zhou ZX, Ma XP, Jin EL, Tang JB, Sui MH, Shen YQ, Van Kirk EA, Murdoch WJ, Radosz M, Linear-dendritic drug conjugates forming long-circulating nanorods for cancer-drug delivery. *Biomaterials*, 2013; 34(22): 5722–5735.
112. Yang Y, Pearson RM, Lee O, Lee C-W, Chatterton RT, Khan SA, Hong S, Dendron-Based Micelles for Topical Delivery of Endoxifen: A Potential Chemo-Preventive Medicine for Breast Cancer. *Adv. Func. Mater,* 2014; 24(17): 2442–2449
113. Tian L, Hammond PT, Comb-dendritic block copolymers as tree-shaped macromolecular amphiphiles for nanoparticle self-assembly. *Chem Mater*, 2006; 18(17): 3976–3984.
114. Tian L, Nguyen P, Hammond PT,. Vesicular self-assembly of comb-dendritic block copolymers. *Chem Commun*, 2006; (33): 3489–3491.
115. Poon, Z., Chen S, Engler AC, Lee HI, Atas E, von Maltzahn G, Bhatia SN and Hammond PT (2010). Ligand-clustered patchy nanoparticles for modulated cellular uptake and *in vivo* tumor targeting. *Angew Chem Int Ed*, 2010; 49(40): 7266–7270.

116. Poon Z, Lee JA, Huang S, Prevost RJ, Hammond PT, Highly stable, ligand-clustered patchy micelle nanocarriers for systemic tumor targeting. *Nanomedicine*, 2011; 7(2): 201–209.
117. Wood KC, Little SR, Langer R, Hammond PT, A Family of hierarchically self-assembling linear-dendritic hybrid polymers for highly efficient targeted gene delivery. *Angew Chem Int Ed*, 2005; 44(41): 6704–6708.
118. Yang B, Sun Y-X,. Yi W-J, Yang J, Liu C-W, Cheng H, Feng J, Zhang X-Z, Zhuo R-X, A linear-dendritic cationic vector for efficient DNA grasp and delivery. *Acta Biomater*, 2012; 8(6): 2121–2132.
119. Percec V, Wilson DA, Leowanawat P, Wilson CJ, Hughes AD, Kaucher MS, Hammer DA, Levine DH, Kim AJ, Bates FS, Davis KP, Lodge TP, Klein ML, DeVane RH, Aqad E, Rosen BM, Argintaru AO, Sienkowska MJ, Rissanen K, Nummelin S, Ropponen J, Self-Assembly of janus dendrimers into uniform dendrimersomes and other complex architectures. *Science*, 2010; 328(5981): 1009–1014.
120. Zhang S, Sun H-J, Hughes AD, Draghici B, Lejnieks J, Leowanawat P, Bertin A, Otero De Leon L, Kulikov OV, Chen Y, Pochan DJ, Heiney PA, Percec V, Single–Single amphiphilic janus dendrimers self-assemble into uniform dendrimersomes with predictable size. *ACS Nano*, 2014, 8(2), 1554–1565.
121. Riehemann K, Schneider SW, Luger TA, Godin B, Ferrari M, Fuchs H, Nanomedicine — challenge and perspectives. *Angew Chemie Int Ed*, 2009; 48(5): 872–897.
122. Pearson RM, Hsu H-J, Bugno J, Hong S, Understanding nano-bio interactions to improve nanocarriers for drug delivery. *MRS Bulletin*, 2014; 39(3), 227–237.

Chapter 6

ORAL DRUG DELIVERY SYSTEMS FOR GASTROINTESTINAL CANCER THERAPY

Yue-Wern Huang and Sutapa Barua*[†,‡]

**Department of Biological Sciences*
[†]Department of Chemical and Biochemical Engineering
Missouri University of Science and Technology, Rolla, MO 65409, USA
[‡]baruas@mst.edu

Abstract

The focus of this review is the oral delivery of anticancer drugs. The oral route is the most convenient method of delivering drugs as it precludes the discomfort of injection, thereby enhancing patient compliance and quality of life. Challenges associated with the oral delivery of anticancer drugs include low solubility of drug molecules, poor absorption through the gastrointestinal (GI) epithelium, inefficient permeability across biological membranes, and the need for site-specific delivery of these toxic compounds. Nanoscale drug delivery systems (DDS) improve protection of drugs from degradation in the GI tract, and drug absorption and diffusion through the intestinal epithelium. This chapter covers recently published approaches to oral DDS with an emphasis on nanoscale DDS, biological barriers to the oral delivery of anticancer drugs, techniques designed to overcome these barriers, and the exploitation of physicochemical properties of DDS to improve the therapeutic efficacy of different types of formulations. The potential for the development of efficient anticancer oral DDS that enables high patient compliance is the focus of this review.

6.1 Introduction

Cancers of the GI tract are found mainly in oral, esophagus, small and large intestines, stomach and colon. These cancers accounted for an estimated 3.1 million new cases and 1.9 million deaths worldwide in 2012.[1] Oral cancer affects the cheeks (buccal mucosa), esophagus, lips, mouth and tongue (Figure 6.1). Each of these regions consists of multiple layers of cells. The basal cells differentiate into membrane coating granules that release lipophilic materials into intercellular spaces. These lipophilic materials and lipid layers slow down the entry of exogenous and endogenous materials across the epithelium. The epithelium also produces an acellular mucus layer of 70–100 μm that covers the surface of esophageal glands, stomach, small and large intestines and colon of the GI tract.[2]

The intercellular tight junctions pose a challenge to molecules larger than 1 nm to penetrate across the physical integrity of intestinal epithelial cells.[3,4] Efflux proteins such as P-glycoprotein (P-gp) limit the intestinal transport further.[5] Moreover, rapid pH changes in GI tract degrade the

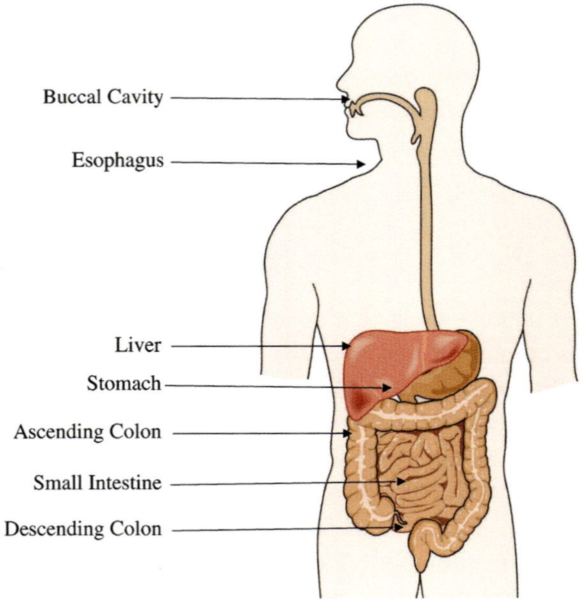

Figure 6.1 Schematic diagram of the GI tract in the human body.

orally administered small molecules, peptides and proteins rapidly by hydrolysis or oxidation. Therapeutic drug molecules are vulnerable to the acidic pH and enzymatic degradation by pepsin in the stomach.[6–8] Further proteases in the intestine and aminopeptidases in the brush-border membrane are also barriers to oral drug delivery.[9] The pre-systemic degradation of drugs in GI tract results in low dose fraction to enter the systemic circulation.

Nanoparticles (NPs) have been investigated as a vector to deliver small drug molecules including small interfering RNA (siRNA), genes, peptides and proteins across the GI tract. Advantages of NPs include enhanced blood circulation,[10] bioavailability,[11] enzymatic tolerance[12] and pH stability.[13] Swallowed NPs pass through the esophagus connecting the back of mouth with stomach.[14–17] The stomach has a low pH (0.8–2.0) due to the production of intrinsic factor and HCl.[18] The therapeutic drugs in the stomach can be degraded by its acidic pH and digestive enzymes.[19]

NPs can encapsulate or chemically modified to prevent degradation of therapeutic agents. Commonly used NPs in oral DDS are polymers,[20–23] liposomes[24–26] and nanoemulsions.[12,27–29] These materials have been developed to deliver small molecule drugs, siRNA, peptides and proteins for the treatment of various diseases, including cancer. For instance, the anticancer drug paclitaxel (PTX) encapsulated in β-casein NPs have shown high efficiency in drug release, with less toxicity to epithelial cells lining the stomach than the free drugs.[30] Doxorubicin (DOX) encapsulated within poly(ethylene glycol) (PEG) immunoliposomes have been shown to be very efficient for the treatment of gastric cancer.[31] Additionally, other chemotherapeutic agents such as peptides, proteins and siRNA are of interests due to their low molecular weights. Section 6.2 discusses the therapeutic agents that have been evaluated for oral DDS. Understanding the biological barriers is important in the design of oral DDS in order to enhance the therapeutic activity, drug release and penetration across the mucosal epithelium. Section 6.3 of this chapter reviews the biological barriers that influence NP permeation.

Commercially available formulations for oral drug delivery include films and tablets. Fentenyl buccal soluble film (Meda Pharmaceuticals Inc., Somerser, NJ, USA) is an oral transmucosal delivery system that contains fentanyl citrate in the bioadhesive layer and a drug-free backing layer.[32,33] This bi-layer design maximizes drug delivery across the buccal mucosa from the adhesive layer while limiting drug diffusion into the saliva. Fentanyl released from the film provides rapid and effective pain relief for the treatment of cancer breakthrough pain. Other DDS for cancer breakthrough pain are Fentora tablets (Cephalon Inc., Frazer, PA, USA) and Actiq lozenges (Teva Pharmaceuticals, Sellersville, PA, USA), which provide drug release and improved patient compliance.[33] Mucoadhesive NPs are used as drug carriers because of their high adhesive properties, solubility, surface area and visco-elastic properties.[34] Details of mucoadhesive patches are discussed in Section 6.4.

The efficiency of NP-mediated drug delivery largely depends on the physicochemical properties of NPs, such as size, shape, and surface properties. NP size of ~50–300 nm with hydrophobic surfaces is optimal for uptake from the GI tract. Carboxylate polystyrene NPs of size 59, 100, 200 and 500 nm bind to mucins, whereas amine-modified polystyrene NPs of similar size penetrate cystic fibrotic sputum more rapidly.[35,36] Coating the surface of NPs with PEG enhances their rate of transport.[37] NP size also plays a role in determining NP ability to cross the mucus gel. NPs of 100 nm size move six and four fold faster than NPs of 200 and 500 nm, respectively.[38] Latex beads of 14 and 415 nm cross the colon mucus gel in 2 and 30 minutes,[39,40] respectively, while 1 μm particles could not permeate the gel layer.[41] The effects of NP physicochemical parameters are described in Section 6.5.

6.2 Types of therapeutic agents

6.2.1 *Small molecule drugs*

In the past ten years, the cancer treatment paradigm has shifted from the one-size-fits-all cytotoxic chemotherapy approach to molecular-based targeted treatment. This change reflects contributions of biotechnology to our understanding of cancer biology. For example, the epidermal growth factor receptor (EGFR) kinase inhibitors gefitinib and lapatinib inhibit EGFR in patients with non-small cell lung cancer (NSCLC).[42] The monoclonal antibody trastuzumab (TTZ) inhibits the growth of Her-2 positive breast cancer cells.[43] Recent developments in nanotechnology promise

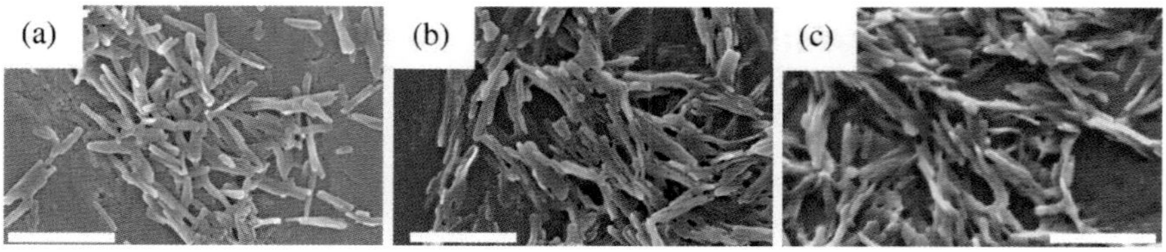

Figure 6.2 Scanning electron microscopy (SEM) images of (a) camptothecin (CPT) drug NPs, (b) CPT coated with trastuzumab (TTZ; Genentech) antibody and (c) CPT-TTZ-doxorubicin (DOX) NPs. Scale bar indicates 500 nm. Reprinted with permission from Ref. 44. Copyright (2013) American Chemical Society.

further advances in cancer treatment. The new approach utilizes nano-scaled drug complexes with specific targeting. For instance, the effects of TTZ coated with camptothecin and doxorubicin (CPT-TTZ-DOX) NPs (Figure 6.2) on breast cancer growth inhibition were synergistic (combination index = 0.17 ± 0.03) with 10–10,000-fold lower inhibitory concentrations (IC_{50}) compared to individual drugs alone.[44]

6.2.2 *RNA interference*

Gene therapy using nucleic acid therapeutics such as antisense oligonucleotides, small hairpin RNA (shRNA) and siRNA have emerged as powerful strategies in cancer treatment.[45,46] Among these, siRNA and shRNA have received the most attention in recent years. RNA interference (RNAi) is an evolutionarily conserved mechanism of gene expression regulation in both animals and plants.[47,48] Using the principals of RNA interference, siRNA or shRNA of 20–25 base pairs in length have been developed to manipulate the expression of specific genes. This siRNA-mediated knockdown strategy represents a new treatment modality in cancer, genetic disorders, infectious diseases, inflammation and metabolic defects. For instance, polyethyleneimine/TNFα siRNA nanocomplexes have shown significant anti-inflammatory effects in inflamed colon.[49] Currently, there are more than 20 siRNAs therapeutics undergoing clinical trials in the United States.[50] One of the major challenges of this technique is the low penetration of uncomplexed siRNA across the cytoplasmic membrane. Accordingly, delivery systems, such as aptamers, cationic polymers, cell-penetrating peptides (e.g., protein transduction domain, PTD) and nanocapsules (e.g., immunoliposomes), have been developed to increase penetration.[51–53] An elegant carrier, that fuses a Tat PTD with a double-stranded RNA-binding domain (dsRBD), allows siRNA to bind to dsRBD while utilizing PTD to increase cellular uptake in primary and transformed cells.[53,54]. Simultaneous delivery of EGFR and Akt2 siRNAs using the carrier system has shown a reduction in tumor mouse models of glioblastoma and improved the longevity of mice. Median survival was improved from 14 days of untreated glioblastoma to 14–31.5 days with PTD-dsRBD EGFR plus Akt2 siRNA ($p < 0.0005$). A similar strategy to target gastrointestinal cancers could be considered.

Although siRNA can down-regulate expression of genes associated with cancer progression, its efficacy is still limited. Consequently, co-delivery of siRNA and small molecule anticancer drugs has been pursued to improve treatment efficacy. The advantages of co-delivery include overcoming multidrug resistance, sensitizing cancer cells to anticancer drugs, and reducing toxicity (i.e., side effects). DOX and PTX along with bcl-2 siRNA or MDR-1 siRNA are commonly used to evaluate co-delivery systems for cancer treatment.[55–57] Additionally, mTERT siRNA (telomerase reverse transcriptase specific siRNA) and siMcl1 siRNA (myeloid cell leukemia sequence-1 specific siRNA) in combination with PTX or DOX have been shown to improve therapeutic outcomes.[58,59]

6.2.3 *Peptides and proteins*

Peptide and protein drugs have become the drugs of choice because of their target specificity and low toxicity. However, improving the oral bioavailability of these molecules is a challenge due to unfavorable properties, including large molecular weight, enzymatic degradation, and poor penetration through the GI tract. Further, biological obstacles such as low permeation, variable pH, and breakdown of peptides and proteins in the stomach lead to the poor bioavailability.[60] Some peptide drugs such as cyclosporin A and desmopressin show enhanced permeability across intestinal epithelium and improved enzymatic stability when delivered orally in NP formulations.[61]

Arginine-Glycine-Aspartic (RGD) peptide binds to integrin receptors overexpressed on the tumor neovasculature and tumor cells, becoming the new target for drug and gene delivery for tumor treatment.[62] A prototypic class of RGD is iRGD (CRGDKGPDC) that provides better tumor homing and tissue penetration than RGD due to its K variant.[9] A recent study has shown that iRGD enhances intratumoral entry of intraperitoneally co-injected doxorubicin by 250% compared to RGD control peptide, thereby inhibiting the growth of peritoneal carcinomatosis which is a common metastasis in colorectal (30%), gastric (50%), and appendiceal cancers (nearly 100%).[63]

Salmon calcitonin (sCT) is a cationic peptide that forms complexes with the cationic amphiphilic polyelectrolyte poly(allyl)amine (QPa) to generate sCT–QPa NPs of 200 nm in diamater.[64] sCT–QPa NPs promote oral delivery of sCT peptide because of their bioactivity, stability, and resistivity to peptidases. sCT–QPa NPs had effective IC_{50} concentrations in the nanomolar range of 3.3 and 7.9 nM at pH 7.4 and 5, respectively in T47D epithelial cells.[64] Chitosan/sulphobutyl ether-b-cyclo-dextrin (CS/SBE7m-b-CD) NPs were used for oral delivery of glutathione (GSH) peptide *in vivo*.[65] The enhanced loading efficiency of glutathione into CS/SBE7m-β-CD NPs was attributed to the nano-formation with cyclodextrin. These NPs had two fold higher absorption efficiency in duodenum than the peptide (GSH) solution alone.

CSK peptide was modified with trimethyl chitosan chloride (TMC) for the oral delivery of insulin *in vivo*.[66] CSK–TMC NPs increased the bioavailability of insulin by 50% relative to the uncomplexed insulin because the modified NPs prevented enzymatic degradation of insulin in the GI tract.[66,67] CSK9 peptide increased the transcytosis of chitosan NPs by 50% in M cells *in vitro* (Figure 6.3).[68] The CSK9-chitosan NPs consistently target at the follicle-associated epithelium (FAE) of Peyer's patches *in vivo*. Quaternized palmitoyl graft polyallylamine (PAA) polymeric NPs have also been used for oral delivery of insulin since these polymeric NPs protect insulin from enzymatic degradation by trypsin and pepsin.[64]

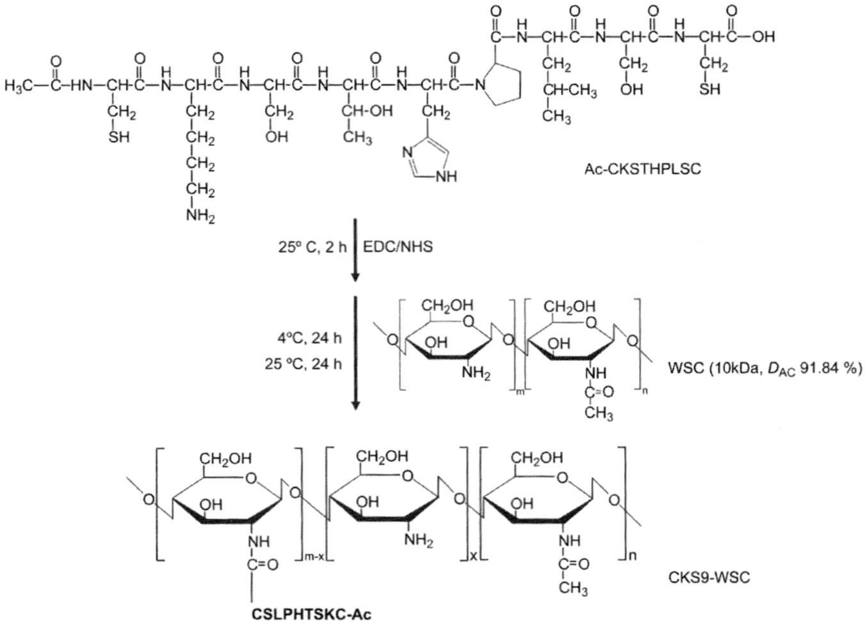

(a)

Figure 6.3 (a) Reaction scheme for the synthesis of CKS9-conjugated water-soluble chitosan (CKS9-WSC).[68] CKS9 peptide was conjugated to water-soluble chitosan (WSC) by coupling through the carboxylic groups of CKS9 peptide with amino groups of WSC using EDC/NHS chemistry. (b) Temperature dependent translocation of FITC-conjugated latex NPs in human M cell model as co-cultured with Caco-2/Raji coculture *versus* Caco-2 monolayer across apical to the basal site.[68] The translocation increased during the first 37°C incubation period, decreased during the 4°C incubation period and further enhanced at 37°C. In contrast, Caco-2 monolayer did not show any changes at given time points. This *in vitro cell* model indicates the translocation of NPs by transcytosis through the cells, and not by simple diffusion through intercellular spaces. TEM images of (c) Caco-2 monolayer[68] and (d) Caco-2/Raji confirm the Raji cell migration in a transwell system.[68] (c) Caco-2 cells showed well-organized brush borders, while (d) disorganized apical microvilli as indicated by arrowheads were found due to Raji cell (R) migration. Scale bars = 2 μm. Reprinted with permission from Elsevier.[68]

Tyrosine kinase has been a favored target that has, in turn, led to inhibition of mammalian target of rapamycin (mTOR) signaling pathway.[69] Imatinib (Gleevec, Novartis) is an mTOR-specific kinase inhibitor that has been tested in GI stromal tumor therapy.[70] Sorafenib (Nexavar) is an

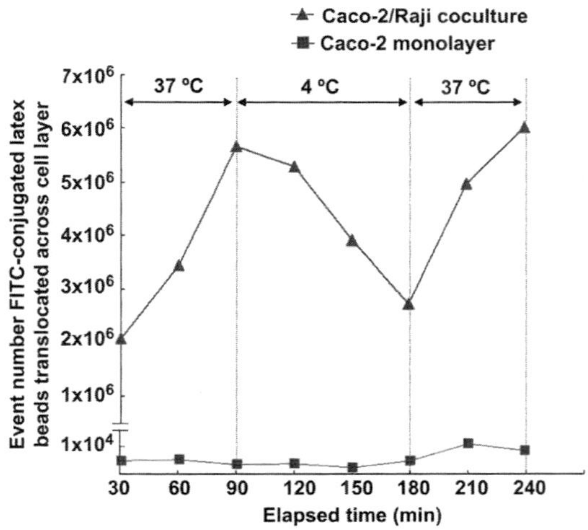

(b)

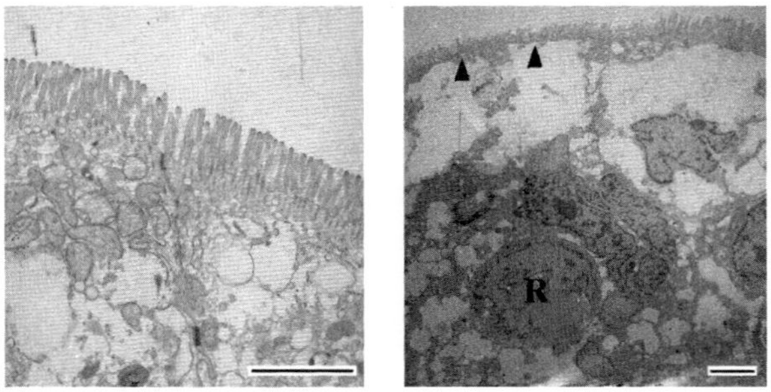

(c) (d)

Figure 6.3 (*Continued*)

orally active multikinase inhibitor that has been used for the treatment of renal cell cancer and hepatocellular cancer.[71,72]

6.3 Challenges for the oral delivery of drugs

When NPs pass through the GI tract, the first barrier of absorption is the mucus followed by the layers of epithelial cells and connective tissues underlying the epithelium (Figure 6.4). The epithelial barrier is lipophilic in nature allowing hydrophobic NPs to pass through the cells (transcellular). Paracellular transport between passages of cells has been reported for small

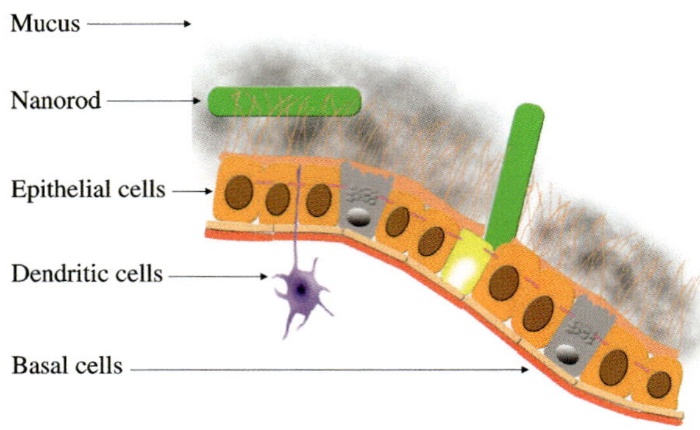

Figure 6.4 NPs travel through mucus and reach the underlying epithelia in the GI tract. The epithelial cells (orange color) are linked together by tight junctions (pink). Immune cells are embedded in the keratinocyte layer. NPs permeate the cells by passage through the cells (transcellular) or between the cells (paracellular).[132] The gray and yellow colored cells represent Goblet and M cell, respectively.

(< 1000 Da) hydrophilic molecules such as dextran. Transport of NPs occurs *via* energy-dependent endocytosis process,[73] or energy-independent diffusion process.[74] In the following sections, biological effects on NP penetration are discussed.

6.3.1 *Mucus layer*

Mucus is a lubricating slippage plane that protects epithelial cell layers from exogenous substances and toxins and enables the exchanges of cellular nutrients, gases and water with the extracellular environment.[75] The mucus layer is composed of mucins, antibodies, bacteria, cellular debris, lipids, proteins, and salts.[17] Mucins can be either secreted or cell-bound.[14] The secreted mucins form a gel-like structure by linking monomers of proteins *via* disulfide bonds. The protein backbone of mucins largely contains serine, threonine and proline residues with a majority (40–80%) of *O*-linkage oligosaccharides and a small fraction (2–3%) of *N*-linked oligosaccharides to the ends of mucin monomers. These oligosaccharides vary in size imposing different levels of resistance to NP diffusion. In addition, enzymes, electrolytes, and water form a thick fibrillar coating on top of the cell bound mucus layer just above the epithelium known as glycocalyx. Glycocalyx also contains lipids and glycoproteins such as MUC1, 3, 4, 12 and 13 mucins. Mucin fibers generally range from 10–40 MDa in size with a diameter of around 4–10 nm.[76] Depending on the body areas, the thickness of the mucus layer varies between 70 to 100 μm.[2,77] The thickness can be as high as 897–1354 μm and 730–1136 μm in stomach and rectum areas, respectively.[78] This complex structures of mucins affect the transport rates of NPs.[15,79] Amine functionalized polystyrene particles of 100, 200 and 500 nm in diameter penetrated more rapidly than the same sizes of carboxylate functionalized particles in cystic fibrosis sputum.[35] Bare polystyrene NPs of 59 nm diffuse the mucus layer faster than carboxylate-coated particles.[36] Covalent binding of a low molecular weight PEG increased diffusion of larger (200 and 500 nm) polystyrene NPs than smaller (100 nm) NPs in cervicovaginal mucus (CVM).[16] This was attributed to decreased interactions of NP surface with proteins and biomacromolecules in the mucus.[80] High PEG density using a short chain (2 kDa) PEG showed better mucus penetration of NPs than that of a heavy chain (10 kDa) PEG.[81] This effect was due to muco-inertness of NPs by shielding the NP surface with short chain PEG, resulting in little NP interactions with mucin fibers. A decrease of 40% PEG surface coverage lead to 700 fold reduction in NP transport rate in human mucus further supports the importance of NP surface modification with PEG.[81]

6.3.2 *Epithelium and tissue barriers*

The epithelial layer beneath the mucus layer is another barrier for NPs. The GI tract has an impressive absorptive surface area of 200–300 m^2. The absorptive area is composed of villi with absorptive cells (enterocytes) and mucus secreting (goblet) cells.[14] These cells are joined together by tight junctions (TJs). NPs need to perforate through these cells *via* paracellular and transcellular pathways. Paracellular transport occurs by diffusion through TJs between epithelial cells; molecules less than 1–2 nm are transported through TJs.[82] Transcellular transport is accomplished by active or passive processes from the apical site of the cells to the basal site; only a small number of drugs are transported *via* active transport systems.[83] The thickness of a simple columnar epithelium is 20–25 μm, whereas it can be 300–500 μm in the esophagus.[84] The epithelial cells reside on a basal membrane where the cells face capillaries, connective tissues, lymph follicles, lymph vessels and peripheral nerves. Drug NPs must cross the epithelial layer and basal tissues to enter the systemic circulation.

6.3.3 *Enzymatic barrier*

Most NPs are degraded by enzymatic reactions of peptidases in the stomach and small intestine. The peptidases include aminopeptidases, carboxypeptidases, chymotrypsin, elastase, pepsin and trypsin. They degrade proteins by cleaving the bonds at different hydrophobic residues of phenylalanine, methionine, leucine and tryptophan.[19] Aminopeptidases and intracellular proteases exist in all mucosal tissues. Pepsin is active only at pH ~2 in the stomach. Carboxypeptidases, chymotrypsin, elastase and trypsin are active in the small intestine to induce proteolysis of proteins into small fractions of amino acids, dipeptides, and tri-peptides. NPs need to be fully protected from enzymatic degradation to enter the target site. Breast cancer cells treated with the oral delivery of DNA vaccines showed a significant reduction in tumor volumes by protecting the encapsulated DNA using alginic acid.[85] Excess alginic acid on the surface of chitosan NPs was insoluble at gastric fluid of pH 1.5, resulting in an amorphous shell surrounding the NPs by self-assembly. Oral administration of the particles carrying legumain-based DNA enhanced the responses of legumain-specific cytotoxic T cells, inhibited regulatory T cells and reduced orthotopic breast tumor in mice by oral vaccination.[85] Poly(lactide-co-glycolide) (PLGA) NPs induced potent mucosal and systemic immunity against hepatitis B surface antigen by coating the NPs with mucoadhesive glycol chitosan.[86]

6.4 Types of oral delivery systems

Commonly used oral delivery systems include, but not limited to, buccal patches, tablets, nanoemulsions and NPs. Bioadhesive DDS are attractive for patients' compliance. To date, only a few bioadhesive formulations for oral drug delivery are in clinical trials or commercially available.[87]

6.4.1 *Mucoadhesive delivery*

Mucoadhesive designs include buccal patches and tablets. Buccal patches are flexible films that interact with the mucin surface of a tissue using bioadhesive polymers.[88] The adhesive layer is made of polymers like chitosan, hydroxypropyl methylcellulose (HPMC), polycarbophil, sodium alginate, etc., which exhibit high mucoadhesion. The increase in the bioadhesion of buccal patches is mainly attributed to interactions of cationic charges of polymers with the negative charges on mucus layer. The presence of carboxyl and hydroxyl groups on the polymer surface also contributes to mucoadhesion through hydrogen bonding with the mucosa.[89,90] Drugs are cast upon an adhesive layer that offers prolonged contact at the administration site. Some examples of mucoadhesive delivery are listed below.

Fenretinide (4-hydroxyphenylretinamide), a hydrophobic therapeutic agent, has been used in formulating mucoadhesive patches for chemoprevention in buccal cancer treatment.[89] The patches were prepared by assembling the fenretinide — Eudragit® based layer encompassed by an adhesive layer which is made using a combination of HPMC and polycarbophil. Eudragit® is an acrylic copolymer with quaternary ammonium groups widely used in drug delivery. The ammonium groups in Eudragit® improve the permeability of the hydrophobic drug. It was used in the preparation of the fenretinide-based patches to improve the mucoadhesion as well as to understand the drug release behavior. The release kinetics of fenretinide from the patch was studied in a simulated saliva environment (pH 6.8) as well as *in vivo* on white rabbits. The patch was weighed after immediate freeze-drying at regular intervals and the cumulative fenretinide release was calculated. A slow continuous release of the drug was observed up to 42.36 μg/g after 6 hours in rabbit buccal mucosa *in vivo*.[89]

Erlotinib hydrochloride (Tarceva®) tablet is an orally administered EGFR inhibitor to treat metastatic NSCLC and pancreatic cancer.[91] Erlotinib treatment showed altered blood vessel morphology and decreased vessel permeability in head and neck cancer xenogratfs in nude mice.[92] A combination treatment of erlotinib and radiation synergistically inhibited the tumor growth of NSCLC xenografts.[93] Combination treatments of erlotinib with another EGFR inhibitor, gefitinib decreased vascular endothelial growth factor (VEGF) mRNA and protein expressions in NSCLC xenografts.[92]

Granisetron hydrochloride is a highly hydrophilic antiemetic drug used to avoid vomiting during cancer chemotherapy. The granisetron formulation was prepared as buccoadhesive tablets by mixing with carbopol, HPMC polymer and/or xantham gum in an optimal ratio to enhance mucoadhesion.[94] The mixture was compressed as a tablet form. *In vivo* and *in vitro* studies were conducted using granisetron with different polymers in rabbits and human volunteers to understand the adhesion time and release behavior. Granisetron with HPMC polymer and carbopol showed longer adhesion time, less bitterness, slow drug release and low irritation compared to other combinations. Collectively, bioadhesive DDS have potential in controlled release, and improved efficacy.

6.4.2 *Nanoparticles*

NP-based carrier platforms have the potential to deliver drugs across mucosal and intestinal barriers. The source materials include polymers, liposomes, micelles and some metals to transport drugs for targeting tumor cells. Targeted delivery can be achieved by conjugating the surface of NPs with functional groups such as peptides and proteins that recognize its complementary binding sites on the cell membrane. Lectins,[95] invasins,[96] and vitamin B12[97] derivatives are natural ligands that bind to sugars on epithelial cells. Dendrimers are branched, tree-like macromolecules that are in the size range (150–200 nm) between liposomes and polymeric nanospheres. Polymeric NPs cross the oral bio-barriers by paracellular and transcellular pathways, thereby enhancing the permeability of drugs. DOX was conjugated to polyamidoamine (PAMAM) dendrimers using a pH sensitive linkage to enhance the permeability of the drugs in Caco-2 cells and transport across rat intestinal segments after oral administration.[98] The pH sensitive linker was the acid labile hydrazine group that was stable at the physiological pH of 7.4 and cleaved at a lower pH of cancer cells releasing DOX. The release of DOX resulted in higher DOX uptake in Caco-2 cells than the free drug. The transport efficiency of the DOX-PMAM complex was also 4–7 times higher than the free drug solution from the mucosal side to the serosal side. A 300-fold enhancement in the bioavailability of DOX was observed for the PAMAM complex than free drugs following a single dose by oral delivery in rats. A significant amount of DOX was also released in blood, most likely, due to its hydrophobic effect. Overall, the pH driven hydrolysis of the drug-dendrimer linker enhanced oral bioavailability of anti-cancer drugs. The following sections discuss in details the major NPs that have been used for oral cancer drug delivery.

6.4.2.1 *Polymers*

The effort has been made to design polymeric NPs for oral delivery. By conjugating drugs to the backbone of polymers, both therapeutic efficacy and water solubility of drugs are significantly improved. Commonly used polymeric NPs for oral drug delivery include PEG, poly(lactide) (PLA) and PLGA.[20] Microspheres (680–850 μm) of two mucoadhesive copolymers, fumaric acid and sebacic acid, showed longer retention time in the rat intestine and 1.5 fold more area under the plasma concentration-time curve than weakly adhesive alginate particles.[21] Tocopheryl polyethylene glycol succinate (TPGS) is a PEGylated vitamin E that suppressed human lung cancer cell growth *in vitro* and *in vivo* both as a single agent treatment and in combination with other chemotherapeutic drugs.[22] TPGS emulsified PLGA NPs showed 10 times higher bioavailability, 9.7-fold higher therapeutic effects of PTX, and 12.6-fold longer therapeutic time than the PTX drug solution alone in breast tumor after oral administration.[23] Oral bioavailability of PTX reached ~80% using cyclodextrin-poly(anhydride) NPs in rats.[11] PTX encapsulated in PLGA NPs with a stabilizer, didodecyl dimethylammonium bromide (DMAB) ammonium salt achieved a 3-fold reduction in tumor growth than the drug solution alone in Sprague–Dawley rats.[99] Docetaxel drug delivery using PLA–TPGS NPs, or PLA–TPGS NPs with montmorillonite (MMT) mucoadhesive (PLA–TPGS/MMT) NPs showed ~2.6- and 4-fold, respectively, more effective than a current clinical dose form of Docetaxel (Taxotere®) alone in MCF-7 breast cancer cells.[100] The IC_{50} of PLA–TPGS, PLA–TPGS/MMT and Taxotere® were ~5.1, 3.7 and 11.3 μg/ml, respectively. *In vivo* pharmacokinetics study revealed that oral administration of

PLA–TPGS and PLA–TPGS/MMT NPs had 20.6 and 26.4 times, respectively, longer half-life than intravenously injected Taxotere® in Sprague Dawley rats.[100] Bioavailability of a single oral administration of PLA–TPGS/MMT NPs revealed a sustained chemotherapy of three weeks compared to only 22 hours for the intravenous administration of Taxotere® (Figure 6.5).

Albumin coated poly(methylvinylether-*co*-maleic anhydride) NPs showed targeting of stomach mucosa and strong bioadhesivity towards the intestine after their oral administration in animals.[101] The NPs were tested to deliver 5-fluorouridine (FURD) that showed ~79% bioavailability compared to 21% by the drug alone.[101] Polymeric micelles are self-assembled amphiphilic macromolecules that are used for the oral delivery of hydrophobic drugs. PTX was encapsulated in the core of poly(2-(4-vinylbenzyloxy)-N,N-diethylnicotinamide) (p-VBODENA) micelle block and PEG shell.[102] The bioavailability of PTX was 12.4% after

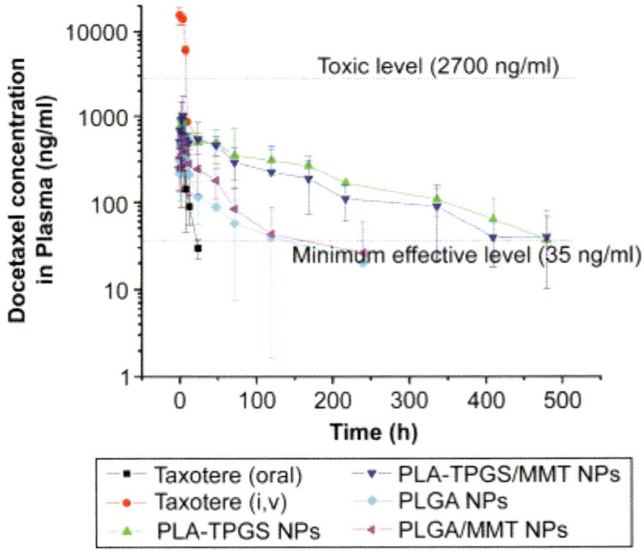

Figure 6.5 *In vivo* pharmacokinetics of Docetaxel concentrations in the plasma *versus* time curve after oral administration with PLGA, PLA-TPGS, PLGA/MMT and PLA-TPGA/MMT NPs and intravenous injection of Taxotere.[100] Reprinted with the permission from Elsevier.

3.8 mg/kg of oral administration in rats.[102] PTX loaded in lipid nanocapsules improved bioavailability 3-fold higher than an orally dosed Taxol.[24]

6.4.2.2 *Liposomes*

Liposomes are concentric bilayered lipid particles encapsulating an aqueous core. The structure is made by dissolving phospholipids in a suitable organic solvent (mostly chloroform or ethanol), then evaporating the solvent in a rotary evaporator under vacuum followed by a series of sizing and separation processes.[103] Liposomes are amphiphilic in structure making it easy to attach to tumor cells and infuse the payload of drugs.[104] Bioavailability of liposomal delivery of cefotaxime was 2.3 to 2.7 times higher than the aqueous solution of the drug itself.[105] Folic acid conjugated liposomes showed 1.2–1.8 times higher oral bioavailability of cefotaxime than unconjugated liposomes in SD rat animal models.[106] Oral administration of liposomal β-sitosterol penetrated the intestinal mucosa and upregulated the host defense of metastatic cancer cells.[25] PTX loaded liposomes enhanced oral bioavailability and avoided P-glycoprotein efflux transports.[24] Folate receptor (FR) targeted liposomes containing PTX ($IC_{50} \approx 0.048$ mM) showed 3.8 fold greater toxicity than untargeted PTX liposomes ($IC_{50} \approx 0.184$ μM) in KB oral cancer cells.[26] FR-targeted PTX liposomes had 3.4-fold higher blood circulation half-life than free PTX in Cremophor EL/ethanol formulation.[107] Collagen polymer was used in the core of liposomes (LSP) to increase oral absorption of 5-fluorouracil (5-FU) and methotrexate (MTX) drug concentrations at the intestinal absorption site (Caco-2 cells).[108]

6.4.2.3 *Nanoemulsions*

Nanoemulsions (NEs) are isotropic mixtures of drugs, oil and surfactant droplets in water that form NPs of less than 100 nm.[27] β-Lactamase (BLM) protein loaded NEs were prepared by mixing lauroglycol FCC oil, cremophor EL surfactant, and transcutol HP co-surfactant in a weight ratio of 5:4:3.[12] The droplet size was less than 50 nm. Oral delivery of this formulation showed 1.3-fold higher mean residence time (12.12 hours), 1.5-fold more relative bioavailability (6.34%) and 2.7-fold more of the maximum plasma concentration (C_{max}=1.9 mU/ml) than free BLM solution in Sprague–Dawley rats. Oral delivery of PTX NEs displayed 4-fold higher bioavailability than Taxol.[109] Mice pretreated with curcumin NEs enhanced the bioavailability and antitumor activity of PTX in ovarian adenocancer.[28] Eugenol oil in colchicine drug formulations enhanced the relative bioavailability of the drug 2.1 fold more than free colchicine solution.[110] NEs improve the water solubility of hydrophobic drugs by protecting from physicochemical and enzymatic degradation. For example, 9-nitrocamptothecin (9-NC) is a poorly soluble topoisomerase-I inhibitor to treat pancreatic cancer. A mixture of NEs was prepared for 9-NC using ethyl oleate, Tween-80 or Cremophor EL, and PEG-400/ethanol. Incorporation of 9-NC in these ~30–40 nm NEs improved the oral drug bioavailability by 2-fold compared to 9-NC suspension.[29]

6.5 Influence of physicochemical parameters of nanoparticles

Targeted delivery has made impressive advancement due to an understanding of unique characteristics of the cancer microenvironment such as enhanced permeability and retention (EPR) effect and over-expressed membrane surface proteins and receptors.[111] Understanding physicochemical properties of NPs that govern biological responses further facilitates the design of safe and effective therapeutic agents. Since the uptake of drug NPs depends on the effect of the shape, size and surface charge of NPs, we discuss these parameters in the following sections.

6.5.1 *Shape*

The shape of a nanoparticle can be controlled to tailor the contact surface area. Techniques to prepare shape-specific NPs include film stretching,[112] jet and flash imprint lithography (J-FIL), [113] particle replication in non-wetting templates (PRINT)[114,115] and other nanofabrication processes [116,117] (for detailed methods, please see Refs.112–117). The greater the contact surface area, the greater the receptor-ligand interactions are for binding. For instance, elongated rod-shaped NPs (nanorods) showed higher receptor ligand binding over spherical NPs due to higher contact surface to the cell membrane.[118] Silica beads with silicon nanowires outgrown on the surface effectively adhered to the microvilli of the epithelial cells of the GI tract, thereby prolonged the drug retention time and promoted the paracellular transport of drug molecules across the epithelium.[119] Irregularly shaped, quasi-planar microparticles (5 × 10 μm diameter) are more effective drug carriers than spherical counterparts for delivery across the epithelium.[120]

6.5.2 *Size*

The size of a nanoparticle is another determinant of penetration, accumulation, and cellular uptake of drug NPs in the tumor. The size effect is tested by preparing particles of various sizes and shape while maintaining the same composition.[121,122] Particles were synthesized by the PRINT method using cross-linked PEG hydrogels.[114,123] The shape and size were analyzed using confocal microscopy and Transmission Electron Microscopy (TEM). The effect of size and shape of these cationic PRINT particles were studied using cervical carcinoma epithelial (HeLa) cells. All particles, ranging from 150 nm to 5 μm in cubic and cylindrical shape, were maintained with a constant surface charge.[114] The cylindrical particles with 150 nm diameter and 450 nm height (aspect ratio = 3) showed higher internalization than those with 100 nm diameter and the same aspect ratio.[114] The cubic particles with 2 μm were internalized more efficiently due to its smaller size than the 3 and 5 μm. The higher internalization rate of non-spherical particles than spherical particles was mainly due to the larger surface area per unit volume of the particles, and thereby higher binding interactions with the cell surface than that of symmetrical shape spherical particles. Elongated particles can also avoid phagocytosis by macrophages depending on their degree of interactions.[124] Non-spherical particles can avoid macrophage engulfment by interacting with the cell surface through its major axis.[124] These findings suggest that the intracellular uptake kinetics is likely a function of size, shape, and aspect ratio.[125]

6.5.3 *Surface*

Surface property of drug NPs influences absorption kinetics in the GI tract. Surface modifications include conjugation of functional groups, PEG moieties, polymers, and targeting ligands such as folic acids,[126] lectins, pectins,[127] and peptides.[128] Conjugation of PEG enhances NP transport by shielding the NP surface, thereby decreasing its interactions with mucin fibers. Diffusion of PEG-modified NPs showed higher mean squared displacement (MSD) than unmodified NPs in human cervicovaginal mucus (CVM).[81,129] A 40% reduction in PEG surface coverage decreased NP transport by 700 folds in human mucus.[81] The molecular weight of PEG also influenced the oral bioavailability of anticancer drugs. For instance, PEG2000 enhanced the bioavailability of PTX 1.7- and 4-fold higher than PEG6000 and PEG10,000, respectively.[130] Peptide ligands such as TAT-conjugated with insulin showed five to 8-fold higher transcytosis than free insulin in a Caco-2 monolayer transwell assay.[128] When conjugated with CSK (CSKSSDYQC) peptide, Goblet cell targeting NPs showed 1.15- to 1.75-fold higher transport ability than unconjugated NPs in a co-culture of mucus producing HT29-MTX cells and adsorptive enterocyte-like Caco-2 cells, respectively.[131]

6.6 Summary

Orally delivered anti-cancer drug NPs have been designed to permeate barriers such as the complex passage in the GI tract, mucus layer, enzymatic degradation, and differential pHs. NP properties directly contribute to overcoming these barriers across the GI tract. NPs with peptidic ligands are promising candidates for oral drug delivery. Efficient DDS depends on the selection of polymers, in particular, mucoadhesive polymers offering prolonged contact area, low enzymatic activity, and patient compliance. Shape, size and surface charge of NP interplay for greater uptake, binding, and delivery of drugs. Receptor-targeted specific interactions are advantageous to target the epithelium. More physiologically relevant *in vitro* models are needed to understand the NP penetration behavior through the GI barriers, which would lead to better understanding of drug delivery efficiency.

Acknowledgements

The authors gratefully acknowledge Professor Robert S. Aronstam for manuscript revising and Humayun Shariff for schematic drawings. S.B.

acknowledges the start-up fund by the department and seed funding from the Center of Biomedical Science and Engineering (CBSE) and Environmental Research Center (ERC) at Missouri S & T.

References

1. Torre LA, Bray F, Siegel RL, Ferlay J, Lortet-Tieulent J, Jemal, A. Global cancer statistics, 2012. *CA: A Cancer J Clini,* 2015; 65: 87–108.
2. Collins LMC, Dawes, C. The surface area of the adult human mouth and thickness of the salivary film covering the teeth and oral mucosa. *J Dental Res,* 1987; 66: 1300–1302.
3. Lemmer HJR, Hamman JH. Paracellular drug absorption enhancement through tight junction modulation. *Exp Opin Drug Deli,* 2013; 10: 103–114.
4. Salama NN, Eddington ND, Fasano, A. Tight junction modulation and its relationship to drug delivery. *Adv Drug Deli Rev,* 2006; 58: 15–28.
5. Witt KA, Huber JD, Egleton RD, Roberts MJ, Bentley MD, Guo L, Wei H, Yamamura HI, Davis, TP. Pharmacodynamic and pharmacokinetic characterization of poly(ethylene glycol) conjugation to met-enkephalin analog [d-Pen2,d-Pen5]-enkephalin (DPDPE). *J Pharmaco Exp Ther,* 2001; 298: 848–856.
6. Liu F, McConnell E, Pygall, S. *Update on Polymers for Oral Drug Delivery*. Smithers Rapra Publishing: Shawbury, US, 2011.
7. Khutoryanskiy VV. *Mucoadhesive Materials and Drug Delivery Systems.* Wiley: Somerset, GB, 2014.
8. *Therapeutic Proteins: Strategies to Modulate Their Plasma Half-lives.* Wiley-Blackwell: Hoboken, DE, 2012.
9. Teesalu T, Sugahara KN, Ruoslahti E. Tumor-penetrating peptides. *Front Oncol,* 2013, 3: 216.
10. Geng Y, Dalhaimer P, Cai S, Tsai R, Tewari M, Minko T, Discher DE. Shape effects of filaments versus spherical particles in flow and drug delivery. *Nat Nano,* 2007; 2: 249–255.
11. Agüeros M, Zabaleta V, Espuelas S, Campanero MA, Irache, JM. Increased oral bioavailability of paclitaxel by its encapsulation through complex formation with cyclodextrins in poly(anhydride) nanoparticles. *J Control Release,* 2010; 145: 2–8.

12. Rao SVR, Yajurvedi K, Shao J. Self-nanoemulsifying drug delivery system (SNEDDS) for oral delivery of protein drugs: III. *In vivo* oral absorption study. *Int J Pharm,* 2008; 362: 16–19.
13. Mitragotri S, Burke PA, Langer R. Overcoming the challenges in administering biopharmaceuticals: formulation and delivery strategies. *Nat Rev Drug Discov,* 2014; 13: 655–672.
14. Ensign LM, Cone R, Hanes J. Oral drug delivery with polymeric nanoparticles: the gastrointestinal mucus barriers. *Adv Drug Deliv Rev,* 2012; 64: 557–570.
15. Ensign LM, Schneider C, Suk JS, Cone R, Hanes J. Mucus penetrating nanoparticles: biophysical tool and method of drug and gene delivery. *Adv Mater,* 2012; 24: 3887–3894.
16. Lai SK, O'Hanlon DE, Harrold S, Man ST, Wang Y-Y, Cone R, Hanes J. Rapid transport of large polymeric nanoparticles in fresh undiluted human mucus. *Proc Natl Acad Sci,* 2007; 104: 1482–1487.
17. Lai SK, Wang YY, Hanes J. Mucus-penetrating nanoparticles for drug and gene delivery to mucosal tissues. *Adv Drug Deliv Rev,* 2009; 61: 158–171.
18. Sawaya RA, Jaffe J, Friedenberg L, Friedenberg FK. Vitamin, mineral, and drug absorption following bariatric surgery. *Curr Drug Metab,* 2012; 13: 1345–1355.
19. Lee VH. Enzymatic barriers to peptide and protein absorption. *Crit Reviews Ther Drug Carrier Syst,* 1988; 5: 69–97.
20. Vicent MJ, Duncan R. Polymer conjugates: nanosized medicines for treating cancer. *Trends Biotechnol,* 2006; 24: 39–47.
21. Chickering Iii DE, Jacob JS, Desai TA, Harrison M, Harris WP, Morrell CN, Chaturvedi P, Mathiowitz E. Bioadhesive microspheres: III. An in vivo transit and bioavailability study of drug-loaded alginate and poly(fumaric-co-sebacic anhydride) microspheres. *J Control Release,* 1997; 48: 35–46.
22. Youk H-J, Lee E, Choi M-K, Lee Y-J, Chung JH, Kim S-H, Lee C-H, Lim S-J. Enhanced anticancer efficacy of α-tocopheryl succinate by conjugation with polyethylene glycol. *J Control Release,* 2005; 107: 43–52.
23. Zhao L, Feng, S.-S.: Enhanced oral bioavailability of paclitaxel formulated in vitamin E-TPGS emulsified nanoparticles of biodegradable polymers: *In vitro* and *in vivo* studies. *J Pharm Sci,* 2010; 99: 3552–3560.

24. Peltier S, Oger J-M, Lagarce F, Couet W, Benoît J-P. Enhanced oral paclitaxel bioavailability after administration of paclitaxel-loaded lipid nanocapsules. *Pharm Res,* 2006; 23: 1243–1250.
25. Imanaka H, Koide H, Shimizu K, Asai T, Kinouchi Shimizu N, Ishikado A, Makino T, Oku N. Chemoprevention of tumor metastasis by liposomal & beta-sitosterol intake. *Biol Pharm Bull,* 2008; 31: 400–404.
26. Wu J, Liu Q, Lee, RJ. A folate receptor-targeted liposomal formulation for paclitaxel. *Int J Pharm,* 2006; 316: 148–153.
27. C'Solans PI, Nolla JN, Azemar M.J. Garcia-Celma: Nano-emulsions. *Curr Opin Colloid,* 2005; 10: 102–110.
28. Ganta S, Devalapally H, Amiji M. Curcumin enhances oral bioavailability and anti-tumor therapeutic efficacy of paclitaxel upon administration in nanoemulsion formulation. *J Phar Sci,* 2010; 99: 4630–4641.
29. Lu J-L, Wang J-C, Zhao S-X, Liu X-Y, Zhao H, Zhang X, Zhou S-F, Zhang Q. Self-microemulsifying drug delivery system (SMEDDS) improves anticancer effect of oral 9-nitrocamptothecin on human cancer xenografts in nude mice. *Eur J Pharm Biopharm,* 2008; 69: 899–907.
30. Bachar M, Mandelbaum A, Portnaya I, Perlstein H, Even-Chen S, Barenholz Y, Danino D. Development and characterization of a novel drug nanocarrier for oral delivery, based on self-assembled beta-casein micelles. *J Control Release,* 2012; 160: 164–171.
31. Matsumura Y, Gotoh M, Muro K, Yamada Y, Shirao K, Shimada Y, Okuwa M, Matsumoto S, Miyata Y, Ohkura H, Chin K, Baba S, Yamao T, Kannami A, Takamatsu Y, Ito K, Takahashi K. Phase I and pharmacokinetic study of MCC-465, a doxorubicin (DXR) encapsulated in PEG immunoliposome, in patients with metastatic stomach cancer. *Ann Oncol,* 2004; 15: 517–525.
32. Davies A, Finn A, Tagarro I. Intra- and Interindividual Variabilities in the Pharmacokinetics of Fentanyl Buccal Soluble Film in Healthy Subjects. *Clin Drug Investig,* 2011; 31: 317–324.
33. Elsner F, Zeppetella G, Porta-Sales J, Tagarro I. Newer Generation Fentanyl Transmucosal Products for Breakthrough Pain in Opioid-Tolerant Cancer Patients. *Clin. Drug Investig,* 2011; 31: 605–618.
34. Kharia AA, Singhai AK. Effective parameters for formulation of gastro adhesive nanoparticles: screening by design-of-experiments approach. *J Microencapsul,* 2014; 31: 399–405.

35. Dawson M, Wirtz D, Hanes J. Enhanced viscoelasticity of human cystic fibrotic sputum correlates with increasing microheterogeneity in particle transport. *J Biol Chem,* 2003; 278: 50393–50401.
36. Olmsted SS, Padgett JL, Yudin AI, Whaley KJ, Moench TR, Cone, R. A. Diffusion of macromolecules and virus-like particles in human cervical mucus. *Biophysical Journal* 2001; 81: 1930–1937.
37. Wang JJ, Sanderson BJS, Wang H. Cyto- and genotoxicity of ultrafine TiO2 particles in cultured human lymphoblastoid cells. *Mut Res — Gen Tox En Mutagenesis,* 2007; 628: 99–106.
38. Lai SK, O'Hanlon DE, Harrold S, Man ST, Wang YY, Cone R, Hanes J. Rapid transport of large polymeric nanoparticles in fresh undiluted human mucus. *Proc Natl Acad Sci USA*, 2007; 104: 1482–1487.
39. Szentkuti L. Light microscopical observations on luminally administered dyes, dextrans, nanospheres and microspheres in the pre-epithelial mucus gel layer of the rat distal colon. *J Control Rel,* 1997; 46: 233–242.
40. Ensign LM, Cone R, Hanes J. Oral drug delivery with polymeric nanoparticles: The gastrointestinal mucus barriers. *Adv Drug Deliv Rev,* 2012; 64: 557–570.
41. Jani P, Halbert GW, Langridge J, Florence, A. T.: The uptake and translocation of latex nanospheres and microspheres after oral administration to rats. *J Pharm Pharmacol,* 1989; 41: 809–812.
42. Yap TA, Workman P. Exploiting the cancer genome: Strategies for the discovery and clinical development of targeted molecular therapeutics. In *Ann Rev of Pharmacol Toxicol,* 2012; 52; 549–573.
43. Slamon DJ, Leyland-Jones B, Shak S, Fuchs H, Paton V, Bajamonde A, Fleming T, Eiermann W, Wolter J, Pegram M, Baselga J, Norton L. Use of chemotherapy plus a monoclonal antibody against her2 for metastatic breast cancer that overexpresses HER2. *New Engl J Med,* 2001; 344: 783–792.
44. Barua S, Mitragotri S. Synergistic targeting of cell membrane, cytoplasm, and nucleus of cancer cells using rod-shaped nanoparticles. *ACS Nano,* 2013; 7: 9558–9570.
45. Wang Z, Rao DD, Senzer N, Nemunaitis J. RNA Interference and Cancer Therapy. *Pharm Res,* 2011; 28: 2983–2995.
46. Izquierdo M. Short interfering RNAs as a tool for cancer gene therapy. *Cancer Gene Ther,* 2004; 12: 217–227.

47. Antonin de Fougerolles H-PV. John Maraganore and Judy Lieberman: Interfering with disease: a progress report on siRNA-based therapeutics. *Nat Rev Drug Discov,* 2007; 6: 443–453.
48. Fire A, Xu S, Montgomery MK, Kostas SA, Driver SE, Mello CC. Potent and specific genetic interference by double-stranded RNA in *Caenorhabditis elegans*. *Nature* 1998; 391: 806–811.
49. Laroui H, Dalmasso G, Nguyen HT, Yan Y, Sitaraman SV, Merlin D. Drug-loaded nanoparticles targeted to the colon with polysaccharide hydrogel reduce colitis in a mouse model. *Gastroenterology,* 2010; 138: 843–853 e1–e2.
50. Kanasty R, Dorkin JR, Vegas A, Anderson D. Delivery materials for siRNA therapeutics. *Nat mater,* 2013; 12: 967–977.
51. Toub N, Bertrand JR, Tamaddon A, Elhamess H, Hillaireau H, Maksimenko A, Maccario J, Malvy C, Fattal E, Couvreur P. Efficacy of siRNA nanocapsules targeted against the EWS-Fli1 oncogene in Ewing sarcoma. *Pharm Res,* 2006; 23: 892–900.
52. Farokhzad OC, Jon S, Khademhosseini A, Tran TN, Lavan DA, Langer R. Nanoparticle-aptamer bioconjugates: a new approach for targeting prostate cancer cells. *Cancer Res,* 2004; 64: 7668–7672.
53. Eguchi A, Dowdy SF. Efficient siRNA delivery by novel PTD-DRBD fusion proteins. *Cell Cycle,* 2010; 9: 424–425.
54. Michiue H, Eguchi A, Scadeng M, Dowdy SF. Induction of *in vivo* synthetic lethal RNAi responses to treat glioblastoma. *Cancer Biol Ther,* 2009; 8: 2306–2313.
55. Cao Y, Yu SL, Wang Y, Guo GY, Ding Q, An RH. MicroRNA-dependent regulation of PTEN after arsenic trioxide treatment in bladder cancer cell line T24. *Tumour Biol,* 2011; 32: 179–188.
56. Beh CW, Seow WY, Wang Y, Zhang Y, Ong ZY, Ee PL, Yang YY. Efficient delivery of Bcl-2-targeted siRNA using cationic polymer nanoparticles: downregulating mRNA expression level and sensitizing cancer cells to anticancer drug. *Biomacromolecules,* 2009; 10: 41–48.
57. Saraswathy M, Gong, S. Q. Recent developments in the co-delivery of siRNA and small molecule anticancer drugs for cancer treatment. *Mater Today* 2014; 17: 298–306.
58. Strebhardt K, Ullrich A. Targeting polo-like kinase 1 for cancer therapy. *Nat Rev Cancer,* 2006; 6: 321–330.

59. Han S, Li M, Liu X, Gao H, Wu Y. Construction of amphiphilic copolymer nanoparticles based on gelatin as drug carriers for doxorubicin delivery. *Colloids Surfaces B,*2013; 102: 833–841.
60. Renukuntla J, Vadlapudi AD, Patel A, Boddu SH, Mitra AK. Approaches for enhancing oral bioavailability of peptides and proteins. *Int J Pharm,* 2013; 447: 75–93.
61. Mahato RI, Narang AS, Thoma L, Miller DD. Emerging trends in oral delivery of peptide and protein drugs. *Crit Rev Ther Drug Carrier Systems,* 2003; 20: 153–214.
62. Ruoslahti E. RGD and other recognition sequences for integrins. *Annu Rev Cell Dev Biol,* 1996; 12: 697–715.
63. Sugahara KN, Scodeller P, Braun GB, de Mendoza TH, Yamazaki CM, Kluger MD, Kitayama J, Alvarez E, Howell SB, Teesalu T, Ruoslahti E, Lowy AM. A tumor-penetrating peptide enhances circulation-independent targeting of peritoneal carcinomatosis. *J Control Release,* 2015; 212: 59–69.
64. Cheng W-P, Thompson C, Ryan SM, Aguirre T, Tetley L, Brayden DJ. *In vitro* and *in vivo* characterisation of a novel peptide delivery system: Amphiphilic polyelectrolyte–salmon calcitonin nanocomplexes. *J Control Release,* 2010; 147: 289–297.
65. Trapani A, Lopedota A, Franco M, Cioffi N, Ieva E, Garcia-Fuentes M, Alonso MJ. A comparative study of chitosan and chitosan/cyclodextrin nanoparticles as potential carriers for the oral delivery of small peptides. *Eur J Pharm Biopharm,* 2010; 75: 26–32.
66. Jin Y, Song Y, Zhu X, Zhou D, Chen C, Zhang Z, Huang Y. Goblet cell-targeting nanoparticles for oral insulin delivery and the influence of mucus on insulin transport. *Biomaterials,* 2012; 33: 1573–1582.
67. Zhang J, Zhu X, Jin Y, Shan W, Huang Y. Mechanism study of cellular uptake and tight junction opening mediated by goblet cell-specific trimethyl chitosan nanoparticles. *Mol Pharm,* 2014; 11: 1520–1532.
68. Yoo M-K, Kang S-K, Choi J-H, Park I-K, Na H-S, Lee H-C, Kim E-B, Lee N-K, Nah J-W, Choi Y-J, Cho C-S. Targeted delivery of chitosan nanoparticles to Peyer's patch using M cell-homing peptide selected by phage display technique. *Biomaterials,* 2010; 31: 7738–7747.
69. Meric-Bernstam F, Gonzalez-Angulo AM. Targeting the mTOR Signaling Network for Cancer Therapy. *J Clin Oncol,* 2009; 27: 2278–2287.

70. Din OS, Woll PJ. Treatment of gastrointestinal stromal tumor: focus on imatinib mesylate. *Ther Clini Risk Manag,* 2008; 4: 149–162.
71. Jäger D, Ma JH, Mardiak J, Ye DW, Korbenfeld E, Zemanova M, Ahn H, Guo J, Leonhartsberger N, Stauch K, Böckenhoff A, Yu J, Escudier B. Sorafenib treatment of advanced renal cell carcinoma patients in daily practice: the large international PREDICT study. *Clini Genitour Can,* 2015; 13: 156–164.e1.
72. Llovet JM, Ricci S, Mazzaferro V, Hilgard P, Gane E, Blanc J-F, de Oliveira AC, Santoro A, Raoul J-L, Forner A, Schwartz M, Porta C, Zeuzem S, Bolondi L, Greten TF, Galle PR, Seitz J-F, Borbath I, Häussinger D, Giannaris T, Shan M, Moscovici M, Voliotis D, Bruix J. Sorafenib in advanced hepatocellular carcinoma. *New Engl J Med,* 2008; 359: 378–390.
73. Sahay G, Alakhova DY, Kabanov AV. Endocytosis of nanomedicines. *J Control Release,* 2010; 145: 182–195.
74. Geiser M, Rothen-Rutishauser B, Kapp N, Schürch S, Kreyling W, Schulz H, Semmler M, Im Hof V, Heyder J, Gehr P. Ultrafine particles cross cellular membranes by nonphagocytic mechanisms in lungs and in cultured cells. *Environ Health Persp,* 2005; 113: 1555–1560.
75. Knowles MR, Boucher RC. Mucus clearance as a primary innate defense mechanism for mammalian airways. *J Clini Invest,* 2002; 109: 571–577.
76. Shogren R, Gerken TA, Jentoft N. Role of glycosylation on the conformation and chain dimensions of O-linked glycoproteins: light-scattering studies of ovine submaxillary mucin. *Biochemistry,* 1989; 28: 5525–5536.
77. Harris D, Robinson JR. Drug delivery *via* the mucous membranes of the oral cavity. *J Pharm Sci,* 1992; 81: 1–10.
78. Huh CH, Bhutani MS, Farfán EB, Bolch WE. Individual variations in mucosa and total wall thickness in the stomach and rectum assessed *via* endoscopic ultrasound. *Physiol Meas,* 2003, 24, N15.
79. Larsson JM, Karlsson H, Sjovall H, Hansson GC. A complex, but uniform O-glycosylation of the human MUC2 mucin from colonic biopsies analyzed by nanoLC/MSn. *Glycobiology,* 2009; 19: 756–766.
80. Hermanson GT. *Bioconjugate Techniques.* Academic Press: Amsterdam (Netherlands); Boston [Massachusetts.], 2008.
81. Wang Y-Y, Lai SK, Suk JS, Pace A, Cone R, Hanes J. Addressing the PEG mucoadhesivity paradox to engineer nanoparticles that "slip" through the human mucus barrier. *Angew Chem Int Ed,* 2008; 47: 9726–9729.

82. Anderson JM, Van Itallie CM. Physiology and Function of the tight junction. *CSH Perspect Biol,* 2009, 1, a002584.
83. Madara JL. Regulation of the movement of solutes across tight junctions. *Annu Review of Physiol,* 1998; 60: 143–159.
84. Diaz del Consuelo I, Falson F, Guy R, Jacques Y. Transport of fentanyl through pig buccal and esophageal epithelia *in vitro*. Influence of concentration and vehicle pH. *Pharm Res,* 2005; 22: 1525–1529.
85. Liu Z, Lv D, Liu S, Gong J, Wang D, Xiong M, Chen X, Xiang R, Tan X. Alginic acid-coated chitosan nanoparticles loaded with legumain DNA vaccine: effect against breast cancer in mice. *PLoS ONE,* 2013, 8, e60190.
86. Pawar D, Mangal S, Goswami R, Jaganathan KS. Development and characterization of surface modified PLGA nanoparticles for nasal vaccine delivery: effect of mucoadhesive coating on antigen uptake and immune adjuvant activity. *Eur J Pharm Biopharm,* 2013; 85: 550–559.
87. Modi P, Mihic M, Lewin A. The evolving role of oral insulin in the treatment of diabetes using a novel RapidMist™ System. *Diabetes/Metabo Res Rev,* 2002; 18: S38–S42.
88. Chinna Reddy P, Chaitanya KS, Madhusudan Rao Y. A review on bioadhesive buccal drug delivery systems: current status of formulation and evaluation methods. *DARU: J Pharm Sci,* 2011; 19: 385–403.
89. Wu X, Desai KG, Mallery SR, Holpuch AS, Phelps MP, Schwendeman SP. Mucoadhesive fenretinide patches for site-specific chemoprevention of oral cancer: enhancement of oral mucosal permeation of fenretinide by coincorporation of propylene glycol and menthol. *Mol Pharm,* 2012; 9: 937–945.
90. Desai KG, Mallery SR, Holpuch AS, Schwendeman SP. Development and *in vitro–in vivo* evaluation of fenretinide-loaded oral mucoadhesive patches for site-specific chemoprevention of oral cancer. *Pharm Res,* 2011; 28: 2599–609.
91. Shepherd FA, Rodrigues Pereira J, Ciuleanu T, Tan EH, Hirsh V, Thongprasert S, Campos D, Maoleekoonpiroj S, Smylie M, Martins R, van Kooten M, Dediu M, Findlay B, Tu D, Johnston D, Bezjak A, Clark G, Santabárbara P, Seymour L. Erlotinib in previously treated non-Small Cell Lung Cancer. *New Engl J Medicine,* 2005; 353: 123–132.
92. Cerniglia GJ, Pore N, Tsai JH, Schultz S, Mick R, Choe R, Xing X, Durduran T, Yodh AG, Evans SM, Koch CJ, Hahn SM, Quon H, Sehgal CM, Lee WMF, Maity A. Epidermal growth factor receptor inhibition

modulates the microenvironment by vascular normalization to improve chemotherapy and radiotherapy efficacy. *PLoS ONE,* 2009; 4: e6539.

93. Pore N, Jiang Z, Gupta A, Cerniglia G, Kao GD, Maity A. EGFR tyrosine kinase inhibitors decrease VEGF expression by both hypoxia-inducible factor (HIF)-1-independent and HIF-1-dependent mechanisms. *Cancer Res,* 2006; 66: 3197–3204.
94. Ahmed S, El-Setouhy DA, El-Latif Badawi AA, El-Nabarawi MA. Provesicular granisetron hydrochloride buccal formulations: in vitro evaluation and preliminary investigation of *in vivo* performance. *Eur J Pharm Sci,* 2014; 60: 10–23.
95. Bies C, Lehr C-M, Woodley JF. Lectin-mediated drug targeting: history and applications. *Adv Drug Deliv Rev,* 2004; 56: 425–435.
96. Dawson G, Halbert G. The in vitro cell association of invasin coated polylactide-co-glycolide nanoparticles. *Pharm Res,* 2000; 17: 1420–1425.
97. Russell-Jones GJ, Arthur L, Walker H. Vitamin B12-mediated transport of nanoparticles across Caco-2 cells. *Int J Pharm,* 1999; 179: 247–255.
98. Ke W, Zhao Y, Huang R, Jiang C, Pei Y. Enhanced oral bioavailability of doxorubicin in a dendrimer drug delivery system. *J Pharm Sci,* 2008; 97: 2208–2216.
99. Bhardwaj V, Ankola DD, Gupta SC, Schneider M, Lehr CM, Kumar MNVR. PLGA nanoparticles stabilized with cationic surfactant: safety studies and application in oral delivery of paclitaxel to treat chemical-induced breast cancer in rat. *Pharm Res,* 2009; 26: 2495–2503.
100. Feng S-S, Mei L, Anitha P, Gan CW, Zhou W. Poly(lactide)-vitamin E derivative/montmorillonite nanoparticle formulations for the oral delivery of Docetaxel. *Biomaterials,* 2009; 30: 3297–3306.
101. Arbós P, Campanero MA, Arangoa MA, Renedo MJ, Irache JM. Influence of the surface characteristics of PVM/MA nanoparticles on their bioadhesive properties. *J Control Release,* 2003; 89: 19–30.
102. Lee SC, Huh KM, Lee J, Cho YW, Galinsky RE, Park K. Hydrotropic polymeric micelles for enhanced paclitaxel solubility: *in vitro* and *in vivo* characterization. *Biomacromolecules,* 2007; 8: 202–208.
103. Thassu D, Deleers M, Pathak Y. *Nanoparticulate Drug Delivery Systems.* Informa Healthcare: New York, 2007.
104. Faraji AH, Wipf P. Nanoparticles in cellular drug delivery. *Bioorg Med Chemi,* 2009; 17: 2950–2962.

105. Kim H, Kim Y, Lee J. Liposomal formulations for enhanced lymphatic drug delivery. *Asian J Pharm Sci,* 2013; 8: 96–103.
106. Ling SSN, Yuen KH, Magosso E, Barker SA. Oral bioavailability enhancement of a hydrophilic drug delivered *via* folic acid-coupled liposomes in rats. *J Pharm Pharmacol,* 2009; 61: 445–449.
107. Cai L, Wang X, Wang W, Qiu N, Wen J, Duan X, Li X, Chen X, Yang L, Qian Z, Wei Y, Chen L. Peptide ligand and PEG-mediated long-circulating liposome targeted to FGFR overexpressing tumor *in vivo*. *Int J Nanomed,* 2012; 7: 4499–4510.
108. Moutardier V, Tosini F, Vlieghe P, Cara L, Delpero JR, Clerc T. Colloidal anticancer drugs bioavailabilities in oral administration models. *Int J Pharm,* 2003; 260: 23–38.
109. Gao P, Rush BD, Pfund WP, Huang T, Bauer JM, Morozowich W, Kuo M-S, Hageman MJ. Development of a supersaturable SEDDS (S-SEDDS) formulation of paclitaxel with improved oral bioavailability. *J Pharm Sci,* 2003; 92: 2386–2398.
110. Shen Q, Wang Y, Zhang Y. Improvement of colchicine oral bioavailability by incorporating eugenol in the nanoemulsion as an oil excipient and enhancer. *Int J Nanomedicine,* 2011; 6: 1237–1243.
111. Agarwal R, Singh V, Jurney P, Shi L, Sreenivasan SV, Roy K. Mammalian cells preferentially internalize hydrogel nanodiscs over nanorods and use shape-specific uptake mechanisms. *Proc Natl Acad Sci USA,* 2013; 110: 17247–17252.
112. Champion JA, Katare YK, Mitragotri S. Making polymeric micro- and nanoparticles of complex shapes. *Proc Natl Acad Sci,* 2007; 104: 11901–11904.
113. Agarwal R, Singh V, Jurney P, Shi L, Sreenivasan SV, Roy K. Scalable imprinting of shape-specific polymeric nanocarriers using a release layer of switchable water solubility. *ACS Nano,* 2012; 6: 2524–2531.
114. Gratton SEA, Ropp PA, Pohlhaus PD, Luft JC, Madden VJ, Napier ME, DeSimone JM. The effect of particle design on cellular internalization pathways. *Proc Natl Acad Sci,* 2008; 105: 11613–11618.
115. Perry JL, Reuter KG, Kai MP, Herlihy KP, Jones SW, Luft JC, Napier M, Bear JE, DeSimone JM. PEGylated PRINT nanoparticles: The impact of PEG density on protein binding, macrophage association, biodistribution, and pharmacokinetics. *Nano Letters,* 2012; 12: 5304–5310.

116. Adriani G, de Tullio MD, Ferrari M, Hussain F, Pascazio G, Liu X, Decuzzi P. The preferential targeting of the diseased microvasculature by disk-like particles. *Biomaterials,* 2012; 33: 5504–5513.
117. Godin B, Chiappini C, Srinivasan S, Alexander JF, Yokoi K, Ferrari M, Decuzzi P, Liu X. Discoidal porous silicon particles: fabrication and biodistribution in breast cancer bearing mice. *Adv Fun Mater,* 2012; 22: 4225–4235.
118. Barua S, Yoo J-W, Kolhar P, Wakankar A, Gokarn YR, Mitragotri S. Particle shape enhances specificity of antibody-displaying nanoparticles. *Proc Natl Acad Sci,* 2013; 110: 3270–3275.
119. Fischer KE, Nagaraj G, Hugh Daniels R, Li E, Cowles VE, Miller JL, Bunger MD, Desai TA. Hierarchical nanoengineered surfaces for enhanced cytoadhesion and drug delivery. *Biomaterials,* 2011; 32: 3499–3506.
120. Uskoković V, Lee K, Lee PP, Fischer KE, Desai TA. Shape effect in the design of nanowire-coated microparticles as transepithelial drug delivery devices. *ACS Nano,* 2012; 6: 7832–7841.
121. Li Y, Xiao K, Luo J, Lee J, Pan S, Lam KS. A novel size-tunable nanocarrier system for targeted anticancer drug delivery. *J Control release,* 2010; 144: 314–323.
122. Tang L, Fan TM, Borst LB, Cheng J. Synthesis and biological response of size-specific, monodisperse drug-silica nanoconjugates. *ACS Nano,* 2012; 6: 3954–3966.
123. Perry JL, Herlihy KP, Napier ME, DeSimone JM. PRINT: a novel platform toward shape and size specific nanoparticle theranostics. *Acco Chem Res,* 2011; 44: 990–998.
124. Champion JA, Mitragotri S. Role of target geometry in phagocytosis. *Proc Natil Acad Sci USA,* 2006; 103: 4930–4934.
125. Gratton SE, Ropp PA, Pohlhaus PD, Luft JC, Madden VJ, Napier ME, DeSimone JM. The effect of particle design on cellular internalization pathways. *Proc Natl Acad Sci USA,* 2008; 105: 11613–11618.
126. Yang S-J, Lin C-F, Kuo M-L, Tan C-T. Photodynamic detection of oral cancers with high-performance chitosan-based nanoparticles. *Biomacromolecules,* 2013; 14: 3183–3191.
127. Bosio V, Machain V, López A, De Berti IP, Marchetti S, Mechetti M, Castro G. Binding and encapsulation of doxorubicin on smart pectin hydrogels for oral delivery. *Appl Biochem Biotechnol* 2012; 167: 1365–1376.

128. Liang JF, Yang VC. Insulin-cell penetrating peptide hybrids with improved intestinal absorption efficiency. *Biochem Biophys Res Comm,* 2005; 335: 734–738.

129. Tang BC, Dawson M, Lai SK, Wang Y-Y, Suk JS, Yang M, Zeitlin P, Boyle MP, Fu J, Hanes J. Biodegradable polymer nanoparticles that rapidly penetrate the human mucus barrier. *Proc Natl Acad Sci,* 2009; 106: 19268–19273.

130. Zabaleta V, Ponchel G, Salman H, Agüeros M, Vauthier C, Irache JM. Oral administration of paclitaxel with pegylated poly(anhydride) nanoparticles: permeability and pharmacokinetic study. *Eur J Pharm Biopharm,* 2012; 81: 514–523.

131. Jin Y, Song Y, Zhu X, Zhou D, Chen C, Zhang Z, Huang Y. Goblet cell-targeting nanoparticles for oral insulin delivery and the influence of mucus on insulin transport. *Biomaterials,* 2012; 33: 1573–1582.

132. Torres-Lugo M, García M, Record R, Peppas NA. Physicochemical behavior and cytotoxic effects of p(methacrylic acid–g-ethylene glycol) nanospheres for oral delivery of proteins. *J Control Release,* 2002; 80: 197–205.

Chapter 7

CANCER THERAPEUTICS WITH LIGHT: ROLE OF NANOSCALE AND TISSUE ENGINEERING IN PHOTODYNAMIC THERAPY

*Joyce Liu**, *Huang-Chiao Huang**, *Emma Briars**, *Girgis Obaid**, *Imran Rizvi** *and Tayyaba Hasan**[†,‡,§]

**Wellman Center for Photomedicine, Massachusetts General Hospital and Harvard Medical School, Boston, MA 02114, USA*

[†]*Department of Dermatology, Massachusetts General Hospital, Boston, MA 02114, USA*

[‡]*Division of Health Sciences and Technology, Harvard University and Massachusetts Institute of Technology, Cambridge, MA 02139, USA*

[§]*thasan@mgh.harvard.edu*

Abstract

The ability of cancer cells to compensate for therapeutic insults by rewiring their growth and survival signaling circuits suggests that combination regimens targeting multiple non-overlapping pathways, ideally where each component of the combination enhances the effect of the others, is most likely to improve treatment outcomes. Light-activated approaches combined with nanotechnology provide a unique opportunity to co-deliver several cooperative agents, targeting multiple key molecular pathways and critical components of the tumor microenvironment. To efficiently and rigorously pinpoint these candidate therapies in pre-clinical settings, an effort has been made to develop high-throughput heterocellular models that replicate the key physiological and architectural cues of cancers to screen potential therapies. This chapter focuses on recent advances in the engineering of light-activatable nanoconstructs for targeted photodynamic therapy, a photochemistry-based modality that is effective against chemo- and radioresistant cancers. This chapter then discusses the current trends and future directions in the engineering of three-dimensional cancer cultures for high-throughput dosimetry analyses, mechanistic studies, and evaluation of treatment response.

7.1 Introduction

Photodynamic therapy (PDT) is a photochemistry-based approach in which a chemical agent, termed a photosensitizer, is energized by light of a specific wavelength to generate cytotoxic molecular species that can kill target cells, modulate tumor microenvironments, or elicit systemic immune responses.[1,2] Upon absorbing a quantum of light, the photosensitizer is electronically excited from its ground state (S_0) to a short-lived singlet excited state (S_1; nanosecond lifetime). From this state, the photosensitizer may relax back to S_0, either by radiative or non-radiative decay. During radiative decay, photons that can be harnessed for fluorescence imaging are emitted. On the other hand, non-radiative decay *via* thermal relaxation leads to local heating and can be exploited for both photothermal therapy and photoacoustic imaging. Alternatively, the S_1 molecule can undergo spin inversion through intersystem crossing to a long-lived triplet excited state (T_1; micro- to milli-second lifetime) (Figure 7.1a).[3] Photodynamic phototoxicity comes primarily from the triplet manifold (T_1), where photochemistry can be initiated *via* electron transfer reactions with a substrate to produce free radicals and reactive oxygen species (type I reaction), or *via* triplet–triplet energy transfer to ground-state molecular oxygen ($^3O_2$) to give rise to singlet excited state oxygen molecules ($^1O_2$, singlet oxygen) that are cytotoxic (type II reaction).[3] This chapter will largely focus on the therapeutic and imaging applications that result from both the radiative decay and energy transfer processes that occur following excitation of the photosensitizer.

Clinically, the photosensitizer is administered systemically (*i.v.*) or topically. Topical application is primarily used in dermatological

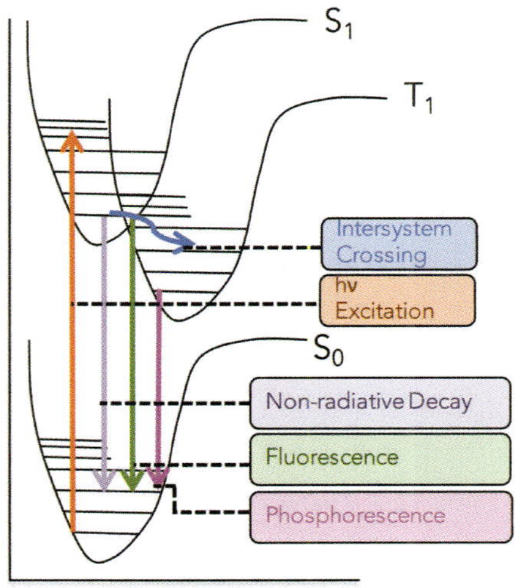

(a)

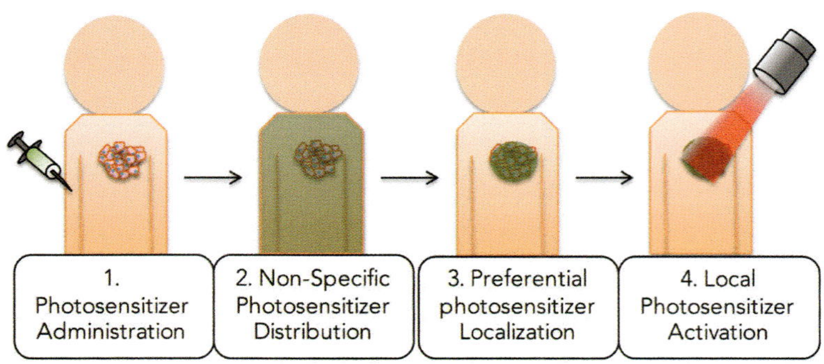

(b)

Figure 7.1 (a) Jablonski energy diagram of a chromophore (e.g. photosensitizer). (b) Idealized clinical process of photodynamic therapy (PDT). Following systemic administration and a period of time, the photosensitizer localizes to the tumor and is locally activated by light irradiation.

purposes, although it has also been documented in the treatment of bladder cancer. When administered systemically, the photosensitizer initially is distributed non-specifically, but localizes with modest selectivity to the tumor after a photosensitizer-specific time-delay. While this low-level inherent selectivity can be achieved from phenomenon such as the enhanced permeability and retention (EPR) effect and metabolic conversion of specific photosensitizers, much ongoing work has been done in photosensitizer design and delivery to increase selectivity in tumors, a field that will be reviewed in this chapter. Following photosensitizer accumulation, light is delivered to the target tissue using an appropriate delivery source for the anatomical site of interest (Figure 7.1b). In combination with photosensitizer localization, site-specific light delivery allows PDT to achieve dual-selectivity by only activating photosensitizers in the irradiated tissue. Importantly, the same equipment can be modified to harness the fluorescence of PDT agents for diagnostic imaging applications.

The historical origins of PDT date back thousands of years, when ancient Egyptians and Indians were able to harness light from the sun to perform light-based therapies, using sunlight to excite psoralens as a treatment for vitiligo.[2,4] More recently in 1901, Oscar Raab, a student of Herman Von Tappeiner, observed that acridine orange was cytotoxic to *Paramecia* after it had been exposed to light.[5] In 1913, the field witnessed the first-in-human study with porphyrins after Friedrich Meyer-Betz observed the effects of self-administered hematoporphyrin in his own skin after light exposure.[6] The first evidence of PDT for cancers was also attributed to Von Tappeiner and A. Jesionek, who topically applied eosin to skin tumors and irradiated with white light in 1903.[7] Although the term "photodynamic" arose from these treatments, it was not until scientists such as Policard, Figge, and Rasmussan-Taxdal observed the localized fluorescence of porphyrins in tumors in the mid-19th century that PDT for cancer began to gain momentum.[8,9] These discoveries paved the way for what would become the development of the first FDA-approved photosensitizer that began when Samuel Schwartz reacted crude hematoporphyrin to yield a derivative compound, which Richard Lipson of the Mayo Clinic found to localize and fluoresce in neoplastic lesions during surgery.[10,11] In the 1970s, T.J. Dougherty and others established the clinical photodynamic efficacy of this compound, hematoporphyrin derivative (HpD), in various tumors. This series of events led to the first approval of the purified version of HpD, called Photofrin, for the treatment of bladder cancer.[12–15] Since then, several other photosensitizers (e.g. Foscan®, Visudyne®, Levulan®, Metvix®) have been approved for various malignant and non-malignant diseases (e.g. esophageal cancer, non-small cell lung cancer, actinic keratosis, age-related macular degeneration).[16] Among these, Visudyne® (nanoliposomal verteporfin) is the only FDA-approved nanoformulation of a photosensitizer.[17] Currently, verteporfin is being tested in clinical studies for patients with cancers that are difficult to treat (e.g. pancreatic cancer), where the majority of therapeutic interventions have failed to improve the outcome.[18]

PDT is a logical adjuvant cancer therapy due to its unique mechanisms of cytotoxicity.[19] Often, chemo- and radio-resistance arise from an increase in the ratio of anti- to pro-apoptotic proteins. PDT has been shown to directly damage these anti-apoptotic proteins, such as Bcl-2 and Bcl-xL, preventing this mechanism of resistance and tipping the balance towards apoptosis.[20] Certain photosensitizers also attack the mitochondria, promoting the release of pro-apoptotic cytochrome c.[21] Amongst other mechanisms, destruction of these targets helps PDT in bypassing the usual mechanisms of resistance, making PDT effective against cancers resistant to chemotherapy and radiation. In preclinical studies, several groups, including ours, have also shown that PDT can reverse treatment resistance by re-sensitizing residual disease to chemotherapy and by synergizing with biologics.[22,23] In addition, several studies have shown that PDT also can modulate the extracellular matrix[24] and permeabilize tumor vasculature to enhance drug delivery into tumors.[25]

To date, PDT is largely used for localized advanced cancers without curative intent.[16] For deep and non-topical tumors, photosensitizers that are administered intravenously must arrive at the tumor vasculature, extravasate through the vessel wall, and finally migrate throughout the tumor interstitium to reach the target cells.[26] Light can be delivered topically, interstitially, intraoperatively or endoscopically to activate photosensitizers.[27] Moreover, the highly complex compensatory signaling pathways in chemo- and radio-resistant cancer cells[28] suggest that combining PDT with treatment modalities that inhibit these pathways would be a more rational and promising therapeutic approach. This chapter will discuss the recent advances in the engineering of sophisticated nanoconstructs that can selectively co-deliver multiple agents to target cells and be photoactivated by visible and near-infrared light as well as alternative energy sources, thereby improving PDT and broadening its application to complex anatomical sites, such as the pleural and peritoneal cavities.[29]

Whilst showing great promise at the benchtop, PDT has not been adopted as a mainstay of cancer management yet. This is likely due to the fact that PDT has mainly been used on patients with very advanced disease, which is refractory to all other treatment options. Other factors hampering the therapeutic translation of PDT from benchtop to bedside include the heterogeneous nature of tumors, lack of sufficiently realistic and complex cancer models for preclinical treatment development, and the "one size fits all" dosimetry approach.[30] Therefore, utilizing biologically relevant platforms that restore the physiological and architectural characteristics of human cancers for the evaluation of treatment response is critical. Moreover, for light-based therapies, spheroid and adherent tumor models have proven to be valuable tools to study photosensitizer distribution and light dosimetry and, with further development, may help to improve the efficacy and predictability of PDT-based regimens.[31,32]

Pre-clinical investigations of nanoscale light-based therapeutic formulations often take place in two-dimensional (2D) *in vitro* models, or monolayer cell culture. However, these models are inherently limited as cells only grow in flat layers and therefore fail to recapitulate the *in vivo* architecture with all its physical and biochemical barriers.[33] Whilst *in vivo* animal models provide a platform that partially restores the architecture and complexities of cancer biology, they are expensive, time-consuming and can be limited by the characteristics of the host biology. Human tissues are also valuable for preclinical research, but tissue acquirement is often challenging due to many legal, ethical, logistical and practical issues. Studies using human subjects are the most informative models; however, they are not a viable model for development and optimization of novel therapeutics. Three-dimensional (3D) tissue culture models have emerged as valuable complementary tools to identify and optimize new candidate therapies prior to animal and human studies. These 3D models capture phenotypic heterogeneity and recapitulate complex signals found in the tumor microenvironment, as their architecture can incorporate multiple subpopulations of cells and can factor in the effects of distances from supplies of nutrients and oxygen.[34] This chapter will review the advances in the engineering of 3D culture models and the quantitative analyses required for the high-throughput evaluation of photosensitizer penetration, light dosimetry and treatment response of PDT, to help bridge the gap between bench and bedside research in light-based cancer therapy.

7.2 Selective photosensitizers as therapeutic and imaging agents

One of the major hurdles encountered with most cancer therapies, including PDT, is the poor tumor selectivity, which can reduce overall tumor control and increase off-site toxicity, therefore limiting maximum tolerated treatment dosages. Poor tumor selectivity can also have a negative impact on the diagnostic and theranostic use of photosensitizers. Multiple mechanisms of targeting have been utilized to enhance the tumor selectivity of photosensitizers and are discussed further in this section. These include both passive and active targeted delivery of PDT agents to tumors, the tumor-specific activation of photosensitizing agents, and pro-drug approaches that are selectively converted to active agents by tumor cells (Figure 7.2). These mechanisms, among others, have shown an enhancement in PDT efficacy and promise to revolutionize many cancer treatment modalities including PDT.

7.2.1 *Site-specific targeted delivery*

The most common modality studied to introduce selectivity for cancer therapy using photosensitizers is site-specific delivery of the photoactive agents to enhance the tumor-to-normal ratio and to achieve effective intratumoral sensitizer threshold levels with lower administered doses. Targeted delivery can either be passive or active, depending on the mechanisms used to transport photosensitizers to the target site. Passive delivery refers to the exploitation of a natural physical tumor phenomenon, which arises from the rapidly developed, poor quality intratumoral vasculature. This phenomenon is known as the EPR effect and is observed in tumors exceeding 2–3 mm in diameter. The tumor vasculature exhibiting the EPR effect is characterized by irregular and dysfunctional vessels with multiple fenestra that enhance the extravasation of large molecular weight sensitizer delivery vehicles.[35] This enhanced permeability towards large molecular weight macromolecule or nanosized complexes is accompanied by poor lymphatic drainage from the tumors resulting in a simultaneous enhancement in tumor retention.[35] By loading hydrophobic or amphiphilic photosensitizers

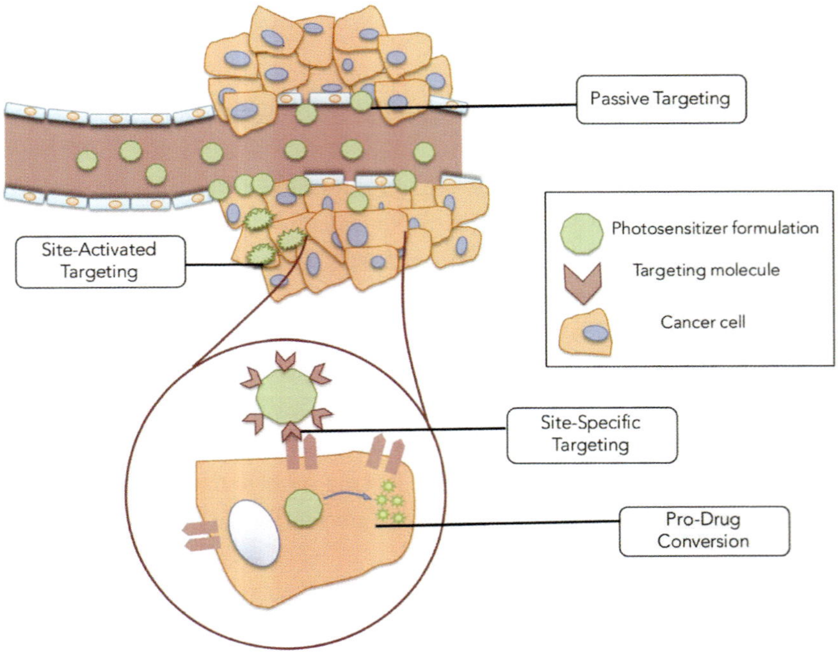

Figure 7.2 Methods of photosensitizer selectivity. Passive targeting takes advantage of leaky tumor vasculature to deliver nanoscale encapsulated photosensitizers or large molecule weight molecules. Site-activated targeting is mediated by biochemical characteristics of the tumor environment, such as low pH and high levels of glutathione and matrix metalloproteinases. Site-specific targeting uses conjugation of specific tumor recognition molecules to the photosensitizer. Pro-drug conversion requires intracellular conversion to become active.

into a variety of nanoparticle constructs,[36,37] liposomes[38] or dendrimers,[39] their photoactivity is retained and the agents are passively targeted to tumors following intravenous administration.

Active targeting of photosensitizers requires the conjugation of sensitizers or sensitizer formulations to bioactive molecules with specific tumor recognition and binding capabilities, whilst retaining the photoactivity of the photosensitizer for imaging and therapy. The earliest studies demonstrating active targeting of sensitizers to tumor tissue utilized sensitizer conjugates of antibodies raised against tumor-specific antigens; these conjugates were termed photoimmunoconjugates (PICs).

The very first study demonstrating the selectivity of uptake and phototoxicity of cancer cells using PICs was by Levy *et al.* who developed a tumor cell-selective anti-M-1 antibody conjugated to hematoporphyrin.[40] The accumulation of the photosensitizer in the M-1 myosarcoma tumors and the tumor inhibiting properties of PDT using the PIC were superior to the free hematoporphyrin.[40] Furthermore, the anti-M-1-specific PICs demonstrated negligible levels of phototoxicity in non-target cell lines. The PIC technology has been adapted to an array of photosensitizers and antibodies to tailor and deliver targeted PDT to a variety of tumor antigens, with the ultimate goal of customizing PDT-based regimens to specific cancers.[41–48] A recent advancement in current PIC therapies is the utilization of tumor-targeted, activable photo-immunotherapy (taPIT), which relies on a PIC conjugated to a statically quenched photosensitizer that requires tumor cell internalization, followed by lysosomal degradation of the conjugate leading to dequenching of the photosensitizer for subsequent light-based activation.[49] This technique markedly enhances tumor selectivity and allows for the use of significantly higher PDT doses, which are often toxic when using conventional "always-on" PICs.[49]

Targeted delivery of photosensitizer conjugates has also been achieved using bioactive molecules that target receptors with elevated expression levels in cancer cells. For example, folic acid-porphyrin conjugates have been prepared to target the folate receptor overexpressed on KB human nasopharyngeal carcinoma cells and these sensitizer conjugates demonstrated an enhanced PDT efficacy as compared to the non-targeted sensitizer.[50] Additionally, holo-transferrin has been conjugated to hematoporphyrin, which facilitated the selective transferrin receptor-mediated uptake by target cells and enhanced their phototoxicity following iron starvation.[51]

Photosensitizer-bearing nanoparticles and large molecular weight structures that naturally exhibit higher accumulation in tumors through passive targeting have also been modified with targeting moieties to allow them to target tumor cells actively. Gold nanoparticles functionalized with a thiolated zinc phthalocyanine photosensitizer have been functionalized with anti-HER-2 antibodies and have demonstrated selective phototoxicity towards HER-2 overexpressing breast cancer cells.[52] The same phthalocyanine gold nanoparticles have been functionalized with jacalin, a member of the lectin family, which enabled the T antigen glycan-specific recognition and targeted phototoxicity of HT-29 colon cancer cells.[53] Liposomes have been commonly used to encapsulate photosensitizers with varying degrees of hydrophilicity and hydrophobicity due to their multi-compartmental nature. Liposomes loaded with a sulfonated aluminum phthalocyanine have been surface-modified with a secondary antibody, which selectively targeted leukocytes pre-labeled with a primary antibody *in vivo*.[54] Similarly, a monoclonal antibody, T-43, was covalently bound to the surface of liposomes entrapping pyropheophorbide, which demonstrated selective phototoxicity of the target human bladder carcinoma cells.[55] Liposomes encapsulating a phthalocyanine photosensitizer were also modified with transferrin to target rat bladder carcinoma cells overexpressing the transferrin receptor.[56] More recently, conventional photoactive lipid-based vesicles have been developed and targeted with folate have been formulated to allow for targeting-triggered activation of liposomes formed of porphyrin-lipid conjugates.[57] Bio-functionalized nanoparticles entrapping photosensitizers have also been used for the targeted delivery of PDT agents. For example, polyacrylamide nanoparticles entrapping Photofrin have been surface modified with the tumor vasculature targeting F3 peptide.[58] Similarly, the tumor vasculature targeting NRP-1 peptide has been used to functionalize chlorin e6-loaded silica nanoparticles.[59] Silica nanoparticles surface-modified with mannose residues have been prepared with a conjugated porphyrin derivative for lectin receptor targeted PDT.[60]

Although actively targeted photosensitizers bear huge potential to enhance the selectivity of PDT treatment, a common barrier for these described technologies is the poor control of the conjugation sites of photosensitizers or photosensitizer-bearing constructs.

7.2.2 *Site-activated targeting*

In addition to the aberrant physiology of tumor vasculature and the overexpression of tumor-associated antigens, other biochemical and biophysical characteristics distinguish tumor tissue from healthy tissue. These characteristics can be exploited for the site-activated targeting of photosensitizing agents to enhance the selectivity of PDT. The extracellular pH levels within a tumor interstitium are significantly lower than those in healthy tissue, giving rise to pH gradients that can be exploited for tumor-specific drug activation.[61] pH responsive micelles encapsulating self-quenching local concentrations of protoporphyrin IX have been chemically designed to destabilize at the pH levels present in the tumor interstitium.[62] Once destabilized, the micelles release the payload allowing the photosensitizers to dequench and become activated for tumor diagnosis and PDT. This approach has been shown to be effective for the imaging and therapy of a subcutaneous SCC7 murine tumor model of squamous cell carcinoma. Intratumoral glutathione levels are also markedly higher than in healthy tissue and high glutathione levels correlate with a poor prognosis.[63] This high intratumoral potential for redox reactions can be exploited to activate photosensitizing agents. For example, redox-responsive nanoparticles conjugated to a statically-quenched photosensitizer, chlorin e6, have been reported.[63] In this construct, the disulfide cross-linked dextran nanoparticles are reduced in the presence of high intratumoral glutathione, and destabilize resulting in the release and de-quenching of the photosensitizer and an enhancement in the PDT targeting ability.[63] Tumors also exhibit higher levels of extracellular matrix metalloproteinases (MMPs), which assist in tumor invasion, angiogenesis and disease progression. MMPs can be targeted to mediate selective tumor activation of photosensitizers to aid in imaging and PDT treatment. A PDT-based molecular beacon has been designed to be activated by MMPs in tumors.[64] A photosensitizer, pyropheophorbide *a*, was covalently conjugated to a black hole quencher through a short peptide spacer that quenched the fluorescence and singlet oxygen production of pyropheophorbide *a* through FRET. The peptide sequence contained a substrate for MPPs, which following MPP cleavage, allowed the photosensitizer to separate from the quencher, hence allowing for tumor selective activation of the photosensitizer's fluorescence and PDT capacity *in vitro* and *in vivo*.[64]

7.2.3 *Prodrug approaches*

In addition to the nanocarriers and photosensitizer conjugate preparations, prodrug approaches offer an alternative strategy to improve selectivity, pharmacokinetic and photo-physicochemical properties of the therapeutic agents. Prodrugs (or photosensitizer precursors) are forerunners of a pharmacologically active agent, which have little or no activity until biochemically modified *in vivo* to generate the active form. An example of such a prodrug approach for PDT is aminolevulinic acid (ALA), which, when applied exogenously bypasses the negative feedback control of the heme biosynthesis pathway to produce the endogenous photosensitizing protoporphyrin IX (PpIX). Intracellular accumulation of PpIX can be influenced by the rates of ALA conversion and efflux of PpIX that is produced. The activity of enzymes along the heme biosynthesis pathway also impacts intracellular concentrations of the photosensitizer. For example there can be negative control by the activity of ferrochelatase, which mediates the insertion of iron into PpIX to produce the non-photosensitizing heme molecule, and an increase in PpIX generation by an up-regulation of coproporphyrinogen oxidase.[65,66] Along these lines, the selective accumulation of PpIX in cancer cells following administration of exogenous ALA is attributed to the characteristic low ferrochelatase activity levels in neoplastic tissues. Some ALA-PDT studies have also utilized ion chelators to deprive iron in cancer cells, thereby reducing heme formation and enhancing PpIX accumulation.[67,68]

ALA (Levulan®) is approved for the topical treatment of actinic keratosis (AK), however the hydrophilic nature of ALA limits its penetration through the skin. A number of hydrophobic ALA derivatives have been developed to improve ALA delivery, among which, Metvix® (methyl ester-ALA) and Hexvix® (hexyl ester-ALA) are approved for the treatment of AK, and imaging of bladder cancer, respectively. Packaging of ALA in nanoparticles has also been demonstrated to improve the delivery of ALA to tumors and to increase PpIX biosynthesis. Casas *et al.* showed that liposomal formulations of ALA and ALA-hexyl ester improved porphyrin synthesis in murine mammary adenocarcinoma organ cultures.[69] Both gold nanoparticles and chitosan nanoparticles have also been successfully employed to facilitate delivery of ALA to tumor cells.[70,71]

Other types of photosensitizer-based prodrugs have also synthesized. Lange *et al.* developed non-phototoxic polymeric photosensitizer prodrugs (PPP), which can be converted to the photoactive forms by urokinase-type plasminogen activator (uPA) — a protease that is overexpressed in prostate cancer cells.[72] Light irradiation of uPA-activated PPP resulted in PC-3M-luc-C6 tumor growth inhibition in mice, while light alone and PPP alone did not show tumoricidal activities.[72] Dolphin *et al.* demonstrated in a *proof-of principle* photosensitizer (porphyrin derivative)-drug complex system that, upon visible light activation, the singlet oxygen generated from the porphyrin derivative could oxidatively cleave the photosensitizer-drug linkage to release the drug.[73] This photodynamic prodrug system allows simultaneous site-specific PDT and the triggered release of the complementary drug, offering a unique opportunity to overcome multiple survival pathways in cancers.

7.3 Alternative light sources for photodynamic therapy and imaging

Delivery of light to the desired treatment site still represents one of the challenges associated with the broad applicability of PDT in cancer treatment. The optimal wavelengths of light for deep tissue penetration lie in the range of 700–1100 nm, where water and biological chromophores such as hemoglobin, fat, and melanin absorb the least. However, most currently available photosensitizers absorb in the visible spectrum below 700 nm. At these shorter wavelengths, penetration of light is only several millimeters below the surface. On the other hand, photosensitizers with longer excitation wavelengths, above 850 nm, yield an inefficient production of toxic species due to lack of sufficient energy.[3] Thus, there is an active effort to develop novel nanomaterials and light sources that allow for deeper penetration, while maintaining efficacy. Additionally, the development of advanced endoscopes and other clinically relevant light generating devices have greatly expanded the range of accessible sites for PDT.[74] This section will focus on the various applications of nanoscale and cellular engineering in providing alternative sources of light for PDT and imaging (Figure 7.3).

7.3.1 *Quantum dots*

Since they were first prepared and described in 1982,[75] quantum dots have been synthesized and investigated by numerous groups due to their

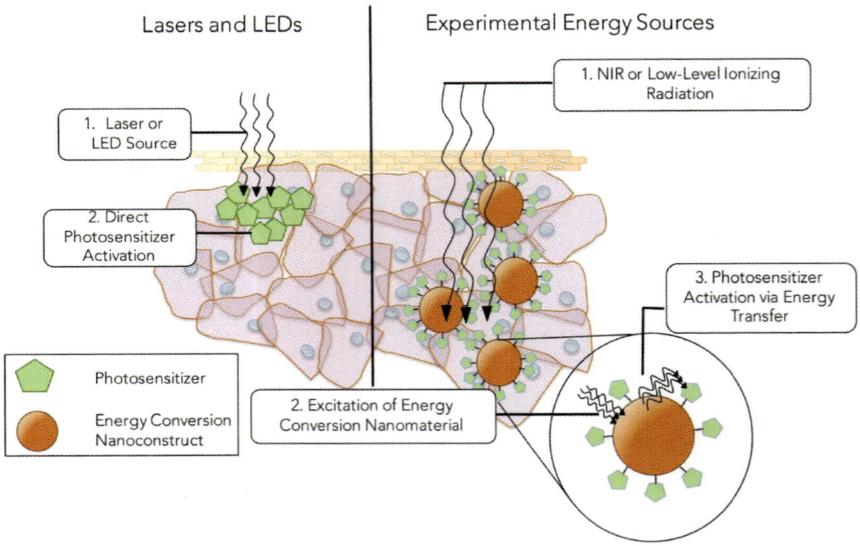

Figure 7.3 Types of light sources for PDT. While traditionally lasers and more recently LEDs have been used, for the future, other innovative approaches are being explored. These approaches aim to activate photosensitizers via energy conversion nanomaterials, such as quantum dots, upconversion nanoparticles, and nanoscintillators and may provide additional penetration depth for deeply seated tumors.

unique photophysical properties. Quantum dots are typically identified as inorganic semiconductor crystals that are smaller than the Bohr radius of their composition material, generally 1–6 nm. Restricting the size of the semiconductor crystal causes the quantum dot to adjust its energy states, a phenomenon known as the quantum confinement effect.[76] Thus, the optical properties of the QDs are highly tunable through size and composition adjustment, allowing for adjustments to the near infrared region and making them potentially attractive for PDT. While the unique optical properties of QDs have been fairly well applied in biological imaging, their use as potential therapeutic agents in PDT has only emerged more recently. These applications include serving as energy donors to traditional molecular photosensitizers or serving as photosensitizers themselves by generating singlet oxygen and other toxic species.

7.3.1.1 *Quantum dot-photosensitizer conjugates*

Several groups have reported the conjugation of conventional photosensitizers to a QD for indirect activation. Under this design, the quantum dot serves to facilitate the excitation of the attached photosensitizer through Forster Resonance Energy Transfer (FRET). In 2003, Samia *et al.* reported the first demonstration of photosensitizer activation through QD-based FRET in a CdSe QD conjugated to a phthalocyanine photosensitizer with a reported 77% FRET efficiency.[77] Another study demonstrated a thiol-capped CdTe QD conjugated to an aluminum tetrasulfophthalocyanine photosensitizer.[78] In both of these cases, the broad absorption spectrum of the QD allowed for flexibility in the excitation wavelength. However, these early and largely proof-of-concept studies suffer from potential QD toxicity and low singlet oxygen yield. New studies from these groups and others have built on these early studies by optimizing the ratio of photosensitizer per QD, as well as FRET pair linker length to improve the energy transfer efficiencies. Alternative materials have also been used to alleviate the toxicity of traditional Cd-based QDs.[79,80]

7.3.1.2 *Quantum dots as photosensitizers*

In the previously discussed study by Samia and colleagues demonstrating the FRET-based activation of CdSe-Pc4, the authors also discovered that the QD alone could transfer energy directly to molecular oxygen, without mediation through a photosensitizer.[77] This direct production of singlet oxygen, the authors believe, was possibly due to a triplet-excited state (T_1) of CdSe and the intercalation of molecular oxygen with the tri-n-octylphosphine oxide layer at the surface of the QD. However, the singlet oxygen quantum yield was low at 5%, as tested in toluene. Still, some suggest that the relative photostability of quantum dots may allow for repeated or prolonged exposure to light irradiation, and thus allow a steady-state level of singlet oxygen to be maintained.[81] A more recent study also reported a new class of QDs, termed graphene quantum dots that produce a singlet oxygen yield of over 100%, which the authors attribute to an additional mechanism that they term multistate sensitization.[82] In this study, the authors also demonstrate the potential of using their construct for *in vivo* imaging and tumor treatment efficacy in a mouse model of breast cancer.

7.3.2 *Nanoscintillators*

Scintillation nanoparticles, sometimes termed nanoscintillators, self-lighting, or self-illuminating nanoparticles, absorb ionizing radiation such as X-rays or gamma rays and subsequently emit visible light. The physical processes that mediate this phenomenon are complex, but include the generation of an electron-hole pair from a high-energy photon, followed by various forms of decay of the excited electron, secondary electron excitations, and ultimately electron-hole recombination.[83] This cascade leads to the emission of a characteristic spectrum of light that is lower in energy than the absorbed energy, often in the visible range. Thus, given the proper spectral overlap between the emission of a nanoscintillator and the absorbance of an attached photosensitizer, these constructs can act as mediators between the deeply penetrating, but highly damaging effects of ionizing radiation and the spatially confined and mechanistically-distinct cytotoxicity of PDT. The combination of radiotherapy and PDT may be advantageous because it not only seeks to alleviate the issue of poor visible light penetration in deeply situated tumors, but it also provides an additional level of treatment selectivity due to tumor-localized PDT toxicity, as radiation does not discriminate between healthy and tumor tissue. Therefore, coupled with PDT, lower clinical doses of ionizing radiation could potentially be required to achieve the same intratumoral efficacy, while damaging less of the surrounding healthy tissue.[84]

Scintillation conversion of ionizing radiation has been reported for several types of nanoparticles. An early study by Liu *et al.* reported the synthesis of lanthanum fluoride nanoparticles doped with terbium ions (LaF_3:Tb^{3+}), which were subsequently conjugated to the photosensitizer *meso*-tetra(4-carboxyphenyl) porphine (MTCP).[85] Following excitation *via* UV irradiation, the LaF_3:Tb^{3+} nanoscintillators generated several blue-green emission bands, where MTCP exhibits strong absorption. The authors further demonstrated that the conjugated nanoscintillator efficiently produces singlet oxygen in water during X-ray irradiation. Importantly, this singlet oxygen production occurred at relatively low-dose irradiation, an important detail since the ultimate clinical goal is safer treatments. Since their earliest report, Liu *et al.* have made several attempts to continue improving several properties of the construct, including luminescence efficiency, spectral overlap, and biocompatibility. A more recent nanoscintillator conjugate from this group, zinc oxide coupled to the photosensitizer *meso*-(o-amino-phenyl) porphine), exhibited 89% energy transfer after UV-excitation, which was used in place of short wavelength radiation for the feasibility of measuring energy transfer.[86] Others have also attempted to optimize the design of nanoscintillators for appropriate biocompatibility and biodistribution. Bulin *et al.* studied a terbium oxide nanoparticle coated in a polysiloxane layer ($Tb_2O_3@SiO_2$), conjugated to Photofrin.[87] With this construct, the authors demonstrate singlet oxygen production following X-ray excitation and further show that the generation of singlet oxygen is radiation dose-dependent.

Quantum dots have also been reported as good candidates for nanoscintillators, although their precise mechanisms of scintillation are still being investigated.[88] It is believed, however, that the electronic band gap change caused by decreasing the QD diameter leads to more electron-hole pairs, and thus more visible photons produced upon electron-hole pair recombination.[89] While quantum dot scintillators have been successfully synthesized to convert energy from gamma, alpha, X-ray, and short wavelength UV radiation to visible light,[89–91] there are still relatively few reports of their conjugation to photosensitizers and their efficacy in biological applications. However, one group reported quantum dots conjugated to Photofrin that, upon dose-dependent X-ray excitation, produced singlet oxygen in a radiation dose-dependent manner and induced apoptosis in H460 human lung cancer cells *in vitro*.[92]

7.3.3 *Upconversion nanomaterials*

In contrast to nanoscintillators, which convert high energy, ionizing radiation into lower energy, visible light, upconversion nanomaterials convert lower energy near-infrared light into visible light, a process known as Anti-Stokes Emission.[84] Nanomaterials that facilitate this process are luminescent and possess a triplet-excited state, also known as phosphors. In their nanoscale form, these particles are referred to as upconverting nanoparticles or upconverting phosphors and their components include a transition metal, lanthanide, or actinide ion doped into a solid-state host.[84]

Many combinations of core materials and dopants have been attempted. The first attempt to use upconversion as a mediator for PDT reported $NaYF_4$ nanoparticles, doped with both ytterbium and erbium ions, and coated in a silica shell containing the photosensitizer Merocyanine 540.[93] Near-infrared excitation of these nanocrystals, followed by emission in the green visible range, ultimately led to the generation of singlet oxygen. The authors further demonstrated the customizability of the nanoparticles by functionalizing their surfaces with a tumor-targeting anti-MUC1 antibody.

Further studies have highlighted the potential translational applications of upconversion nanoparticles for PDT by testing them *in vitro* and *in vivo*. In 2006, other examples of $NaYF_4$; Yb^{3+}, Er^{3+} nanoparticles were reported. These modified constructs, which posses a polyethyleneimine coating for solubilization in physiological buffers, had zinc phthalocyanine (ZnPC) photosensitizer molecules adsorbed to their surface with a high encapsulation efficiency.[94] ZnPC was chosen due to its overlapping 670 nm excitation peak with an emission peak of the upconversion particle, following excitation with a 980 nm near-infrared laser. The authors tested these photosensitizer-nanoparticle conjugates *in vitro* in HT29 human colonic carcinoma cells and demonstrated a significant phototoxic effect following irradiation at 980 nm for 30 minutes.

The applications of upconversion nanoconstructs as agents for PDT have also been tested *in vivo* as well. PEGylated $NaYF_4$; Yb^{3+}, Er^{3+} nanoparticles conjugated to chlorin e6 photosensitizer molecules were synthesized by Wang *et al.*[95] After *in vitro* cell viability assays in 4T1 mouse breast cancer showed significant and dose-dependent reductions in cell viability, the constructs were directly injected into 4T1 tumors in mice, followed by irradiation with a NIR 980 nm laser. The authors observed effective tumor reduction and a survival benefit in the mice that received the upconversion-mediated PDT. Further modifications to upconversion nanoconstructs have been reported, which include a mesoporous silica shell to allow for more stable encapsulation of photosensitizers and protection from degradation in harsh environments.[96] These authors also utilized the multi-wavelength distinct luminescence emission profile of $NaYF_4$; Yb^{3+}, Er^{3+} nanoparticles to their advantage, encapsulating both merocyanine 540 and zinc phthalocyanine as wavelength-specific photosensitizers. While *in vivo* results from this study showed tumor growth reduction after intravenous injection and 980 nm irradiation for 2 hours, the authors point out that treatment with laser alone also resulted in tumor destruction, likely due to a local thermal effect.

7.3.4 *Emerging alternatives*

7.3.4.1 *Cherenkov radiation*

Charged particles, such as beta particles, are emitted during the decay of radioactive isotopes that are currently used *in vivo* for diagnostic and therapeutic purposes, such as positron emission tomography (PET). When these charged particles travel faster than the speed of light in a given medium, the molecules of the surrounding medium are polarized into excited states and Cherenkov radiation spanning the entire UV-visible spectrum is emitted upon relaxation.[97] Because the intensity of the emitted Cherenkov radiation is proportional to the frequency of light, maximal intensity occurs in the UV region.[98]

Recently, researchers have explored the idea of harnessing this luminescent phenomenon for PDT, essentially allowing for the placement of a light source within the tissue. Importantly, the luminescence of Cherenkov radiation is of low intensity. Thus, simulations of ultra low (20 $\mu W/cm^2$) irradiances over an extended time period that showed effective phototoxicity in ALA-PDT have been important proof-of-concept studies in applying Cherenkov radiation for PDT.[99] To maintain long exposure durations, radionuclides with long half-lives are ideal, such as Yttrium-90 (90Y), with a half-life of 64 hours.[99] Although Cherenkov radiation has not been directly used to activate molecular photosensitizers, initial proof-of-concept studies have shown promise. One study reports the use of luminescence from PET-imaging agent, FDG, to photocleave a covalent ester bond.[97]

7.3.4.2 *Biological sources of light*

As another alternative, light emitted from biological sources are being explored for photosensitizer activation. Although proteins capable of generating bioluminescence, such as luciferases, have been used widely in biological research as reporters, they are only recently being studied in the context of PDT. Like FRET, bioluminescence resonance energy transfer (BRET) can harness the emission of bioluminescent molecules to activate photosensitizers with appropriate spectral overlap. Yuan *et al.* first reported a BRET-based PDT system in which the molecule luminol is catalyzed to its bioluminescent form in the presence of an oxidizing agent and the enzyme horseradish peroxidase.[100] Emission of luminol at 425 nm overlaps with the broad absorption of photosensitizer oligo (*p*-phenylene vinylene), leading to the production of toxic species. In their study, the authors show that this system leads to both increased *in vitro* cytotoxicity in HeLa cells and tumor growth inhibition following intratumoral injection. However, they admit that their largely proof-of-concept system still lacks tumor cell specificity and possesses inherent toxicity, therefore requiring further engineering and appropriate drug delivery. Hsu *et al.* have also reported a BRET-based PDT system.[101] Their system consisted of quantum dots conjugated to the enzyme Renilla luciferase. Upon exposure of the constructs to coelenterazine, the luciferase substrate, energy is transferred *via* BRET to the QDs, which subsequently emit at 655 nm and are able to activate the photosensitizer *meta*-tetra-hydroxyphenyl-chlorin (*m*-THPC). *In vitro* testing of this BRET-mediated QD system showed coelenterazine dose-dependent cell killing and ROS production. The authors further tested the constructs in a subcutaneous tumor mouse model and observed tumor growth inhibition following intratumoral injection of the QD constructs and intravenous injection of coelenterazine. They attribute the effect to decreased cell proliferation, increased cell apoptosis, and decreased microvessel density.

Unlike the other sources of light for PDT, cell lasers attempt to make cells a living light source, as opposed to exogenous nanomaterials or luminescent molecules. In the first report of a biological cell laser, Gather and Yun transfected mammalian cells with a plasmid encoding for recombinant GFP to test its applicability as a gain medium.[102] After single GFP-transfected cells were placed in a microcavity resonator and pumped with pulses of nanojoule scale light, they produced a bright, directional coherent green beam, demonstrating that GFP can be used as a viable gain medium for optical amplification.

7.3.4.3 *Organic light emitting diodes*

It should be noted that while light penetration to deeply seated tumors represents one of the major hurdles to the applicability of PDT, other challenges, such as the portability of treatment devices and pain from laser irradiation, also are being addressed for treatments in which accessibility is not the main problem. While efforts have been made to create new portable laser devices,[103] organic light emitting diodes (OLEDs) have also emerged as potentially attractive alternatives to these challenges. OLEDs generally consist of a thinly deposited organic material between two electrodes.[104] When an electrical current is applied, light is emitted from the organic layer. Because of their small size and low irradiances, OLEDs have been implicated as good candidates for portable PDT and decreasing PDT-associated pain. A clinical pilot study demonstrated in 2008 that ALA-PDT for Bowen's disease and superficial basal cell carcinoma could be performed with a 2 cm diameter OLED.[105] They reported that seven of the twelve patients remained clear of cancer after 12 months and four patients exhibited peripheral margin failure, perhaps due to improper treatment area. While further work is required, this initial study suggests that the OLEDs may hold promise in decreasing treatment discomfort and increasing the compactness of light-generating devices for PDT.

7.4 Engineering three-dimensional culture models for light-based therapy

The following sections highlight some of the advances in engineering three-dimensional (3D) tumor models and in developing quantitative analysis routines for the evaluation of photosensitizer penetration, light dosimetry, PDT treatment response and the efficacy of PDT-based combinations.

7.4.1 *3D models for investigating photosensitizer penetration and corresponding photoactivity*

In both clinical and preclinical studies investigating PDT treatment response, it has been shown that there are significant differences in the rates and mechanisms of photosensitizer uptake.[106–108] Tumors will exhibit varying patterns of photosensitizer uptake and distribution when treated with a given photosensitizer dose, a fixed sensitizer-light interval, and a predetermined light dose.[30,106] This heterogeneity is an important contributing factor to the variation seen in treatment response, and the subsequent inconsistencies observed in many promising clinical and pre-clinical studies. Understanding the heterogeneity of photosensitizer uptake in tumor models *in vivo* is cumbersome and may not be practical, particularly in the early stages of therapeutic development. Thus, bioengineered 3D tumor models may serve as complementary research tools to characterize and address a number of challenges associated with the variability in PDT response, such as uneven intratumoral sensitizer distribution.

In clinical manifestations of deeply residing tumors or lesions that exceed a certain size threshold, PDT faces new challenges including poor photosensitizer penetration and low intratumoral oxygen saturation levels, both of which lead to multiple physical and biochemical complications. A number of solutions have been proposed to address these limitations that have been described in part in this chapter. These include nanoparticle formulations to enhance photosensitizer delivery throughout the tumor, enhancing tumor selectivity through targeting, and the utilization of mechanistically distinct photosensitizers that are independent of molecular oxygen. Typically, the *in vitro* screening and assessment of the efficiency of PDT agents by measuring photosensitizer fluorescence and photobleaching are performed in *in vitro* 2D monolayer cell cultures.[109,110] However, much of the time, the simplicity of the monolayer studies renders the findings almost irrelevant compared to a more complex three-dimensional tumor model. *In vitro* 3D models provide a unique platform for testing multiple parameters of the treatment efficacy of photosensitizers, formulations and conjugates in a physiologically relevant system that recapitulates some of the architecture and biochemical complexities that lead to the clinical challenges.

In an early example, Georgakoudi *et al.* used *in vitro* 3D EMT6 murine mammary carcinoma spheroids, 500 μm in diameter, to monitor the oxygen dependence of photosensitizer photobleaching during PDT. The spheroids were treated with Photofrin®-PDT, and oxygen saturation was monitored in real time using microelectrode measurements. It was found that photobleaching is oxygen dependent, indicating a predominance of type II reactions, and this dependence varied among other photosensitizers investigated.[111] A more recent study directly investigated the therapeutic potential of micellar formulations of the photosensitizer pheophorbide on HCT-116 human colorectal adenocarcinoma cells grown in *in vitro* 2D or 3D cultures. Differential uptake within the tumor nodule was also investigated, allowing for the identification of one photosensitizer formulation that resulted in 96% of cells to uptake the photosensitizer. However, phototoxicity reached a maximum of 55% in 3D, unlike in 2D where more than 95% phototoxicity was achieved.[112] This study exemplifies that 3D models are powerful tools in the selection of optimal photosensitizer preparations. However, it also demonstrates that sensitizer penetration is not sufficient to guarantee a complete PDT response. Additionally, oxygen consumption by type II photochemical reactions can prove to be self-limiting as a result of generating hypoxia during irradiation.

Porphyrin-based photosensitizers such as Photofrin® and benzoporphyrin derivative (BPD) are examples of the more common type II photosensitizers. Like most therapeutic agents their efficiency can be limited by their penetration depth; however, their dependence on oxygen to mediate cytotoxicity is an additional limitation in completely eradicating tumors that may contain hypoxic regions. EtNBS (5-ethylamino-9-diethyl-aminobenzo[a]phenothiazinium, a small cationic non-porphyrin sensitizer) is a type I photosensitizer that has been commonly used to mediate cytotoxicity through both oxygen-dependent and oxygen-independent pathways, thereby inducing phototoxicity in hypoxic tumor regions that would usually not respond to type II photosensitizer-mediated PDT.[113,114] In early work, Georgakoudi and Foster used *in vitro* 2D and 3D cultures of EMT6 murine mammary cells to study the oxygen consumption and sensitizer redistribution during irradiation of EtNBS.[115] In more recent work, Evans *et al.* utilized an *in vitro* 3D model of micro-metastatic ovarian cancer to investigate the efficacy of EtNBS in eradicating nodule regions with suboptimal oxygen saturation. Using a physiologically tolerated dose of the photosensitizer, it was found that EtNBS rapidly diffuses throughout the entire nodule, and more importantly, can impart cytotoxicity even at the hypoxic core of the nodules.[116,117]

These findings highlight the utility of *in vitro* 3D models for appropriately assessing photobleaching and photosensitizer localization, as well as the utility of these metrics for dose determinants.

7.4.2 *3D models in light dosimetry*

Globally, PDT is in clinical trial phases or clinically approved for the treatment of numerous solid tumors, including breast, lung, ovarian, skin, and the gastrointestinal tract.[14] As with more traditional therapies, such as chemotherapy or radiation therapy, PDT treatment response is patient-specific, and the clinical response does not always correlate with the response observed in pre-clinical studies.[30] One explanation for this phenomenon is the complexity of patient-specific factors and dosimetry, as well as the actual light delivery that compose the total dose. Here, the clinical evidence that has motivated the investigation of light dosimetry in *in vitro* 3D systems is described.

7.4.2.1 *Understanding in vivo PDT dose parameters in 3D models*

In the clinic, dose parameters for PDT treatments are usually pre-determined dosages of administered photosensitizers (per unity body weight), the incident light fluence (J/cm^2) or total energy (J), and the time duration of the photosensitizer-light interval. The sensitizer-light interval is typically selected based on the time of maximal sensitizer uptake in the target tissue, or on maximal target tissue selectivity. However, the rigidity in this dosing approach overlooks the nuances of the body that are known to affect the photobiological response, including the optical properties of the tissue.[118,119] More specifically, the effects of these individual dose parameters on the overall PDT response are hugely variable.

The correlation between light dose and surviving fractions has been reported *in vivo* by Gibson *et al.*, who showed an increase in tumor destruction with lower irradiances in an *in vivo* mouse model of mammary tumors.[120] Foster *et al.* have also shown the same relationship between PDT response and the incident fluence rate in *in vivo* mouse models of breast cancer and mesothelioma.[31,121] Foster *et al.* utilized an *in vitro* 3D system to determine the cellular mechanisms that drive this phenomenon.[31] To do this, EMT6/Ro multi-cell tumor spheroids were treated with the same dose of Photofrin-PDT (10 μg/mL; 60 J/cm^2), but irradiated at varying fluences from 25 mW/cm^2 to 200 mW/cm^2. As seen *in vivo*, as the fluence increased, the amount of surviving fractions also increased. Using a mathematical model to calculate the oxygen concentration in the spheroids during PDT, it was shown that at lower fluences, spheroids had better overall access to oxygen concentrations that were sufficient for singlet oxygen generation, and thus PDT-mediated cytotoxicity. Here, by using both *in vivo* and 3D *in vitro* models as complementary tools, these studies identified effective PDT dose parameters. This may serve as a useful model for similar investigations and for establishing appropriate treatment parameters for maximal clinical effectiveness.

A second example is the reported reciprocity between photosensitizer and light dose reported in both *in vitro* 3D models, and *in vivo* systems. An early *in vivo* study done by Bown *et al.* compared multiple photosensitizers in an *in vivo* rat model of normal livers, to determine which PDT dose parameters control PDT-mediated cell death. In addition to gathering pharmacokinetic data of different photosensitizers in the liver, Bown *et al.* demonstrated that there was a threshold of minimum photosensitizer concentration in which necrosis was achieved, and that maximum necrosis was achieved with a lower dose of photosensitizer and a higher dose of light.[122] A later study done by Patterson *et al.* built upon the work done by Bown, demonstrated in an *in vivo* model of normal rat liver that PDT-mediated necrosis will occur if the amount of photons absorbed by photosensitizer per unit volume of tissue is within a certain range of concentrations.[123] Further *in vivo* studies have reported these phenomena in animal models of brain, and pancreatic cancers as well.[124,125] This early work *in vivo* found that reciprocity between photosensitizer and light dose held over different ranges of doses within certain thresholds, and much work has been dedicated to studying the relationship between drug (PS) and light dose in 2D *in vitro* models, as well as *in vivo* animal models.[126] More recent work reported by Rizvi *et al.* utilized an *in vitro* 3D model of ovarian cancer to investigate this reciprocity using a matrix of PDT dose parameters.[127] Here, using a 3D model provides flexibility in terms of testing numerous combinations of PDT dose parameters in one experimental model. Testing both low and high doses of the photosensitizer BPD and with varying fluences, it was found that lower doses of BPD and a high fluence yields the greatest tumor destruction of 3D ovarian cancer nodules.

7.4.3 *Assessing treatment response in 3D models*

7.4.3.1 *Evaluating the efficacy of PDT and PDT combinations*

There is increased evidence that using combination therapies, which utilize distinct and non-overlapping mechanisms of actions, is advantageous over monotherapies.[128,129] Although it is most relevant to test combination therapies in animal tumor models or on human subjects, these can be cumbersome, expensive and inappropriate when trying to assess treatment response over a variety of therapeutic regimens and doses. There is an emerging body of work using *in vitro* 3D models to evaluate chemotherapy treatment response in various disease models, including breast, colon and prostate cancer.[130,131] As one example, Gurski *et al.* utilize a 3D hydrogel overlaid with poorly adherent prostate cancer cells to model the prostate tumor microenvironment, and test different chemotherapy combinations to determine the specificity, dose and time responses of each drug.[132] However, it is evident that chemotherapy alone is rarely successful in effectively treating the disease. As a result, some groups are looking to combine PDT and chemotherapy, and are evaluating the efficacy of these combinations in *in vitro* 3D models.

PDT efficacy is dependent on the architectural, photophysical, and biological characteristics of the tumor. It has been demonstrated *in vivo* that PDT is a potential adjuvant therapy to other traditional therapeutic modalities.[133] In an early study, Nahabedian *et al.* combined PDT with either doxorubicin or cisplatin in *in vivo* mouse models of RIF-1 and EMT-6 tumors. The tumors were insensitive or mildly sensitive to each therapy on its own, but the combination of PDT and doxorubicin saw an enhanced tumoricidal effect.[134] Although these early studies were crucial in becoming aware of the potential PDT-chemotherapy combinations, these *in vivo* studies are limited in their ability to measure the treatment efficacy of these combinatorial therapies in a high-throughput manner. Conversely, high-throughput *in vitro* 2D models are routinely utilized to assess PDT and chemotherapy combination regimens in a variety of disease models.[135–137] These 2D models, although high-throughput, give limited insight into evaluating treatment response of PDT and PDT combination therapies.

As described in a later section, an added advantage of *in vitro* 3D models is the facilitated quantification of treatment response on an individual nodule basis, in a fashion analogous to high-throughput 2D cell assays. Rizvi *et al.* have developed an *in vitro* 3D model of micrometastatic ovarian cancer to screen PDT and chemotherapy combination regimens.[138] This model provides an efficient high-throughput platform to simultaneously evaluate multiple treatment regimens. The 3D model and image analysis routines demonstrated that low-dose PDT (1.25 μmol/L J/cm^2 BPD-PDT) followed by low-dose carboplatin (400 mg/m^2) has a synergistic effect on reducing the viable fraction. The model also provided insight into the regional differences of a nodule's response to PDT and chemotherapy, which imparted distinctive patterns of cytotoxicity to the 3D nodules. Carboplatin was consistently toxic to cells at the nodule periphery, whereas PDT disrupted the nodule structure. This information uniquely obtained by the 3D platform enabled the rational selection of combination sequences to synergistically reduce ovarian cancer nodule volume. More recently, Yang *et al.* have developed an *in vitro* heterocellular 3D microfluidic model of breast cancer for evaluating PDT.[139] This model involved culturing MCF-7 human breast cancer cells with ASC adipose-derived stromal cells in a microfluidic chip, which formed 3D nodules that display morphologies similar to those observed *in vivo*. The 3D cultures were treated with ALA-PDT assisted by gold nanoparticle delivery vectors, and it was found that a decrease in irradiation time resulted in greater nodule cell viability. Furthermore, it was established that the nanoparticle assisted ALA was markedly more effective in killing the nodule cell populations, as compared to ALA-PDT alone, confirming that the gold nanoparticles enhanced the intratumoral delivery of the sensitizer precursor ALA. Correlating studies performed by Yang in an *in vitro* 2D system confirmed previously reported findings, that 3D nodules are significantly less responsive to treatment than *in vitro* 2D cell monolayers. With the advancement of nano-therapeutics, bio-conjugates and next generation therapies, it is envisaged that the use of 3D models will grow as they become more sophisticated and capture more physiological attributes of cancers *in vitro*.

3D models composed of A549 human lung adenocarcinoma 3D spheroids have also been used to assess the treatment efficacy of photoactive, strained ruthenium (II) complexes. Upon irradiation at 450 nm, intracellular ruthenium (II) complexes cross-linked DNA and were significantly more potent at reducing spheroid cell viability than traditional cisplatin-mediated DNA crosslinking.[140]

7.4.3.2 *Analysis framework for evaluating treatment response*

In order to maximize the potential of 3D tumor models, and to obtain in-depth insight into PDT-based therapeutics, sophisticated imaging-based methods are being developed to quantify multiple parameters that contribute to overall treatment response (Figure 7.4).[141]

Conventional, routine analysis methods for quantifying treatment efficacy in 2D *in vitro* and *in vivo* studies exist, and obtaining metrics for

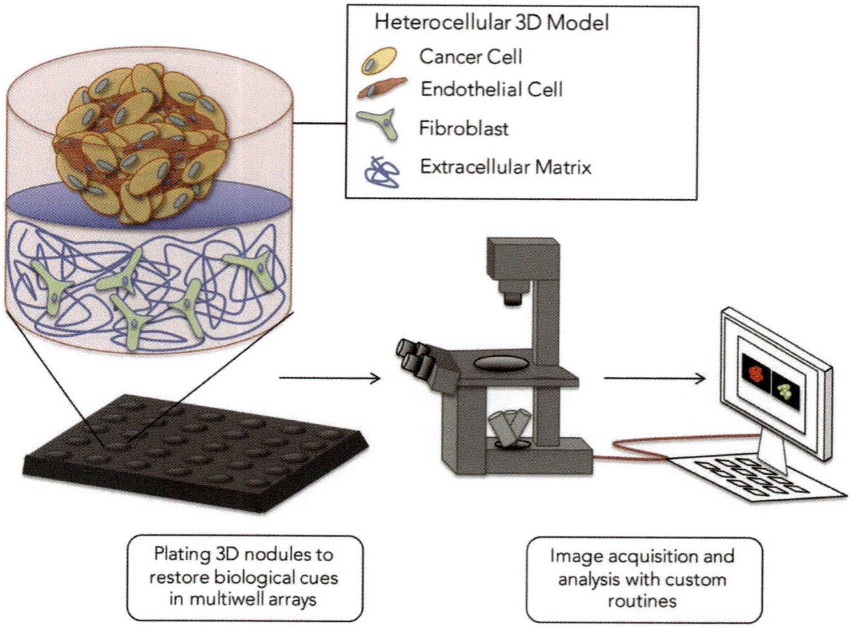

Figure 7.4 Utilizing 3D models to evaluate treatment parameters for PDT. Cancer cells can be plated in a heterocellular 3D model to recapitulate the tumor microenvironment and restore biological cues that may be found *in vivo*. Following plating and treatment of heterocellular 3D models in multiwell plates, the nodules can be evaluated through multi-channel image acquisition and analyzed with custom routines for quantitative readouts.

these are well-established. Assessing treatment response in a 3D *in vitro* system is non-trivial and for the most part, unconventional. Therefore, it is critical to develop appropriate and specialized tools and statistical methods to evaluate characteristics of 3D nodules *in vitro* in order to obtain a relevant insight into the mechanisms occurring *in vivo*.[142] Eliceiri *et al.* provided a comprehensive review of the imaging software tools available for a variety of biological analyses.[143] Methods for image acquisition are presented, as well as the availability of numerous open-source image-analysis algorithms. These algorithms cover a broad spectrum of biological challenges, but are limited by being tailored specifically for the individual studies performed by the developers. Common frameworks such as ImageJ or Fiji, can be utilized by researchers who require basic quantification from images of *in vitro* 3D models. One example of an unconventional and specialized custom analysis framework is that developed by Celli *et al.* This unique workflow allows for the quantitative imaging and subsequent batch-processing of 3D pancreatic and ovarian spheroids in models of the respective micro-metastatic disease.[142,144] This methodology allows an accurate analysis of 3D nodule viability, in addition to both global and individual nodule based size and structural information. More specifically, this framework enabled the quantification of the shifts in frequencies of nodule viabilities and volumes in response to PDT or chemotherapy. This analysis provided information into the heterogeneity of treatment response within an individual disease models in a *in vitro* 3D system. Looking ahead, the ability to quantify this body of information is extremely valuable as it allows insight on how to intelligently combine different therapies, taking advantage of their distinct effects on nodule disruption.

7.5 Perspective and future directions

As nanoscale delivery systems become increasingly sophisticated they may be critical to overcoming the major challenges currently associated with treating advanced stage cancers including poor selectivity and heterogeneity in the biology and therapeutic response of metastatic disease. Within the context of a broader strategy in oncology that is increasingly focused on identifying and targeting patient-specific molecular signatures, the customizable, multi-compartment nanocarriers reviewed here may be essential to delivering tailored therapies designed to interrupt the multiple evasion pathways employed by tumors. To contend with the complexities of not only selecting the appropriate payload, but also establishing the optimal spatial and temporal delivery of the candidate agents, bioengineered tumor models could play an increasingly important role in the high-throughput evaluation of nanoscale carriers. Cancer models that account for the critical biological cues including crosstalk with heterocellular signaling partners and physical influences in the tumor microenvironment may be necessary to restore the architectural barriers and evasion pathways that multi-agent nanocarriers are designed to overcome. Simultaneous efforts to develop quantitative multiplexed imaging systems will be essential to characterizing the genetic, molecular and architectural factors that contribute to treatment failure. These advancements on multiple fronts including targeted multi-compartment delivery systems, bioengineered tumor models and quantitative imaging will be essential to improving outcomes from the complex set of disease that fall under the umbrella of translational oncology.

Acknowledgements

This work was supported by NIH grants P01CA084203, R01CA156177, R01CA160998, S10OD01232601 (T.H.), K99CA194269 (H-C.H.), and K99CA175292 (I.R.), and the MGH-Tosteson-FMD Fellowship # 224889 (H-C.H.), and the Bullock-Wellman Postdoctoral Fellowship (G.O.). The authors thank Dr. Anne-Laure Bulin for helpful discussions.

References

1. Dolmans DEJGJ, Fukumura D, Jain RK, Photodynamic therapy for cancer. *Nat Rev Cancer*, 2003; 3(5): 380–387.
2. Celli JP, Spring BQ, Rizvi I, Evans CL, Samkoe KS, Verma S, *et al.*, Imaging and photodynamic therapy: mechanisms, monitoring, and optimization. *Chem Rev*, 2010; 110(5): 2795–2838.

3. Plaetzer K, Krammer B, Berlanda J, Berr F, Kiesslich T, Photophysics and photochemistry of photodynamic therapy: fundamental aspects. *Laser Med Sci*, 2009; 24(2): 259–268.
4. Hasan T, Ortel B, Solban N, Pogue B, Photodynamic therapy of cancer. In: Kufe DW, Bast RC Jr, Hait WN, Hong WK, Pollock RE, Weichselbaum RR, Holland JF, and Frei E III, editors. *Cancer Medicine*, 7th ed. Hamilton, Ontario: B.C. Decker, Inc., 2006, pp. 537–548.
5. Raab O, Über die Wirkung fluoreszierender Stoffe auf Infusion. *Z Biol*, 1900; 39: 524–546.
6. Meyer-Betz F, Untersuchungen uber die biologische photodynamische Wirkung des Hematoporphyrins und anderer Derivative des Blut und Galenafarbstoffs. *Dtsch Arch Klin*, 1913; 112: 476–503.
7. Von Tappeiner H, Jesionek A, Therapeutische versuche mit fluoreszierenden stoffen. *Muench Med Wochenschr*, 1903; 47: 2042–2044.
8. Rassmussan-Taxdal DS, Ward GE, Figge FH, Fluorescence of human lymphatic and cancer tissues following high doses of intravenous hematoporphyrin. *Cancer*, 1955; 8: 78–81.
9. Policard A, Etudes sur les aspects offerts par des tumeurs experimentales examines a la lumiere de Wood. *CR Soc Biol*, 1924; 91: 1423–1424.
10. Lipson RL, Baldes EJ, Olsen AM, The use of an derivative of hematoporphyrin in tumor detection. *J Natl Cancer Inst*, 1961; 26: 1–11.
11. Schwartz SK, Abolon K, Vermund H, Some relationships of porphyrins, X-rays and tumors. *Univ Minn Med Bull*, 1955; 27: 7–8.
12. Dougherty TJ, Gomer CJ, Henderson BW, Jori G, Kessel D, Korbelik M, *et al.*, Photodynamic therapy. *J Natl Cancer Inst*, 1998; 90(12): 889–905.
13. Dolmans DE, Fukumura D, Jain RK, Photodynamic therapy for cancer. *Nat Rev Cancer* 2003; 3(5): 380–387.
14. Agostinis P, Berg K, Cengel KA, Foster TH, Girotti AW, Gollnick SO, *et al.*, Photodynamic therapy of cancer: an update. CA *Cancer J Clin*, 2011; 61(4): 250–281.
15. Roozeboom MH, Arits AH, Nelemans PJ, Kelleners-Smeets NW, Overall treatment success after treatment of primary superficial basal cell carcinoma: a systematic review and meta-analysis of randomized and nonrandomized trials. *Br J Derma*, 2012; 167(4): 733–756.

16. Allison RR, Downie GH, Cuenca R, Hu XH, Childs CJ, Sibata CH, Photosensitizers in clinical PDT. *Photodiag Photodyn Ther*, 2004; 1(1): 27–42.
17. Bressler NM, Bressler SB, Photodynamic therapy with verteporfin (Visudyne): impact on ophthalmology and visual sciences. *Invest Ophthalmol Visual Sci*, 2000; 41(3): 624–628.
18. Huggett MT, Jermyn M, Gillams A, Illing R, Mosse S, Novelli M, *et al.*, Phase I/II study of verteporfin photodynamic therapy in locally advanced pancreatic cancer. *Br J Cancer*, 2014; 110(7): 1698–1704.
19. Oleinick NL, Evans HH, The photobiology of photodynamic therapy: cellular targets and mechanisms. *Radiat Res*, 1998; 150(5 Suppl): S146–S156.
20. Kessel D, Castelli M, Evidence that bcl-2 is the target of three photosensitizers that induce a rapid apoptotic response. *Photochem Photobiol*, 2001; 74(2): 318–322.
21. Granville DJ, Carthy CM, Jiang H, Shore GC, McManus BM, Hunt DW, Rapid cytochrome c release, activation of caspases 3, 6, 7 and 8 followed by Bap31 cleavage in HeLa cells treated with photodynamic therapy. *FEBS Lett*, 1998; 437(1–2): 5–10.
22. Duska LR, Hamblin MR, Miller JL, Hasan T, Combination photoimmunotherapy and cisplatin: effects on human ovarian cancer *ex vivo*. *J Natl Cancer Inst*, 1999; 91(18): 1557–1563.
23. Bhuvaneswari R, Yuen GY, Chee SK, Olivo M, Hypericin-mediated photodynamic therapy in combination with Avastin (bevacizumab) improves tumor response by downregulating angiogenic proteins. *Photochem Photobiol Sciences*, 2007; 6(12): 1275–1283.
24. Pazos M, Nader HB, Effect of photodynamic therapy on the extracellular matrix and associated components. *Brazilian J Med Biol Res* = Revista brasileira de pesquisas medicas e biologicas 2007; 40(8): 1025–1035.
25. Chen B, Pogue BW, Luna JM, Hardman RL, Hoopes PJ, Hasan T, Tumor vascular permeabilization by vascular-targeting photosensitization: effects, mechanism, and therapeutic implications. *Clin Cancer Res*, 2006; 12(3 Pt 1): 917–923.
26. Chauhan VP, Stylianopoulos T, Boucher Y, Jain RK, Delivery of molecular and nanoscale medicine to tumors: transport barriers and strategies. *An Rev Chem Biomol Engin*, 2011; 2: 281–298.

27. Yoon I, Li JZ, Shim YK, Advance in photosensitizers and light delivery for photodynamic therapy. *Clin Endosc*, 2013; 46(1): 7–23.
28. Hanahan D, Weinberg RA, Hallmarks of cancer: the next generation. *Cell*, 2011; 144(5): 646–674.
29. Spring BQ, Abu-Yousif AO, Palanisami A, Rizvi I, Zheng X, Mai Z, *et al.*, Selective treatment and monitoring of disseminated cancer micrometastases in vivo using dual-function, activatable immunoconjugates. *Proc Natl Acad Sci*, 2014; 111(10): E933–E942.
30. Hopper C, Photodynamic therapy: a clinical reality in the treatment of cancer. *Lancet Oncol*, 2000; 1: 212–219.
31. Foster TH, Hartley DF, Nichols MG, Hilf R, Fluence rate effects in photodynamic therapy of multicell tumor spheroids. *Cancer Res*, 1993; 53(6): 1249–1254.
32. Jarvi MT, Patterson MS, Wilson BC, Insights into photodynamic therapy dosimetry: simultaneous singlet oxygen luminescence and photosensitizer photobleaching measurements. *Biophy J*, 2012; 102(3): 661–671.
33. Abbott A, Cell culture: biology's new dimension. *Nature*, 2003; 424(6951): 870–872.
34. Sutherland RM, McCredie JA, Inch WR, Growth of multicell spheroids in tissue culture as a model of nodular carcinomas. *J Natl Cancer Inst*, 1971; 46(1): 113–120.
35. Iyer AK, Khaled G, Fang J, Maeda H, Exploiting the enhanced permeability and retention effect for tumor targeting. *Drug Discov Today*, 2006; 11(17–18): 812–818.
36. Huang H-C, Hasan T, The "nano" world in photodynamic therapy. *Austin J Nanomed Nanotechnol*, 2014; 2(3).
37. Chatterjee DK, Fong LS, Zhang Y, Nanoparticles in photodynamic therapy: An emerging paradigm. *Adv Drug Delivery Rev*, 2008; 60(15): 1627–1637.
38. Derycke AS, de Witte PA, Liposomes for photodynamic therapy. *Adv Drug Deliv Rev*, 2004; 56(1): 17–30.
39. Wolinsky JB, Grinstaff MW, Therapeutic and diagnostic applications of dendrimers for cancer treatment. *Adv Drug Deliv Rev*, 2008; 60(9): 1037–1055.
40. Mew D, Wat CK, Towers GH, Levy JG, Photoimmunotherapy: treatment of animal tumors with tumor-specific monoclonal antibody-hematoporphyrin conjugates. *J Immunol*, 1983; 130(3): 1473–1477.

41. Oseroff AR, Ohuoha D, Hasan T, Bommer JC, Yarmush ML, Antibody-targeted photolysis: selective photodestruction of human T-cell leukemia cells using monoclonal antibody-chlorin e6 conjugates. *Proc Natl Acad Sci USA*, 1986; 83(22): 8744–8748.
42. Goff BA, Hermanto U, Rumbaugh J, Blake J, Bamberg M, Hasan T, Photoimmunotherapy and biodistribution with an OC125-chlorin immunoconjugate in an in vivo murine ovarian cancer model. *Br J Cancer*, 1994; 70(3): 474–480.
43. Molpus KL, Hamblin MR, Rizvi I, Hasan T, Intraperitoneal photoimmunotherapy of ovarian carcinoma xenografts in nude mice using charged photoimmunoconjugates. *Gynecol Oncol*, 2000; 76(3): 397–404.
44. Soukos NS, Hamblin MR, Keel S, Fabian RL, Deutsch TF, Hasan T, Epidermal growth factor receptor-targeted immunophotodiagnosis and photoimmunotherapy of oral precancer *in vivo*. *Cancer Res,* 2001; 61(11): 4490–4496.
45. van Dongen GA, Visser GW, Vrouenraets MB, Photosensitizer-antibody conjugates for detection and therapy of cancer. *Adv Drug Deliv Rev*, 2004; 56(1): 31–52.
46. Savellano MD, Pogue BW, Hoopes PJ, Vitetta ES, Paulsen KD, Multiepitope HER2 targeting enhances photoimmunotherapy of HER2-overexpressing cancer cells with pyropheophorbide-a immunoconjugates. *Cancer Res,* 2005; 65(14): 6371–6379.
47. Mitsunaga M, Ogawa M, Kosaka N, Rosenblum LT, Choyke PL, Kobayashi H, Cancer cell-selective in vivo near infrared photoimmunotherapy targeting specific membrane molecules. *Nat Med*, 2011; 17(12): 1685–1691.
48. Rizvi I, Dinh TA, Yu W, Chang Y, Sherwood ME, Hasan T, Photoimmunotherapy and irradiance modulation reduce chemotherapy cycles and toxicity in a murine model for ovarian carcinomatosis: perspective and results. *Israel J Chem*, 2012; 52(8–9): 776–787.
49. Spring BQ, Abu-Yousif AO, Palanisami A, Rizvi I, Zheng X, Mai Z, *et al.*, Selective treatment and monitoring of disseminated cancer micrometastases in vivo using dual-function, activatable immunoconjugates. *Proc Natl Acad Sci USA*, 2014; 111(10): E933–E942.
50. Schneider R, Schmitt F, Frochot C, Fort Y, Lourette N, Guillemin F, *et al.*, Design, synthesis, and biological evaluation of folic acid targeted tetraphenylporphyrin as novel photosensitizers for selective photodynamic therapy. *Bioorg Med Chem*, 2005; 13(8): 2799–2808.

51. Hamblin MR, Newman EL, Photosensitizer targeting in photodynamic therapy. I. Conjugates of haematoporphyrin with albumin and transferrin. *J Photochem Photobiol B,* 1994; 26(1): 45–56.
52. Stuchinskaya T, Moreno M, Cook MJ, Edwards DR, Russell DA, Targeted photodynamic therapy of breast cancer cells using antibody-phthalocyanine-gold nanoparticle conjugates. *Photochem Photobiol Sci,* 2011; 10(5): 822–831.
53. Obaid G, Chambrier I, Cook MJ, Russell DA, Targeting the oncofetal Thomsen-Friedenreich disaccharide using jacalin-PEG phthalocyanine gold nanoparticles for photodynamic cancer therapy. *Angew Chem Int Ed Engl*, 2012; 51(25): 6158–6162.
54. Morgan J, MacRobert AJ, Gray AG, Huehns ER, Use of photosensitive, antibody directed liposomes to destroy target populations of cells in bone marrow: a potential purging method for autologous bone marrow transplantation. *Br J Cancer*, 1992; 65(1): 58–64.
55. Bergstrom LC, Vucenik I, Hagen IK, Chernomorsky SA, Poretz RD, *In-vitro* photocytotoxicity of lysosomotropic immunoliposomes containing pheophorbide a with human bladder carcinoma cells. *J Photochem Photobiol B Biol*, 1994; 24(1): 17–23.
56. Derycke AS, Kamuhabwa A, Gijsens A, Roskams T, De Vos D, Kasran A, *et al.*, Transferrin-conjugated liposome targeting of photosensitizer AlPcS4 to rat bladder carcinoma cells. *J Natl Cancer Inst*, 2004; 96(21): 1620–1630.
57. Jin CS, Cui L, Wang F, Chen J, Zheng G, Targeting-triggered porphysome nanostructure disruption for activatable photodynamic therapy. *Adv Healthcare Mat*, 2014; 3(8): 1240–1249.
58. Reddy GR, Bhojani MS, McConville P, Moody J, Moffat BA, Hall DE, *et al.*, Vascular targeted nanoparticles for imaging and treatment of brain tumors. *Clin Cancer Res,* 2006; 12(22): 6677–6686.
59. Benachour H, Seve A, Bastogne T, Frochot C, Vanderesse R, Jasniewski J, *et al.*, Multifunctional peptide-conjugated hybrid silica nanoparticles for photodynamic therapy and MRI. *Theranostics,* 2012; 2(9): 889–904.
60. Brevet D, Gary-Bobo M, Raehm L, Richeter S, Hocine O, Amro K, *et al.*, Mannose-targeted mesoporous silica nanoparticles for photodynamic therapy. *Chem Comm,* 2009(12):1475–1477.
61. Gerweck LE, Seetharaman K, Cellular pH gradient in tumor *versus* normal tissue: potential exploitation for the treatment of cancer. *Cancer Res*, 1996; 56(6): 1194–1198.

62. Koo H, Lee H, Lee S, Min KH, Kim MS, Lee DS, *et al.*, *In vivo* tumor diagnosis and photodynamic therapy *via* tumoral pH-responsive polymeric micelles. *Chem Comm*, 2010; 46(31): 5668–5670.
63. Liu P, Yue C, Sheng Z, Gao G, Li M, Yi H, *et al.*, Photosensitizer-conjugated redox-responsive dextran theranostic nanoparticles for near-infrared cancer imaging and photodynamic therapy. *Polym Chem*, 2014; 5(3): 874–881.
64. Zheng G, Chen J, Stefflova K, Jarvi M, Li H, Wilson BC, Photodynamic molecular beacon as an activatable photosensitizer based on protease-controlled singlet oxygen quenching and activation. *Proc Natil Aca Sci USA,* 2007; 104(21): 8989–8994.
65. Anand S, Hasan T, Maytin EV, Mechanism of differentiation-enhanced photodynamic therapy for cancer: upregulation of coproporphyrinogen oxidase by C/EBP transcription factors. *Mol Cancer Ther*, 2013; 12(8): 1638–1650.
66. Sinha AK, Anand S, Ortel BJ, Chang Y, Mai Z, Hasan T, *et al.*, Methotrexate used in combination with aminolaevulinic acid for photodynamic killing of prostate cancer cells. *Br J Cancer* 2006; 95(4): 485–495.
67. Choudry K, Brooke RC, Farrar W, Rhodes LE, The effect of an iron chelating agent on protoporphyrin IX levels and phototoxicity in topical 5-aminolaevulinic acid photodynamic therapy. *Br J Derm,* 2003; 149(1): 124–130.
68. Campbell SM, Morton CA, Alyahya R, Horton S, Pye A, Curnow A, Clinical investigation of the novel iron-chelating agent, CP94, to enhance topical photodynamic therapy of nodular basal cell carcinoma. *Br J Derm*, 2008; 159(2): 387–393.
69. Casas A, Perotti C, Saccoliti M, Sacca P, Fukuda H, Batlle AM, ALA and ALA hexyl ester in free and liposomal formulations for the photosensitisation of tumour organ cultures. *Br J Cancer* 2002; 86(5): 837–842.
70. Oo MK, Yang X, Du H, Wang H, 5-aminolevulinic acid-conjugated gold nanoparticles for photodynamic therapy of cancer. *Nanomedicine* (Lond), 2008; 3(6): 777–786.
71. Yang SJ, Shieh MJ, Lin FH, Lou PJ, Peng CL, Wei MF, *et al.*, Colorectal cancer cell detection by 5-aminolaevulinic acid-loaded chitosan nanoparticles. *Cancer Lett*, 2009; 273(2): 210–220.
72. Zuluaga MF, Sekkat N, Gabriel D, van den Bergh H, Lange N, Selective photodetection and photodynamic therapy for prostate cancer through targeting of proteolytic activity. *Mol Cancer Ther*, 2013; 12(3): 306–313.

73. Jiang MY, Dolphin D, Site-specific prodrug release using visible light. *J Am Chem Soc*, 2008; 130(13): 4236–4237.
74. Messmann H, Endlicher E, Freunek G, Rummele P, Scholmerich J, Knuchel R, Fluorescence endoscopy for the detection of low an high grade dysplasia in ulcerative colitis using systemic or local 5-aminolaevulinic acid sensitisation. Inflammatory Bowel Disease 2003(52):1003–1007.
75. Rossetti R, Brus L, Electron-hole recombination emission as a probe of surface chemistry in aqueous CdS colloids. *J Phys Chem,* 1982; 86: 4470–4472.
76. Samia ACS, Dayal S, Burda C, Quantum dot-based energy transfer: perspectives and potenital for applications in photodynamic therapy. *Photochem Photobiol*, 2006; 82: 617–625.
77. Samia ACS, Chen X, Burda C, Semiconductor quantum dots for photodynamic therapy. *J Am Chem Soc Comm*, 2003; 125: 15736–15737.
78. Ma J, Chen J-Y, Idowu M, Nyokong T, Generation of singlet oxygen *via* the composites of water-soluble thiol-capped CdTe quantum dots-sulfonated aluminum phthalocyanines. *J Phys Chem Lett B,* 2008(112): 4465–4469.
79. Cao L, Wang X, Meziani MJ, Lu F, Wang H, Luo PG, *et al.*, Carbon dots for multiphoton bioimaging. *J Am Chem Soc*, 2007; 129(37): 11318–11319.
80. Kovalev D, Fujii M, Silicon Nanocrystals: Photosensitizers for oxygen molecules. *Adv Mater,* 2005; 17(21): 2531–2544.
81. Bakalova R, Ohba H, Zhelev Z, Ishikawa M, Baba Y, Quantum dots as photosensitizers? *Nat Biotechnol*, 2004; 22(11): 2.
82. Ge J, Lan M, Zhou B, Liu W, Guo L, Wang H, *et al.*, A graphene quantum dot photodynamic therapy agent with high singlet oxygen generation. *Nat Comm*, 2014; 5: 4596.
83. Weber MJ, Scintillation: mechanisms and new crystals. *Nucl Instrum Methods Phys Res A*, 2004; 527(1–2): 9–14.
84. Chatterjee DK, Fong LS, Zhang Y, Nanoparticles in photodynamic therapy: an emerging paradigm. *Adv Drug Deliv Rev*, 2008; 60(15): 1627–1637.
85. Liu Y, Chen W, Wang S, Joly AG, Investigation of water-soluble x-ray luminescence nanoparticles for photodynamic activation. *Appl Phys Lett*, 2008; 92(4): 043901.
86. Liu Y, Zhang Y, Wang S, Pope C, Chen W, Optical behaviors of ZnO-porphyrin conjugates and their potential applications for cancer treatment. *Appl Phys Lett*, 2008; 92(14): 143901.

87. Bulin A-L, Truillet C, Chouikrat R, Lux F, Frochot C, Amans D, *et al.*, X-ray-induced singlet oxygen activation with nanoscintillator-coupled porphyrins. *J Phys Chem, C* 2013; 117(41): 21583–21589.
88. Juzenas P, Chen W, Sun YP, Coelho MA, Generalov R, Generalova N, *et al.*, Quantum dots and nanoparticles for photodynamic and radiation therapies of cancer. *Adv Drug Deliv Rev*, 2008; 60(15): 1600–1614.
89. Letant SE, Wang T-F, Semiconductor quantum dot scintillation under gamma-ray irradiation. *Nano Lett*, 2006; 6(12): 2877–2880.
90. Létant SE, Wang TF, Study of porous glass doped with quantum dots or laser dyes under alpha irradiation. *Appl Phys Lett*, 2006; 88(10): 103110.
91. Mutlugun E, Soganci IM, Demir HV, Nanocrystal hybridized scintillators for enhanced etection and imaging on Si platforms in UV, *Opt Express*, 2007; 15(3): 1128–1134.
92. Wang L, Yang W, Read P, Larner J, Sheng K, Tumor cell apoptosis induced by nanoparticle conjugate in combination with radiation therapy. *Nanotechnology* 2010; 21(47): 475103.
93. Zhang P, Steelant W, Kumar M, Scholfield M, Versatile photosensitizers for photodynamic therapy at infrared excitation. *J Am Chem Soc*, 2007; 129(15): 4526–4527.
94. Wang F, Chatterjee DK, Li Z, Zhang Y, Fan X, Wang M, Synthesis of polyethylenimine/NaYF4 nanoparticles with upconversion fluorescence. *Nanotechnology* 2006; 17(23): 5786–5791.
95. Wang C, Tao H, Cheng L, Liu Z, Near-infrared light induced *in vivo* photodynamic therapy of cancer based on upconversion nanoparticles. *Biomaterials,* 2011; 32(26): 6145–6154.
96. Idris NM, Gnanasammandhan MK, Zhang J, Ho PC, Mahendran R, Zhang Y, *In vivo* photodynamic therapy using upconversion nanoparticles as remote-controlled nanotransducers. *Nat Med*, 2012; 18(10): 1580–1585.
97. Ran C, Zhang Z, Hooker J, Moore A, *In vivo* photoactivation without "light": use of Cherenkov radiation to overcome the penetration limit of light. *Mol Imag Biol*, 2012; 14(2): 156–162.
98. Jang KW, Shin SH, Kim SG, Kim JS, Yoo WJ, Ji YH, *et al.*, Measurement of Cerenkov radiation induced by the gamma-rays of Co-60 therapy units using wavelength shifting fiber. *Sensors*, 2014; 14(4): 7013–7025.
99. Hirschberg H, Madsen SJ, Jansen ED, Luo Q, Mohanty SK, Thakor NV, *et al.*, Ultra low fluence rate photodynamic therapy: simulation of light emitted by the Cerenkov effect. *SPIE Proceedings*, 2014; 8928: 89280F.

100. Yuan H, Chong H, Wang B, Zhu C, Liu L, Yang Q, *et al.*, Chemical molecule-induced light-activated system for anticancer and antifungal activities. *J Am Chem Soc*, 2012; 134(32): 13184–13187.
101. Hsu CY, Chen CW, Yu HP, Lin YF, Lai PS, Bioluminescence resonance energy transfer using luciferase-immobilized quantum dots for self-illuminated photodynamic therapy. *Biomaterials*, 2013; 34(4): 1204–1212.
102. Gather MC, Yun SH, Single-cell biological lasers. *Nat Photonics,* 2011; 5(7): 406–410.
103. Huang Z, A Review of progress in clinical photodynamic therapy. *Technol Cancer Res Treat,* 2005; 4(3): 283–293.
104. Geffroy B, Le Roy P, Prat C, Organic light-emitting diode (OLED) technology: materials, devices and display technologies. *Polym Int*, 2006; 55(6): 572–582.
105. Attili SK, Lesar A, McNeill A, Camacho-Lopez M, Moseley H, Ibbotson S, *et al.*, An open pilot study of ambulatory photodynamic therapy using a wearable low-irradiance organic light-emitting diode light source in the treatment of nonmelanoma skin cancer. *Br J Dermatol*, 2009; 161(1): 170–173.
106. Zhou X, Pogue BW, Chen B, Demidenko E, Joshi R, Hoopes J, *et al.*, Pretreatment photosensitizer dosimetry reduces variation in tumor response. *Int J Radiat Oncol Biol Phys*, 2006; 64(4): 1211–1220.
107. Pogue BW, Chen B, Zhou X, Hoopes PJ, Analysis of sampling volume and tissue heterogeneity on the *in vivo* detection of fluorescence. *J Biomed Optics,* 2005; 10(4): 41206.
108. Busch TM, Hahn SM, Wileyto EP, Koch CJ, Fraker DL, Zhang P, *et al.*, Hypoxia and Photofrin uptake in the intraperitoneal carcinomatosis and sarcomatosis of photodynamic therapy patients. *Clin Cancer Res*, 2004; 10(14): 4630–4638.
109. Dysart JS, Singh G, Patterson MS, Calculation of singlet oxygen dose from photosensitizer fluorescence and photobleaching during mTHPC photodynamic therapy of MLL cells. *Photochem Photobiol*, 2005; 81(1): 196–205.
110. Dysart JS, Patterson MS, Characterization of Photofrin photobleaching for singlet oxygen dose estimation during photodynamic therapy of MLL cells *in vitro*. *Phys Med Biol*, 2005; 50(11): 2597–2616.
111. Georgakoudi I, Nichols MG, Foster TH, The mechanism of Photofrin photobleaching and its consequences for photodynamic dosimetry. *Photochem Photobiol*, 1997; 65(1): 135–144.

112. Gibot L, Lemelle A, Till U, Moukarzel B, Mingotaud AF, Pimienta V, *et al.*, Polymeric micelles encapsulating photosensitizer: structure/photodynamic therapy efficiency relation. *Biomacromolecules*, 2014; 15(4): 1443–1455.
113. Cincotta L, Foley JW, MacEachern T, Lampros E, Cincotta AH, Novel photodynamic effects of a benzophenothiazine on two different murine sarcomas. *Cancer Res,* 1994; 54(5): 1249–1258.
114. Verma S, Sallum UW, Athar H, Rosenblum L, Foley JW, Hasan T, Antimicrobial photodynamic efficacy of side-chain functionalized benzo[a] phenothiazinium dyes. *Photochem Photobiol*, 2009; 85(1): 111–118.
115. Georgakoudi I, Foster TH, Effects of the subcellular redistribution of two nile blue derivatives on photodynamic oxygen consumption. *Photochem Photobiol* 1998; 68(1): 115–122.
116. Evans CL, Abu-Yousif AO, Park YJ, Klein OJ, Celli JP, Rizvi I, *et al.*, Killing hypoxic cell populations in a 3D tumor model with EtNBS-PDT, *PloS One*, 2011; 6(8): e23434.
117. Shannon AM, Bouchier-Hayes DJ, Condron CM, Toomey D, Tumour hypoxia, chemotherapeutic resistance and hypoxia-related therapies. *Cancer Treat Rev* 2003; 29(4): 297–307.
118. McIlroy BW, Mann TS, Dysart JS, Wilson BC, The effects of oxygenation and photosensitizer substrate binding on the use of fluorescence photobleaching as a dose metric for photodynamic therapy. *Vib Spectros*, 2002; 28(1): 25–35.
119. Wilson BC, Patterson MS, Lilge L, Implicit and explicit dosimetry in photodynamic therapy: a new paradigm. *Lasers Med Sci*, 1997; 12(3): 182–199.
120. Gibson SL, VanDerMeid KR, Murant RS, Raubertas RF, Hilf R, Effects of various photoradiation regimens on the antitumor efficacy of photodynamic therapy for R3230AC mammary carcinomas. *Cancer Res*, 1990; 50(22): 7236–7241.
121. Feins RH, Hilf R, Ross H, Gibson SL, Photodynamic therapy for human malignant mesothelioma in the nude mouse. *J Surg Res*, 1990; 49(4): 311–314.
122. Bown SG, Tralau CJ, Smith PD, Akdemir D, Wieman TJ, Photodynamic therapy with porphyrin and phthalocyanine sensitisation: quantitative studies in normal rat liver. *Br J Cancer*, 1986; 54(1): 43–52.
123. Patterson MS, Wilson BC, Graff R, *In vivo* tests of the concept of photodynamic threshold dose in normal rat liver photosensitized by aluminum

chlorosulphonated phthalocyanine. *Photochem Photobiol*, 1990; 51(3): 343–349.

124. Moesta KT, Greco WR, Nurse-Finlay SO, Parsons JC, Mang TS, Lack of reciprocity in drug and light dose dependence of photodynamic therapy of pancreatic adenocarcinoma *in vitro*. *Cancer Res*, 1995; 55(14): 3078–3084.
125. Farrell TJ, Wilson BC, Patterson MS, Olivo MC, Comparison of the *in vivo* photodynamic threshold dose for photofrin, mono- and tetrasulfonated aluminum phthalocyanine using a rat liver model. *Photochem Photobiol*, 1998; 68(3): 394–399.
126. Fingar VH, Henderson BW, Drug and light dose dependence of photodynamic therapy: a study of tumor and normal tissue response. *Photochem Photobiol*, 1987; 46(5): 837–841.
127. Rizvi I, Anbil S, Alagic N, Celli J, Zheng LZ, Palanisami A, *et al.*, PDT dose parameters impact tumoricidal durability and cell death pathways in a 3D ovarian cancer model. *Photochem Photobiol*, 2013; 89(4): 942–952.
128. Zuluaga MF, Lange N, Combination of photodynamic therapy with anticancer agents. *Curr Med Chem*, 2008; 15(17): 1655–1673.
129. Greco WR, Bravo G, Parsons JC, The search for synergy: a critical review from a response surface perspective. *Pharma Rev*, 1995; 47(2): 331–385.
130. Pickl M, Ries CH, Comparison of 3D and 2D tumor models reveals enhanced HER2 activation in 3D associated with an increased response to trastuzumab. *Oncogene*, 2009; 28(3): 461–468.
131. Hicks KO, Pruijn FB, Secomb TW, Hay MP, Hsu R, Brown JM, *et al.*, Use of three-dimensional tissue cultures to model extravascular transport and predict in vivo activity of hypoxia-targeted anticancer drugs. *J Natl Cancer Inst*, 2006; 98(16): 1118–1128.
132. Gurski LA, Jha AK, Zhang C, Jia X, Farach-Carson MC, Hyaluronic acid-based hydrogels as 3D matrices for *in vitro* evaluation of chemotherapeutic drugs using poorly adherent prostate cancer cells. *Biomaterials*, 2009; 30(30): 6076–6085.
133. Canti G, Nicolin A, Cubeddu R, Taroni P, Bandieramonte G, Valentini G, Antitumor efficacy of the combination of photodynamic therapy and chemotherapy in murine tumors. *Cancer Lett*, 1998; 125(1–2): 39–44.
134. Nahabedian MY, Cohen RA, Contino MF, Terem TM, Wright WH, Berns MW, *et al.*, Combination cytotoxic chemotherapy with cisplatin or doxoru-

bicin and photodynamic therapy in murine tumors1. *J Natl Cancer Inst,* 1988; 80(10): 739–743.

135. Datta SN, Allman R, Loh C, Mason M, Matthews PN, Effect of photodynamic therapy in combination with mitomycin C on a mitomycin-resistant bladder cancer cell line. *Br J Cancer*, 1997; 76(3): 312–317.
136. Ma LW, Steen HB, Moan J, Berg K, Peng Q, Saether H, *et al.*, Cytotoxicity and cytokinetic effects of mitomycin C and/or photochemotherapy in a human colon adenocarcinoma cell line. *Int J Biochem,* 1992; 24(11): 1807–1813.
137. Nonaka M, Ikeda H, Inokuchi T, Effect of combined photodynamic and chemotherapeutic treatment on lymphoma cells *in vitro*. *Cancer Lett,* 2002; 184(2): 171–178.
138. Rizvi I, Celli JP, Evans CL, Abu-Yousif AO, Muzikansky A, Pogue BW, *et al.*, Synergistic enhancement of carboplatin efficacy with photodynamic therapy in a three-dimensional model for micrometastatic ovarian cancer. *Cancer Res*, 2010; 70(22): 9319–9328.
139. Yang Y, Yang X, Zou J, Jia C, Hu Y, Du H, *et al.*, Evaluation of photodynamic therapy efficiency using an in vitro three-dimensional microfluidic breast cancer tissue model. *Lab Chip*, 2015 Feb 7; 15(3): 735–744.
140. Howerton BS, Heidary DK, Glazer EC, Strained ruthenium complexes are potent light-activated anticancer agents. *J Am Chem Soc*, 2012; 134(20): 8324–8327.
141. Myers G, Why bioimage informatics matters. *Nat Methods*, 2012; 9(7): 659–660.
142. Celli JP, Rizvi I, Evans CL, Abu-Yousif AO, Hasan T, Quantitative imaging reveals heterogeneous growth dynamics and treatment-dependent residual tumor distributions in a three-dimensional ovarian cancer model. *J Biomed Opt*, 2010; 15(5): 051603.
143. Eliceiri KW, Berthold MR, Goldberg IG, Ibanez L, Manjunath BS, Martone ME, *et al.*, Biological imaging software tools. *Nat Meth*, 2012; 9(7): 697–710.
144. Celli JP, Rizvi I, Blanden AR, Massodi I, Glidden MD, Pogue BW, *et al.*, An imaging-based platform for high-content, quantitative evaluation of therapeutic response in 3D tumour models. Scientific Reports 4, Article number: 3751 (2014).

Chapter 8

TARGETED CONTRAST AGENTS FOR ^{1}H MRI OF TUMOR MICROENVIRONMENT

*Shubhangi Agarwal and Vikram D. Kodibagkar**

School of Biological and Health Systems Engineering,
Arizona State University, Tempe, AZ 85287, USA
**Vikram.Kodibagkar@asu.edu*

Abstract

In-vivo imaging of cancer can readily provide anatomical and functional information about the tumor, non-invasively, and assist in guiding treatment choices. Proton (^{1}H) magnetic resonance imaging (MRI) is a powerful imaging modality that is particularly useful for studying the dynamic processes in the body with anatomical co-registration. MRI can safely provide diagnostic information to the clinicians repeatedly over the course of therapy. This information on the temporal evolution of tumors can greatly assist the clinicians and researchers in prognosis as well as in selection of better therapeutic protocols and assessment of novel therapies. MRI can be enhanced by the use of contrast agents that enable better delineation of the tumor from its surroundings and these are now routinely used in the clinic for examination of pathologies, particularly cancer. Development of novel contrast agents with specific targeting abilities to the tumor has the potential to greatly improve the diagnosis of the cancer. It can also assist in further improving our understanding of the cancer progression/treatment and aid in designing better therapies. This chapter discusses targeted ^{1}H MRI contrast agents for application in cancer imaging. We begin with an introduction to MRI and the mechanism of MRI contrast agents. We then classify the contrast agents for ^{1}H MRI of tumor microenvironment into two categories, namely lanthanide-based and iron-based, respectively, and discuss the general properties that affect the sensitivity. Targeting agents are also classified based on size and structure (small-molecular, macromolecular, micellar, liposomal and nanoparticulate) and their target in the tumor (vascular *versus* extravascular). We discuss examples of each category of targeted agent and highlight *in vivo* applications. Finally we outline the future prospects for this important field of research.

8.1 Introduction

Cancer is a collective organization of malignant tumor cells and non-malignant tumor promoting cells. Together these constituents form the tumor microenvironment which differs significantly from the surrounding host tissue as it contains several modified cellular and molecular signatures.[1] These signatures or "hallmarks"[2] define the properties of a tumor and distinguish it from the host both morphologically and functionally. These properties as well as the underlying signatures can be exploited for diagnosis and treatment of a tumor by developing novel targeted drugs and drug delivery agents for tumor specific accumulation and action.[3–6] A major benefit of characterizing these hallmarks has been the improvement in *in vivo* imaging of cancer by development of tumor targeted imaging agents. Targeted imaging significantly improves the contrast of tumors from surrounding tissue and provides a better delineation of the tumor margins and possible metastasis.[7,8] Targeted imaging can also improve the prognosis of cancer by helping identify biomarkers associated with either more or less malignant subtypes. There are several imaging modalities that have been used for cancer diagnosis as well as prognosis such as ultrasound,[9] positron emission tomography (PET),[10] single-photon emission computed tomography (SPECT),[11] and magnetic resonance imaging (MRI).[12–14] Amongst these techniques MRI provides an excellent platform for non-invasive *in vivo* molecular imaging of cancer, providing superior soft tissue contrast and high resolution with moderate sensitivity compared to other imaging modalities.[15] MRI can generate images depicting anatomical, biochemical and functional information while offering a good balance between image resolution and imaging time.[16] Additionally, MRI can also help in identifying the hallmarks of cancer and lead to development, analysis and therapeutic response assessment of improved targeted therapies on tumors. In this chapter we will focus on the design and application of some ground breaking MRI agents for cancer imaging which strengthen the path to the goal of personalized therapeutics. Our goal in this chapter is to provide a broad overview of various strategies and approaches used for targeted imaging of cancer by MRI. We do not aim to provide an exhaustive list of targeting agents and the reader is encouraged to refer to excellent reviews on the subject that have been recently published.[17–19]

8.1.1 *Magnetic resonance imaging: An introduction*

MRI is a non-invasive tomographic imaging technique based on the phenomenon of nuclear magnetic resonance (NMR). MRI utilizes the fundamental property of nuclear spin to detect measurable signals arising from bulk nuclear spin polarization in the presence of an external magnetic field (nuclear paramagnetism). Two and three-dimensional images of the body with high spatial resolution and contrast can be generated using MRI by acquiring these signals in the presence of position encoding gradients. Advances in the last three decades have led to a substantial improvement in the quality of the MR images generated with higher signal to noise ratio and sensitivity.

Signal in MRI is generated from the currents induced in a resonant loop when nuclear spin ensembles are exposed to static and oscillating magnetic fields. Hydrogen atom contains a single proton with +1/2 and –1/2 spin states which results in a bulk magnetic moment in presence of a strong external magnetic field. This forms the basis for proton MRI that is widely prevalent in clinical settings. Human body is ~70% water by body weight and (along with hydrogen from fat and other biochemicals) provides plenty of hydrogen protons that can be used as NMR active nuclei. Image contrast is defined as the ability to discriminate tissues based on their relative brightness and darkness on an image. Contrast obtained in the MRI images depends on several intrinsic properties of the tissue such as proton density (ρ), diffusion, spin-lattice relaxation time (T_1), spin-spin relaxation time (T_2), observed spin de-phasing time (T_2^*) and chemical exchange effects. The images are generated by using specific "weighting" (e.g. T_1 or T_2 weighting), i.e. choice of scanner settings, that exploit the differences in the above mentioned properties between the pathology and healthy tissue and results in image contrast that reflects the variation in the chosen property. For example the T_1 and T_2 of water at 1.5 T are ~ 4000 ms and 2000 ms, respectively.[20] Cerebrospinal fluid has similar T_1 and T_2 as water because the hydrogen protons are less restricted while fat protons have restricted mobility, resulting in shorter T_1 (280–340 ms) and T_2 (60–100 ms). In comparison, the T_1 and T_2 of white matter are ~ 800 ms and 80 ms, respectively, while those for gray matter are slightly longer with values ~1100 ms and 90 ms, respectively. Abnormalities such as edema, tumor, inflammation, infarction and infection have increased water content (in case of tumors, lower cell density as well) and can be detected by using T_2 weighting during the MRI acquisition as they appear bright compared to their surrounding normal tissue. In a routine clinical examination, T_2 weighted scan is often the first indicator of presence of an abnormality.

The first MR measurements of a tumor were performed in 1971 and paved the path for use of MR for discrimination of tumors from surrounding tissues based on differences in relaxation times.[21] While this has been generally quite useful, the contrast generated by the intrinsic property of tissues is sometimes not enough to efficiently discriminate the diseased tissue because of small difference in the tissue relaxation times. Contrast agents and reporter agents are thus used to enhance the contrast in MRI. With the development of high relaxivity and targeted MR agents, the image quality (contrast between tissue of interest and its surroundings) has improved considerably.[22,23]

8.1.2 *MRI contrast agents*

MRI contrast agents are biocompatible, small-molecular or nanoscale chemical entities introduced exogenously into the body in order to improve the contrast between the area of interest and the surrounding tissue. The properties and advantages of MR contrast agents were first outlined by Lauffer in 1987[22] and since then tremendous improvements have been achieved in the field and several new agents are developed each year.[15,24–29] MRI based contrast agents typically use paramagnetic molecules (e.g.

chelates of gadolinium or iron oxide) to enhance the contrast of the region of their accumulation by shortening the relaxation times of the water protons around them via a water exchange mechanism.[22] Majority of gadolinium based contrast agents are T_1 relaxation agents and improve the contrast of the region of accumulation in a T_1 weighted image. On the other hand, iron oxide based contrast agents are predominantly T_2 relaxation agents and make the accumulation region appear dark on a T_2 or T_2^* weighted image. The efficiency of a contrast agent is based on its relaxivity (r_1 or r_2), which is the change in spin-lattice relaxation rate R_1 (= $1/T_1$) or spin-spin relaxation rate R_2 (= $1/T_2$), respectively, of water per unit concentration of contrast agent. The relaxivity is typically calculated by measuring the relaxation rate (R_1 or R_2) as a function of agent concentration and by determining the slope of the graph (R_1 or R_2 vs concentration). The relationship between the relaxation times, the relaxivity and the concentration of the contrast agent is given by the following equation:

$$1/T_{i\,(\text{measured})} = 1/T_{i\,(\text{water})} + r_i\,[\text{Agent}]; \quad i = 1 \quad \text{or} \quad 2 \qquad (8.1)$$

The detailed description of mechanisms that relate the various properties of contrast agents, such as lanthanide (Ln) chelates, to their relaxation effect is outside the scope of this chapter and reader is encouraged to refer to detailed reviews.[22,23,30] To summarize (see Figure 8.1), the important properties that influence contrast agent relaxivity are (a) spin S and the electronic relaxation times T_{ie} of the paramagnetic ion (b) the number of inner (directly coordinated to the Ln atom) and outer (hydrogen-bonded) coordination sphere water molecules (q and q′ respectively), (c) the residence lifetime of the inner and outer sphere water molecule(s) (τ_m and $\tau_{m'}$, respectively), (d) the tumbling rate of the agent in solution (t_R, rotational correlation time) and (e) the longitudinal relaxation time of the water molecule(s) protons present in the inner and outer coordination spheres (T_{im} and T_{im}' respectively). Here i = 1 or 2 and refers to the T_1 and T_2 relaxation times, respectively. The major contribution to the relaxation properties of the contrast agents is due to the inner sphere mechanisms. This contribution is given by:

$$r_i^{IS} = q/([H_2O](T_{im} + \tau_m)); \quad i = 1 \quad \text{or} \quad 2 \qquad (8.2)$$

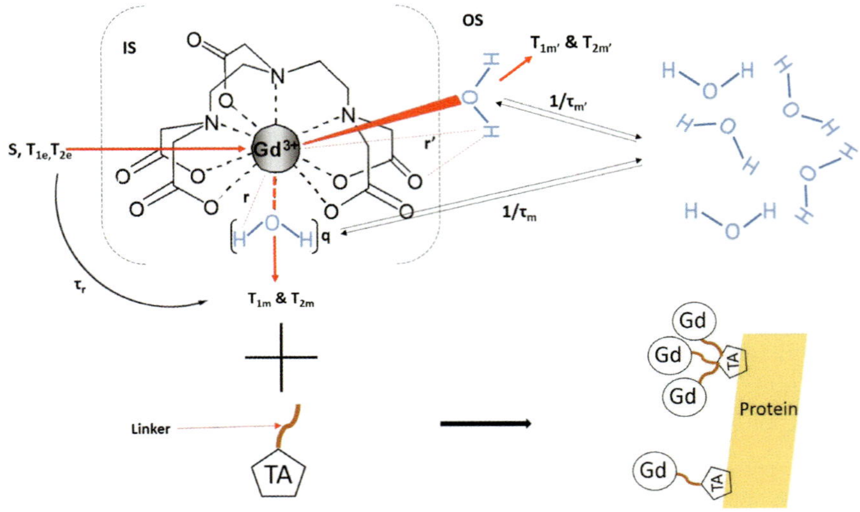

Figure 8.1 Factors influencing the solvent water relaxation and strategies for controlling rotational dynamics of multimeric, targeted contrast agents. IS: inner sphere, OS: outer sphere, T_{1e}/T_{2e}: electronic relaxation times, S: electron spin, τ_r: rotation tumbling time, τ_m: residence lifetime of water molecule, T_{1m}/T_{2m}: relaxation times of bound water, r: Ln-H distance for IS water molecules, q: number of IS water molecules, TA: targeting agent/ligand). Corresponding parameters for OS water molecules are denoted with a prime (′) e.g r′: Ln–H distance for OS water molecules.

where q is the number of bound water nuclei per metal ion; τ_m is the life time of water in the inner sphere and T_{im} is the relaxation time of bound water. Magnetic field strength alters the relaxation rate of the contrast agent which typically decreases with increasing field strength.[31] The relaxation rate at clinically relevant field depends mainly on the τ_m and τ_R properties of the agent. Reduction in the rotational correlation time of the agent, which can be achieved by increasing the molecular weight of the agent, causes an increase in the relaxivity of the agent but the increases in r_2 dominate over r_1 at higher fields, as discussed by Caravan *et al.* in detail.[30] If the residence time of water protons in the inner sphere is long then the corresponding exchange of relaxed water with the bulk water is limited, thus resulting in lower relaxation rate. On the other hand, very short residence time also results in lower relaxation rates due to an inefficient relaxation of the bound water. Thus there is an optimum value of residence time for which the relaxation of the agent is highest. Many of these properties can be altered in order to develop improved contrast agents for cancer imaging by rational design of the chelate ligand.[25]

Another class of MR contrast agents are chemical exchange saturation transfer (CEST) contrast agents which generate contrast based on exchange between bulk water protons (say pool A) and molecules that have exchangeable "bound" protons (say pool B) with a chemical shift difference. The protons in pool B are irradiated by a low power radio frequency pulse with the frequency to match the resonance frequency of the bound protons to saturate their magnetization. Chemical exchange leads to a continuous transfer of saturated nuclei from pool B to pool A and results in signal intensity decrease. The equation given below describes the steady state intensity of pool A protons:

$$\frac{M_{A\infty}}{M_{AO}} = \frac{1}{(1+k_2T_{1B})}$$

where $M_{A\infty}$ is the signal intensity of pool A after prolonged pre-saturation of pool B protons, M_{AO} is the initial intensity of pool A, k_2 is the unidirectional rate constant for the protons moving from pool A to pool B, and T_{1B} is the longitudinal relaxation time of the pool B protons. The CEST phenomenon was first demonstrated by Forsen and Hoffman.[32] CEST agents have a distinct quality of being turned "on" only via application of saturating RF pulses at the right resonant frequencies of the bound water protons and form an important class of "smart" contrast agents. There are several biomolecules with such exchangeable proton pools and these have been exploited as endogenous and exogenous CEST agents, particularly in cancer. The CEST phenomenon has been extended to paramagnetic lanthanide chelate molecules that are similar to Gd based contrast agents but designed to have longer bound water residence times (i.e. slow exchange of water). Also the lanthanide chosen for chelation typically has an asymmetric outer electron cloud (e.g. Eu) giving rise to a large chemical shift for the bound water compared to the bulk water. These agents are referred to as PARACEST agents.[33] While this is a relatively new class of agents and lacks the sensitivity of the T_1 and T_2 based probes, the ability to turn on the contrast effect makes this class of agents very promising for molecular imaging applications. This chapter will include examples of targeted cancer imaging PARACEST probes as well.

8.1.3 *Targeted imaging of cancer*

The need for accurate and early diagnosis of tumors gave rise to the development of targeted contrast agents. Targeted imaging of tumors usually refers to the case when the probe specifically binds to a tumor microenvironmental target to a far greater extent than surrounding tissue and/or is cleared less rapidly than from the surrounding tissue, resulting in image contrast. Targeted imaging also includes the scenario where the relaxivity of the agent is selectively modulated by tumor microenvironmental parameters, also leading to image contrast compared to surrounding tissue. There are several unique characteristics of a tumor, which makes it distinguishable from the rest of the tissue, and these can be exploited in development of the targeted therapeutics and imaging probes. For example, targeted agents are often prepared by conjugating the contrast agents with proteins and growth factors that bind to receptors overexpressed in cancer cells or vasculature. Targeted contrast agents may then accumulate in the tumor cells or vascular endothelium due to the receptor mediated internalization. Alternatively, several microenvironmental properties of the target, e.g. tumor pH, can be assessed based on the modification of probe relaxivities upon accumulation. Targeted agents are broadly subdivided into three groups: small molecular, macromolecular and nano-platform (as an extension of macromolecular agents). In this review, we discuss the various probes that fall into each category as well as the biomarkers selected by scientists to develop new agents.

8.2 Lanthanide based agents

Lanthanides are transition metals with +3 or +2 oxidation state and are diamagnetic or paramagnetic in nature. The paramagnetic elements that have been used for MRI contrast agents are gadolinium (III), europium (II), dysprosium (III), thulium (III) and ytterbium (II). The most prominent element used in MRI is the gadolinium (Gd^{3+}) ion. The ion has seven unpaired electrons whose high spin gives rise to a rapid electron spin relaxation of surrounding water protons. It is a toxic heavy metal and needs to be chelated to proper ligands in order to be used as a contrast agent for *in vivo* imaging. Diethylene triamine pentaacetic acid (DTPA) and 1,4,7,10-tetraazacyclododecane tetraacetic acid (DOTA) are types of ligands that have been extensively used.[23] Gadolinium toxicity can be reduced by use of stable chelates and is then systemically excreted from the body. There are various commercially available contrast agents which have been routinely administered clinically for imaging. The first commercially available contrast agent was Magnevist (Gd-DTPA, Gadopentetate Dimeglumine) in 1988.[34] Several contrast agents were later developed for clinical use amongst which Ablavar (gadofosveset trisodium), Eovist (gadoxetate disodium), Multihance (Gadobenate Dimeglumine), Omniscan (Gadodiamide), Optimark (Gadoversetamide), Prohance (Gadoteridol) are FDA approved.[23]

Gd^{3+} based agents can be categorized into several types: extracellular agents, which are the typical clinical contrast agents, hepabobiliary agents, blood pool agents and targeted agents. These agents can be further categorized into small molecular and macromolecular. Small molecular agents are the typical Gd^{3+} chelates which are either acyclic or macrocyclic. Among the FDA approved agents, Magnevist, Dotarem, Prohance and Omniscan are extracellular agents and Multihance, Eovist and Ablavar are protein binding agents. The above-mentioned agents have spin-lattice relaxivities (r_1) in the range 3.5–3.8 $mM^{-1}s^{-1}$ (at 20 MHz and 37°C) and are lower than that of the protein-binding agents (6.3–19 $mM^{-1}s^{-1}$ at 1.5 T and 37°C[31,35]). The latter agents tend to bind non-covalently with the albumin, thus slowing down the molecular rotation and increasing the relaxivity[22,30] as discussed in Section 8.2. For example, Ablavar (MS-325), which reversibly binds to serum albumin and results in high retention time and better contrast, has been used as a blood pool agent for assessment of microvasculature in tumors.[36,37]

The pharmacokinetics and route of excretion of these agents can be modified by chemical alteration of the chelates. Small molecular agents (linear or macrocyclic) are usually injected intravenously and tend to distribute in the plasma and interstitial spaces. The agents have a half-life of 1–2 hours and are cleared through the renal route. Conjugation of the small molecular Gd^{3+} chelates with macromolecules such as polymers, dendrimers, micelles, proteins and vesicles improves the relaxivity of contrast agents by various mechanisms, though principally by slowing the overall molecular tumbling time τ_r.[38] Macro-molecular contrast agents have been shown to have higher relaxivities, longer blood circulation, better accumulation in tumor tissues, longer retention times (due to slower clearance) and lower toxicities than small molecular agents. Several such agents have been developed and shown to be better than the small molecular contrast agents[38] in specific applications.

8.2.1 *Small molecular targeted contrast agents*

A successful contrast agent needs to have ideal pharmacokinetic, physiochemical and radiological properties. The agent should be (a) thermodynamically and kinetically stable (b) have high water solubility and retention time and (c) non-toxic. Conjugation with biocompatible macromolecules improves these properties of the Gd^{3+} chelates. An important concern while designing Gd^{3+} based agents in particular is the link of Gd^{3+} to nephrotoxicity and nephrogenic systemic fibrosis (NSF). Gd^{3+}, when not completely cleared out, accumulates in the internal organs and results in fibrosis of the organs in patients with compromised renal clearance. Gd^{3+} chelates of DOTA and DTPA can be further conjugated with moieties that bind to cancer specific biomarkers or microenvironmental "hallmarks".[2] Direct conjugations of these contrast agents (CAs) to the targeting moieties increases their relaxivities, specificity, and clearance time although the clearance route is still predominantly renal.

8.2.1.1 *Vascular targeting small molecular probes*

Angiogenesis, which is the formation of new blood vessels, is a marker of invasive tumor growth and metastasis and its inhibition at an early stage is crucial for control of cancer progression. The RGD (Arg–Gly–Asp) peptide binds to the $a_v\beta_3$-integrin, which is known to be over-expressed in nascent endothelial cells during angiogenesis in various tumor types.[39] An early cancer targeting imaging agent was prepared by conjugation of (DOTA) with the cyclic pentapeptide c(RGDyK) and subsequent Gd chelation (Gd-DOTA-RGD)[40] (Figure 8.2). Conjugation of this peptide

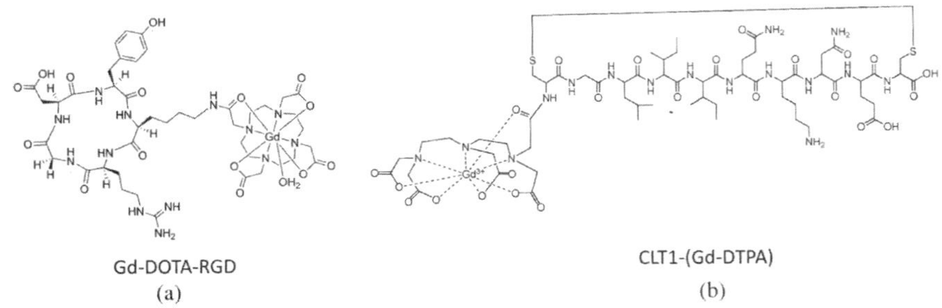

Figure 8.2 (a) Structure of Gd(DOTA) conjugate of RGD and (b) Structure of CLT1-(Gd-DTPA) Reproduced with permission from Refs. 40 and 42.

with the Gd-DOTA chelate resulted in a contrast agent with higher relaxivity than the Gd-DOTA because of the slower molecular tumbling.[40] The agent demonstrated an r_1 and r_2 of 7.4 ± 0.20 $mM^{-1}s^{-1}$and 4.0 ± 0.24 $mM^{-1}s^{-1}$, respectively at 298 K and 64 MHz (1.5 T). T_1-weighted MR imaging of hepatocellular carcinoma in H-*ras*12V transgenic mice injected with this agent confirmed that it binds to the $a_v\beta_3$-integrin in tumors and results in enhancement of contrast. It also has very low toxicity and is cleared through the body mainly *via* glomerular filtration. This agent demonstrated that targeted CAs could be used for better detection of tumors and their properties. Recently, Park *et al.* developed integrin targeting peptide by conjugating cycloalkane-based cyclic RGD peptides that show higher binding efficiency as compared to c(RGDYK) with Gd-DOTA. *In vivo* comparative analysis of the agent with Gd-DOTA-RGD in U87MG glioblastoma bearing mice showed that it had higher signal enhancement and better contrast.[41]

Fibronectin and fibrin clots have also been associated with tumor angiogenesis and targeted agents can also be used to characterize tumor angiogenesis. Fibronectin-fibrin complexes have been reported to be present in tumor extracellular matrix and not in normal tissues and are associated with increased microvessel permeability and angiogenesis. In order to exploit this, in 2008, Ye *et al.* developed a vascular targeting agent CLT1-(Gd-DTPA),[42] consisting of GdDTPA conjugated to a cyclic decapeptide CGLIIQKNEC (CLT1)[43] that targets the fibronectin-fibrin complexes in the extracellular matrix of different tumors (Figure 8.2). The r_1 and r_2

relaxivities of CLT1-(Gd-DTPA) were 4.22 and 4.45 $mM^{-1}s^{-1}$, respectively, at 3 T.[42] Female athymic nu/nu mice bearing HT-29 human colon carcinoma xenografts were injected with 0.1 mmole/kg dose of the agent and T_1-weighted images were collected over a period of 60 minutes. Enhancement was seen after 10 minutes post injection and it increased substantially by 60 minutes post injection with greater uptake in central regions of tumor over time. Similar results were seen in a later study with MDA-MB-231 breast tumor xenografts.[44] In 2013, Wu *et al.* further improved this fibronectin-fibrin targeting contrast agent by conjugating CLT1 with four Gd-DOTA monoamide chelates to form the final product CLT1-dL-$(Gd\text{-}DOTA)_4$.[45] The agent had a relaxivity r_2 of 10.1 ± 1. $mM^{-1}s^{-1}$ and r_1 of 46 ± 5 $mM^{-1}s^{-1}$ at 1.5 T. Orthotopic prostate cancer models (PC3-GFP) grown in NIH athymic nude mice were imaged and analysis of the pre and post injection T_1 weighted images showed an increased contrast in the tumor tissue as compared to the control agent, sCLT1-dL-$(Gd\text{-}DOTA)_4$, which had a scrambled peptide (sCLT1) instead of the targeting moiety. The agent had lower retention in the main organs and tissues of the mouse after two days of injection and was excreted through the kidneys. Zhou *et al.* in 2013 reported a novel breast cancer imaging agent by conjugating Gd-DOTA with a penta-peptide CREKA (Cys–Arg–Glu–Lys–Ala) that targets the fibrin-fibronectin complexes.[46] The agent demonstrated r_1 relaxivity of 8.06 $mM^{-1}s^{-1}$ at 3 T. *In vivo* analysis of the agent in mice bearing orthotopic breast tumors showed that the targeted agent had higher contrast enhancement in the tumors as compared to control agent. The agent also showed the capability to detect micro metastasis (a size less than 0.5 mm).[47]

8.2.1.2 *Extravascular targeted small molecular probes*

A prominent characteristic of malignant tumor microenvironment is low extracellular pH (6.5–6.9), caused due to the excess accumulation of lactic acid.[48,49] Garcia-Martin *et al.*[50] reported the *in vivo* application of a novel pH sensitive contrast agent $GdDOTA\text{-}4AmP^{5-}$ and a pH-independent analog, $GdDOTP^{5-}$ in rat gliomas. The agent $GdDOTA\text{-}4Amp^{5-}$ has been previously demonstrated *in vivo* by Raghunand *et al.* in 2002 and has r_1 relaxivity that is pH dependent.[51] The r_1 relaxivity ranges from 6.8 $mM^{-1}s^{-1}$

at pH 5.75 to 3.6 $mM^{-1}s^{-1}$ at pH 8 (4.7 T) as per the curve-fit to the experimental data.[50] Dual contrast agent imaging was used to measure pH in well perfused rat C6 glioma bearing female Wistar rats by injecting pH sensitive agent (GdDOTA-4AmP^{5-}) and pH insensitive agent (GdDOTP^{5-}) at an infusion rate of 0.24 $mmol \cdot kg^{-1} \cdot min^{-1}$ over a total injection time of 25 seconds. The dual contrast agent technique assumes the same pharmacokinetics for both agents *in vivo*, i.e. both pH independent and dependent contrast agents were assumed to have the same concentration in each voxel at a given time post injection. Pre and post T_1 weighted images were used to determine the concentration of GdDOTP^{5-}, based on its relaxivity *in vitro*, and the same concentrations were assumed for GdDOTA-4AmP^{5-} for calculation of pH. The time to maximal intensity (TMI) in each pixel was compared with the calculated pH values for the entire tumor. The mean pH of the tumors was calculated to be 6.87 ± 0.01. The TMI and pH analysis showed that the regions with TMI of 12.8 minutes or lower had a pH of 6.9, regions with TMI of 12.8–27.65 minutes had a pH of 6.84 and regions with TMI of 27.65 min or higher had a pH of 6.7, implying that regions with poor perfusion had lower pH.

In 2012, Liu *et al.* developed a pH imaging PARACEST agent Yb-DO3A-oAA containing an aryl amine and an amide to measure the pH at a wider range.[52] The r_1 relaxivity of the agent is similar to amide-containing Yb-DOTA i.e. 0.0106 $mm^{-1}s^{-1}$ at 7 T. *In Vivo* analysis of the agent in MCF-7 subcutaneous tumor bearing nude mice was conducted at 9.4 T. Significant change in contrast was seen post injection of the agent and the pH of the tumor was measured to be at or below 6.5. The agent was able to report the pH heterogeneity within the tumor. Nwe *et al.* in 2013 reported a Gd-based MR contrast agent (GC-NH_2-GdDOTA) bearing a glycol chitosan polymer containing pH titratable primary amines, with pH responsive retention in acidic tumor microenvironment.[53] The relaxivity of the agent at 1.41 T was determined to be 9.1 $mm^{-1}s^{-1}$. The gycol chitosan promotes electrostatic interaction with the cell-membranes only at acidic pH while staying inert otherwise. T_1 weighted images of T6–17 flank tumor models administered with 0.03 mmol Gd/kg of the agent showed a significant enhancement in the contrast 2 hours post injection while no enhancement was seen in tumors injected with the control agent.

The tumor microenvironment is often characterized by high oxidative stress due to high amounts of reactive oxygen species generated compared to normal tissue and a direct measurement of the redox status of the tumor can be important in determining tumor invasiveness and predicting resistance to therapies.[54,55] Reducing thiols such as glutathione (GSH) help maintain the redox status of the cell surface. In 2007, Carrera *et al.* developed a redox imaging probe, Gd-DO3AS-Act, containing a 2-pyridyl-dithio group that can form disulfide bonds with the free thiols.[56,57] Gd-DO3AS-Act reacts with free thiols from GSH resulting in a decrease in the relaxivity of the contrast agent. The relaxivity of Gd-DO3AS-Act is 8.1 $mM^{-1}s^{-1}$ (20 MHz and pH 7.4) while for the reduced agent it is 4.1 $mM^{-1}s^{-1}$. In 2011, Menchise *et al.* reported the *in vivo* capability of Gd-DO3AS-Act in reporting extracellular redox status in tumors[58] (referred to as Gd-DO3A-PDP in this article). Gd-DO3A-PDP bears a 2-pyridinedithio group that forms disulfide bridges with the exofacial protein thiols that are over expressed in certain cancers. B16-F10 tumor bearing C57Bl/6 mice were given intratumor injection of the agent and T_1 weighted images were acquired before and after injection. Image analysis showed that Gd-DO3A-PDP was retained in the tumor and showed significant signal enhancement even till 48 hr post injection. This novel agent can label redox cells expressing high levels of thiols and enable spatial mapping of redox status.

Raghunand *et al.* developed a series of thiol-bearing MRI probes and reported their plasma albumin binding capabilities *in vitro* and *in vivo* in mice.[59,60] Guntle *et al.* reported the correlation between the saturating dose of one of these probes, Gd-LC6-SH, and concentration of thiols in tumors.[61] The relaxivity of the free agent was 3.2 $mM^{-1}s^{-1}$ while for the albumin bound agent was 4.7 $mM^{-1}s^{-1}$ at 7 T and 37 °C. Mia-PaCa-2 pancreatic carcinoma cells and NCI-N87 gastric carcinoma xenografts were implanted on the flanks of female SCID mice and treated with glutathione synthesis inhibitor buthionine sulfoximine or the thiol-oxidizing anticancer drug Imexon, respectively (with saline as control). Analysis of time-course T_1-weighted showed that there was a significant enhancement 60 minutes post injection in the R_1 of tumors treated with the drug as compared to saline treated tumors (Mia-PaCa-2 $p < 0.034$ and NCI-N87 $p < 0.001$). An *ex vivo* analysis of thiol levels in both the tumor types after the treatment showed that the drugs had significant oxidizing effect in both Mia-PaCa-2 ($p < 0.05$) and NCI-N87 ($p < 0.01$) tumors as compared to controls. A more oxidizing environment would lead to a larger ΔR_1, thus the results obtained in this study indicate that this agent can be used to examine the redox status of the tumors and changes with treatment. The range of saturating dose for Gd-LC6-SH was found to be between 0.016 – 0.05 mmol Gd/kg.

Rapidly proliferating tumor cells require high levels of oxygen, which drives the cancer cells to promote angiogenesis. The abnormal signaling pathways in the tumor leads to formation of underdeveloped blood vessels which are chaotic in nature, thus leading to regions of hypoxia ($pO_2 < 10$–15 mmHg). Hypoxia has a significant influence on the progression of cancer and is considered as a marker for poor prognosis and its importance in tumor development and survival has presented itself as a target for development of novel hypoxia targeted therapeutics or novel hypoxia-regulated gene therapy strategies.[4,62–68] Rojas-Quijano *et al.* reported a novel hypoxia-targeted contrast agent GdDO3NI.[69] The agent was prepared by conjugation of DO3A-monamide with 2-nitroimidazole moiety, which has been used to sensitize the hypoxic tumors to radiotherapies. GdDO3NI gets accumulated in the hypoxic regions due to nitroreductase-mediated reduction[70,71] and subsequent binding. The agent has r_1 relaxivity of 5.21 ± 0.04 $mm^{-1}s^{-1}$ at 4.7 T. *In Vivo* and *in vitro* analysis confirmed the agent's ability to bind to hypoxic regions.[72] Dunning prostate R3327-AT1 rat tumor bearing animals were injected with GdDO3NI and non-targeted GdDO3ABA and T_1 weighted images were acquired. Clear distinction could be seen in the images 145 minutes post-injection between the control and GdDO3NI (Figure 8.3). This agent shows the capability for studying 3D distribution of the tumor oxygenation that could improve the microenvironmental analysis of the tumor. Iwaki *et al.*[73] developed a hypoxia targeting contrast agent based on the property of nitrobenzenesulfonamide moiety to be reduced into an amino group under low oxygen environment The agent shows a pH dependent r_1 relaxivity between pH 4 to 9. *In Vitro* analysis of the agent ($4NO_22MeOSAGd$) showed a 1.8 fold increase in the r_1 relaxivity under hypoxic conditions when tested in rat liver microsomes. Table 8.1 summarizes all the small molecular Ln based agents discussed above.

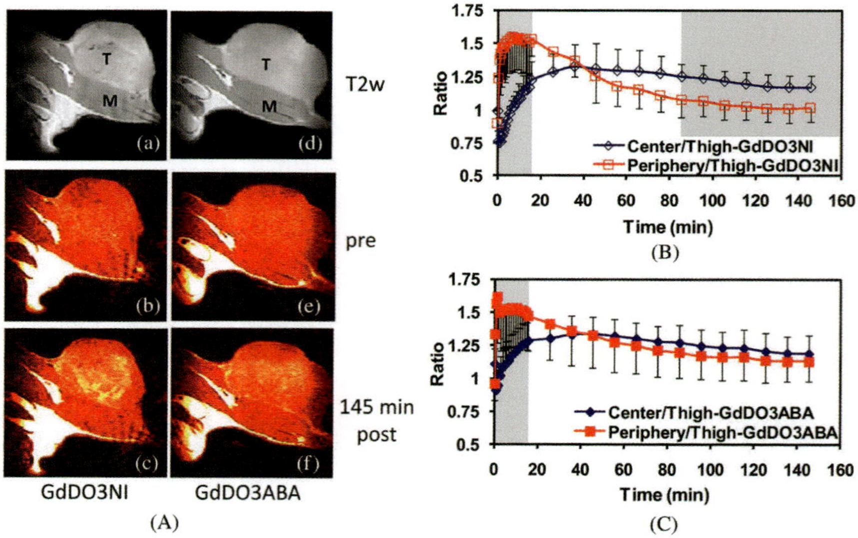

Figure 8.3 (A) Grayscale magnetic resonance T_2-weighted (a, d) and color T_1-weighted (b, c, e, f) images of R3327-AT1 tumors following injection of 0.1 mmol GdDO3NI (a–c) or 0.1 mmol GdDO3ABA (d–f) per kilogram body weight before injection (b, e) and 145 min after injection (c, f). The tumor and thigh muscle are labeled T and M, respectively. Comparison of the time course of normalized mean intensities (n = 5) for well-perfused (periphery) and poorly perfused (center) regions of tumors injected with GdDO3NI (B) and GdDO3ABA (C). The y-axis refers to the ratio of signal intensity compared to muscle. Reproduced with permission from Ref. 72.

8.2.2 *Macromolecular targeted contrast agents*

Small molecular contrast agents showed tremendous improvements over the last few decades, but they have their limitations. The small size and hydrophilic nature of various Gd^{3+} chelates cause their accumulation in the extracellular interstitial and vascular compartment. However, they suffer from rapid renal clearance rate and lower relaxivities due to limited Gd^{3+} payload. Both macromolecular and nanoscale contrast agents provide the capability of surface modification. Several target peptides, proteins, antibodies, etc., can be tethered on the surface of these agents to make them target specific. These surface modifications would also allow for tunable biodistribution with different degrees of biocompatibility and development of multimodality contrast agents. As mentioned in the Section 8.2, the

Table 8.1 Applications of small molecular Ln based agents as actively targeted MRI contrast agents.

Contrast Agent	Target	Targeting moiety	Agent class	Application	Per Ln relaxivity ($mM^{-1}s^{-1}$)
Gd-DOTA-RGD (Ref. 40)	$\alpha_v\beta_3$-integrin	c(RGDYK)	Peptide	Angiogenesis	$r_1 = 7.4 \pm 0.20$; $r_2 = 4.0 \pm 0.24$ (1.5 T)
CLT1-(Gd-DTPA) (Ref. 42)	Fibronectin-fibrin	CGLIIQKNEC (CLT1)	Peptide	Angiogenesis	$r_1 = 4.22$; $r_2 = 4.45$ (3 T)
CLT1-dL-(Gd-DOTA)$_4$ (Ref. 45)	Fibronectin-fibrin	CLT1	Peptide	Angiogenesis	$r_1 = 46 \pm 5$; $r_2 = 10.1 \pm 1$ (1.5 T)
CREKA-Tris(Gd-DOTA)3 (Ref. 46)	Fibronectin-fibrin	Cys–Arg–Glu–Lys–Ala (CREKA)	Peptide	Angiogenesis	$r_1 = 7.34$; $r_2 = 8.36$ (1.5 T)
GdDOTA-4AmP^{5-} (Ref. 50)	Changes in pH	Phosphonate	Functional group	pH	r_1= 6.8 at pH 5.75 and 3.6 at pH 8 (4.7 T)
Yb-DO3A-oAA (Ref. 52)	Changes in pH	Amide	Functional group	pH	$r_1 = 0.0106$ (7 T)*
GC-NH$_2$GdDOTA (Ref. 53)	Changes in pH	Glycol chitosan (GC)	Polymer	pH	r_1= 9.1 (1.41 T)
Gd-DO3AS-Act / Gd-DO3A-PDP (Refs. 56 and 58)	Thiols	2-Pyridyl-dithio	Functional group	Redox	$r_1 = 8.1$ at pH 7.4 and $r_1 = 4.1$ on reduction (0.5 T)
Gd-LC6-SH (Ref. 61)	Albumin	Thiols	Functional group	Redox	r_1 free= 3.23, r_1 bound = 4.74 (7 T)
GdDO3NI (Ref. 72)	Proteins	2-Nitroimidazole	Functional group	Hypoxia	$r_1 = 5.21 \pm 0.04$ (4.7 T)
4NO$_2$2MeOSAGd (Ref. 73)	Hypoxia	Nitrobenzene-sulfonamide	Functional group	Hypoxia	$r_1 = 3.6$; $r_2 = 4.1$ (0.47 T)

*Unlike other agents listed here, Yb-DO3A-oAA is a PARACEST agent.

slower tumbling time of the agent increases the relaxivity of these agents. Molecular weight of these agents is higher than 30 kDa, which facilitates higher retention time in the blood as their size prevents them from diffusing through the vessel walls. Furthermore, the macromolecular moiety allows for more number of chelates and metal ions to be attached per molecule thus improving the functionality per molecule.[38] Table 8.2 summarizes all the macromolecular Ln based agents discussed below.

8.2.2.1 *Vascular targeting macromolecular probes*

The first macromolecular agent albumin-(GD-DPTA) was developed in 1987 by Ogan and colleagues and was specific to albumin which is the most abundant protein in blood.[74] It was prepared by covalent attachment of albumin to GD-DPTA chelate. The average molecular weight of the agent is 92 kDa with r_1 relaxivity of 14.9 $mM^{-1}s^{-1}$ per Gd^{3+} at 0.25 T. The agent was successful in characterizing the vascularity of various tumors models and changes in response to anti-angiogenic therapy.[75–79] However, it suffered from slow clearance rate and immunogenicity.

In 2013, Ye *et al.* reported an extracellular degradable blood pool agent (GODC) based on polydisulphide Gd^{3+} complexes with improved vascular imaging, higher clearance rate and reduced tissue accumulation.[80] The agent does not have a target specific moiety but acts as a blood pool agent for cancer imaging. The GODC had an apparent molecular weight of 26.4 kDa and r_1 relaxivity of 8.25 mM^{-1} s^{-1} per Gd^{3+} at 1.5 T. *In vivo* contrast enhancement was evaluated in mice with breast cancer tumor models. GODC (0.1 mmol Gd/kg^{-1}) was injected intravenously and images were acquired for up to 30 minutes with Magnevist as control (Figure 8.4). Significant and prolonged blood pool enhancement was observed as compared to Magnevist. The agent shows faster clearance rate as compared to other macromolecular blood pool agents and has a potential to be translated clinically.

8.2.2.2 *Extravascular targeted macromolecular probes*

Many biocompatible proteins, native or engineered, are used as conjugates for MRI probes which minimizes the contribution towards toxicity of the

Table 8.2 Applications of macro-molecular Ln based agents as actively targeted MRI contrast agents.

Contrast Agent	**Target**	**Targeting agent**	**Agent class**	**Application**	**Per Ln relaxivity ($mM^{-1}s^{-1}$)**
Albumin-Gd-DTPA (Ref. 74)	Blood	Albumin	Protein	Blood pool agent	r_1 = 14.9 (0.25 T)
GODC (Ref. 80)	Blood	None	Polymer	Blood pool agent	r_1 = 8.25 (1.5 T)
avidin-GdDTPA (Ref. 81)	Her-2/neu receptor	Avidin-biotin	Protein	Tumor progression	r_1 = 128 (4.7 T)
Gd.ProCA1.GRP (Ref. 89)	Gastrin release peptide receptors	Gastrin releasing peptide	Peptide	GR targeted therapies	r_1 ~35; r_2 ~ 50 (3 T)
Gd^{3+}+ ProCA1-affi-m (Ref. 83)	Her-2/neu receptor	Her2/neu	affibody	Tumor progression	r_1 = 21; r_2 = 30 (1.41 T)

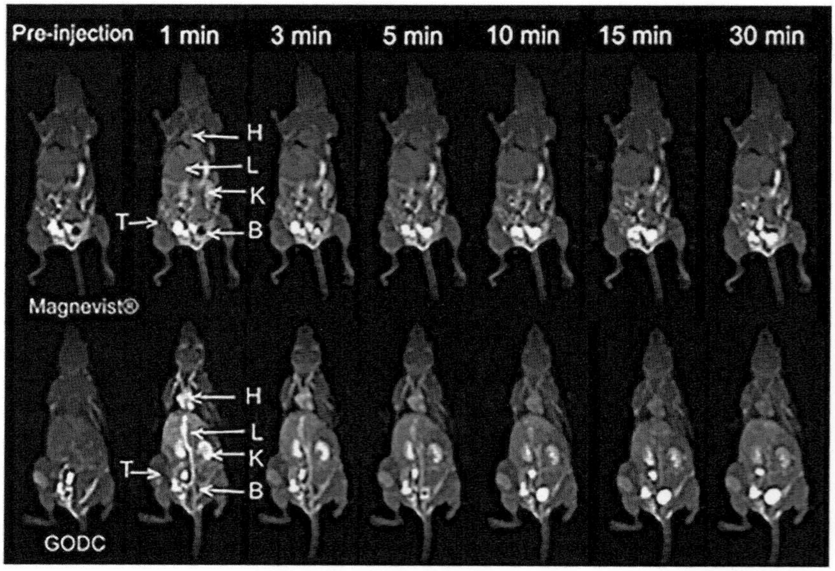

Figure 8.4 Three-dimensional maximum intensity projection images (coronal view) of mice bearing 4 T1 breast tumor xenografts at 1.5 T, before and at 1, 3, 5, 10, 15 and 30 minutes post-injection of Magnevist and GODC at a dose of 0.1 mmol Gd/kg. The major organs are indicated as H (heart), L (liver), K (kidney), B (bladder) and T (tumor). Significant and prolonged blood pool contrast enhancement was observed with GODC in comparison with Magnevist. Reproduced with permission from Ref. 80.

Gd-based probe. Their size also allows for prolonged circulation time and selective uptake of the agents. Several proteins have been studied over the past years for their expressions in tumors and a library has been generated which can be scavenged to develop novel agents with improved properties. Tumor targeted agents are prepared by attaching tumor specific ligands as antibodies to the surface of these agents.[81–83] Multiple antibodies can be attached to a single molecule by improving the specificity of these agents towards tumor. These agents can also act as dual agents by loading therapeutic agents inside the core thus acting as both imaging and therapeutic agents.

HER-2/neu receptor is member of the epidermal growth factor family which is overexpressed in approximately 15–20% of breast cancers.[84,85] HER-2/neu promotes cellular growth, proliferation and is a marker for poor prognosis. Several HER-2/neu targeted therapies that inhibit the

HER2 dependent signaling, such as trastuzumab, pertuzumab, neratinib, etc., have shown better clinical outcomes as compared to non-targeted therapies in clinical settings. In 2003, Artemov *et al.* developed a contrast agent that targeted the HER-2/neu receptor[81] by conjugation of purified avidin to DTPA and labeled with Gd^{3+} ions (avidin-GdDTPA). The r_1 relaxivity of the avidin-GdDTPA was assumed to be same as that of albumin-GdDTPA, (128 mM^{-1} s^{-1} per protein concentration at 4.7 T[86]). For MR imaging of HER-2/neu receptors, the extracellular domain of HER-2/neu receptor was pre-labeled by a biotinylated monoclonal antibody and targeted using avidin-GdDTPA. NT-5 (overexpressing HER-2/*neu* receptors) and EMT-6 (control/does not express HER-2/*neu*) murine mammary carcinomas were implanted in SCID mice and on the day of imaging were injected with biotinylated anti-HER-2/neu monoclonal antibody 12 hours before intravenous injection of 0.2 ml of 60 mg/ml avidin-GdDTPA conjugate. T_1 weighted images were acquired before and 1, 8, 24 and 48 hours after injection. The contrast enhancement between the tumor and muscle was retained till 24 hours after the injection in NT-5 tumors while the contrast reached baseline values in EMT-6 tumors. The two-component approach allows for use of two small molecular size agents rather than a single large molecule making the delivery more efficient to the tumor interstitium. This agent is limited by the immune response that can be induced by the avidin-biotin systems and stability of the agent.

Gastrin releasing peptide receptors are overexpressed in prostate cancers, small cell lung cancer and pancreatic cancer cells and are used as the target for this contrast agent design.[87,88] A novel biomarker gastrin-releasing peptide receptor was used by Wei *et al.* to develop a protein based MRI contrast agent Gd.ProCA1.GRP(52).[89] The agent was prepared by grafting a fragment of gastrin releasing peptide between Ser52 and Gly53 residues of the Gd-binding protein contrast agent ProCA1.[90] Gd.ProCA1.GRP(52) exhibited r_1 and r_2 relaxivity of ~35 $mM^{-1}s^{-1}$ and ~50 $mM^{-1}s^{-1}$, respectively, at 3 T. *In vivo* MR imaging was conducted on immunodeficient mice injected with PC3 and H441 cells. Gd-ProCA1.GRP was directly injected into the tumors and T_1- and T_2-weighted image analysis showed that H441 tumors reached baseline contrast levels within 24 hours post injection while PC3 maintained a strong contrast enhancement until 48 hours post injection. This agent displays a significant contrast enhancement in GRPR overexpressing tumors and can be utilized to assess the level of GRP in tumors noninvasively which is carried out only by taking biopsies. Thus this agent has a great potential in assessing the levels of GR in tumors and helps predict GR targeted therapies.

In 2011, Qiao *et al.* developed a novel HER2 targeting contrast agent, Gd-ProCA1-affi-m, using the above mentioned ProCA1 as the metal binding moiety.[83] The agent expressed higher level of specificity for HER2 and could report different levels of HER2 expression. It was developed by introduction of a high affinity HER2 affibody $Z_{HER2:342}$ into the C terminal of the designed Gd^{3+} binding protein by a flexible linker, followed by PEGylation. *In vivo* MR imaging was carried out on nude mice injected with MDA-MB-231 breast cancer and SKOV3 ovarian cancer cell line, under and overexpressing HER2, respectively (Figure 8.5). 3 mM Gd-ProCA1-affi-m was injected intravenously and pre and post contrast images were acquired using T_1 and T_2 weighted sequences. Contrast enhancement was observed in SKOV3 tumors after 3 hours with no

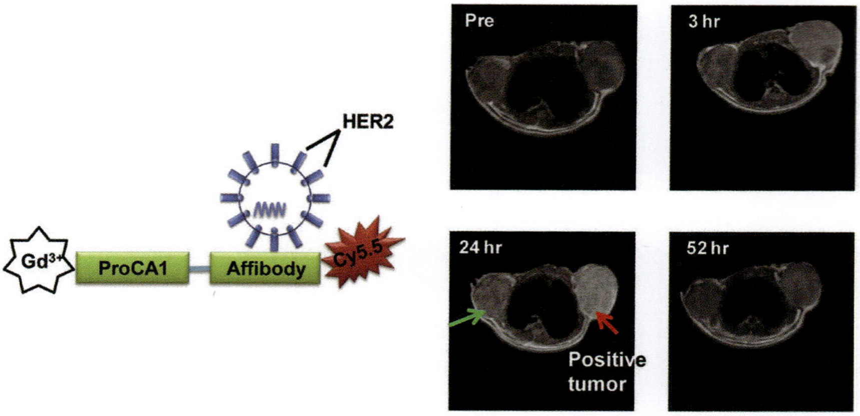

(a) (b)

Figure 8.5 (a) The model structure of multimodal HER2 targeted MR imaging probe created by connecting a high affinity HER2 affibody $Z_{HER2:342}$ at the C-terminal of a de novo designed protein contrast agent ProCA1.CD2 with a designed Gd^{3+} binding site. (b) *In vivo* cancer imaging of contrast agent uptake, retention and distribution in multiple different normal host organs and in both SKOV-3 ovarian cancer (HER2 positive, right) and MDA-MB-231 breast cancer (HER2 negative, left) xenografts. Reproduced with permission from Ref. 83.

significant change in the contrast of MDA-MB-231 tumors. Similar results were seen *via* fluorescence imaging. The agent has reduced toxicity and higher dosages can be used in comparison to commercially available contrast agents. ProCA1-affi-m exhibited r_1 and r_2 relaxivities of 21 $mM^{-1}s^{-1}$ and 30 $mM^{-1}s^{-1}$ at 1.41 T, respectively.

8.2.3 *Nano-platform based contrast agents*

Nano-platform based contrast agents primarily consist of dendrimers, micelles, liposomes and other solid core nanoscale carrier systems. The hyperpermeability of tumors vasculature allows for site specific extravasation and accumulation of nanoparticle agents by the enhanced permeability and retention (EPR) effect[7] which can be further improved *via* targeting the tumor microenvironment. Nano-platforms such as liposomes allow for large loading of the Ln-chelates as they have large surface to volume ratio. Increased loading improves the relaxivity of the agent and also maintains lower toxicity due to encapsulation. Several properties of nanoparticle size agents can be modulated such as size, surface charge, structure and mechanical properties. This property makes them suitable for developing contrast agents with optimum biodegradability, pharmacokinetics, biodistribution and target specificity. This section is sub-divided into the specific platforms that have been used for the formation of Ln-based contrast agents. Table 8.3 summarizes the nano-platform Ln based agents discussed below.

Dendrimer based

Dendrimers are self-assembling nanoscale compounds with size ranging from 1 to 100 nm. They can be synthesized layer-by-layer and can be designed to have various core sizes and different number of layers. The general nomenclature is based on the number of layers, also called as generations (e.g., G1, G2, etc.).[91] The size and properties of the dendrimers can be fine-tuned to meet any purpose, making them very versatile. The surface consists of various functional groups, which can be exploited to attach ligands, antibodies, peptides, etc., to make it target specific. The core is generally hollow and drugs/Gd^{3+} agents can be encapsulated

Table 8.3 Applications of nano-platform Ln based agents as actively targeted MRI contrast agents.

Contrast agent	Target	Targeting agent	Agent class	Application	Per Ln relaxivity ($mM^{-1}s^{-1}$)
Antibody conjugated paramagnetic liposomes (Ref. 104)	$\alpha_v\beta_3$ integrin	Biotinylated LM609	Antibody (liposome)	Angiogenesis	Not reported
T-Gd-LP (Ref. 106)	$\alpha_v\beta_3$ integrin	RGD (arginine-glycine-aspartic acid)	Peptide (liposome)	Angiogenesis	Not reported
CLT1-G2-(Gd-DTPA) (Ref. 107)	fibronectin-fibrin	Cyclic decapeptide CGLIIQKNEC (CLT1)	Peptide (dendrimer)	Angiogenesis	r_1 = 7.92; r_2 = 10.34 (3 T)
TPTN (Ref. 116)	VEGF receptors	Vacsular endothelial growth factor	Antibody (micelle)	Tumor progression, angiogenesis and therapy	r_1 = 17.3 (3.0 T)
G5-FA-(Gd-DOTA) (Ref. 121)	Folic-acid receptors	Folic acid	Ligand (dendrimer)	Cell proliferation	r_1 = 26.0 (2 T)
Folate-receptor targeted DNCs (Ref 122)	Folic-acid receptors	Folic acid	Ligand (dendrimer)	Cell proliferation	r_1 = 12.3 (1.41 T)
F(5000)-NP (Ref. 123)	Folate receptors	Folate-PEG-DSPE	Ligand (Micelle)	Cell proliferation	r_1 =3.7 (9.4 T)
Gd.DOTA.DSA (Ref. 125)	α–FR isoform	Folate-PEG-DSPE	Ligand (liposome)	Cell proliferation	r_1 = 1.30; r_2 = 5.38 (4.7 T)
Eu-G5 (Refs. 126 and 127)	None	None	Dendrimer	Tumor imaging	r_1 = 0.107 (9.4 T)
PEG-p(L-LA)-DTPA-Gd:PEG-p(L-His) (Ref. 128)	Negatively charged cell membrane	Imidazole	Ligand (micelle)	Acidic environment	r_1 = 8.56; r_2 = 26.61 at pH 7.5 (4.7 T)
Met-MSNs-Gd^{3+} (Ref 132)	Methionine transport system	Methionine	Amino-acids (micelle)	Tumor growth	r_1 = 11.65 r_2 = 22.33 (3 T)

making it multifunctional. Since their initial development, significant progress has been made in their functionalization and development into carrier agents.[92] For biological functions the agents need to be nontoxic, permeable and have longer retention times. These properties can be modified *via* various techniques. The dendrimer based contrast agents show higher relaxivity owing to their rigid structure and higher loading capacity.[93,94] The first dendrimer based MRI contrast agent studied *in vivo* was reported by Wiener *et al.* in 1994, which was the pioneering work in the field of dendrimer-based contrast agents.[93] Polyamidoamine (PAMAM) dendrimers were conjugated to DTPA ligands for Gd^{3+} binding. The agents had improved relaxivity and showed enhanced contrast in angiography. PAMAM dendrimers are biocompatible macro-mamolecules and contain multiple amino groups on their surface that can be utilized for conjugation of various molecules. Significant advancements have been made since and various other dendrimers such as poly (propylene imine) (PPI) and polylysine (PLL) have been used for synthesis of contrast agents.[95–98]

Micelle based

Micelles are another platform for delivery of imaging probes and therapeutics. Micelles are supramolecular nanoparticles of 10–100 nm diameters resulting from self-assembly of amphiphilic polymers with typically a hydrophilic shell and hydrophobic core.[99] Amphiphilic macromolecules have the tendency to self-assemble to form nano-sized colloidal micelles in water at a concentration greater than the critical micelle concentration (CMC) making the formulation of micelles straightforward. These self-assemblies are oriented in such a way that the hydrophilic part of the amphiphile is exposed to water while the hydrophobic part forms the core. The enhanced interaction among the hydrophobic polymer chains in the micelle core may contribute to increased stability and allows for incorporation of hydrophobic drugs or imaging probes. Surface functionalization of micelle for cancer targeting requires the modification of individual co-polymer molecules with cancer biomarkers such as small molecules, monoclonal antibodies, aptamers or peptides. Density of targeting moiety can be controlled *via* formulation using the appropriate ratio of targeted and untargeted co-polymer molecules. Biocompatibility and long circulation time is often ensured by the choice of polyethelene glycol based hydrophilic component in the amphiphile.[100]

Liposome based

Liposomes are highly biocompatible spherical vesicles with phospholipid bilayer consisting of at least a single continuous bilayer that separates the inner core from the outer medium. The inner core is typically aqueous, which allows the loading of wide range of conventional imaging agents such as Gd^{3+}-chelates or iron oxide nanoparticles. The ease of surface modification facilitates conjugation of chelates targeted proteins, peptides, antibodies or other molecules.[38] As in case of micelles, the size, charge, mechanical and physical properties of liposomes can be altered to develop improved drug delivery and contrast agents. The liposomes used for encapsulation of Gd^{3+}-chelates are mostly within the range of 50–250 nm in size.[101–103]

8.2.3.1 *Vascular targeting nano-platform probes*

In 1998, Sipkins *et al.* developed paramagnetic liposomes conjugated with $\alpha_v\beta_3$ integrin specific monoclonal antibody.[104] Biotinylated LM609 antibodies were used as the targeting ligands for $\alpha_v\beta_3$ integrin to form $\alpha_v\beta_3$-targeted antibody-conjugated paramagnetic liposomes (ACPLs). Avidin and biotinylated antibodies were added to the paramagnetic polymerized liposome (PPLs)[105] to form the final ACPLs. ACPLs had a mean size of 300–350 nm containing 30% of Gd^{3+} chelate-labeled lipid. Rabbits with sub-cutaneous squamous cell carcinoma tumors were injected intravenously with targeted or control ACPLs. Signal enhancement was seen with targeted ACPLs ($p < 0.02$) at 24 hours post injection compared to non-targeted ACPLs. Also the $\alpha_v\beta_3$-positive vessel density was correlated with the degree of the MR signal enhancement. A receptor negative control experiment to test the specificity of targeted ACPLs was conducted in mice bearing subcutaneous, $\alpha_v\beta_3$ negative, squamous cell carcinomas. In these, no signal enhancement was seen in the MR images. Thus LM609 conjugated APCLs were seen to improve the visualization of the tumors and could identify angiogenic hot spots in the tumor.

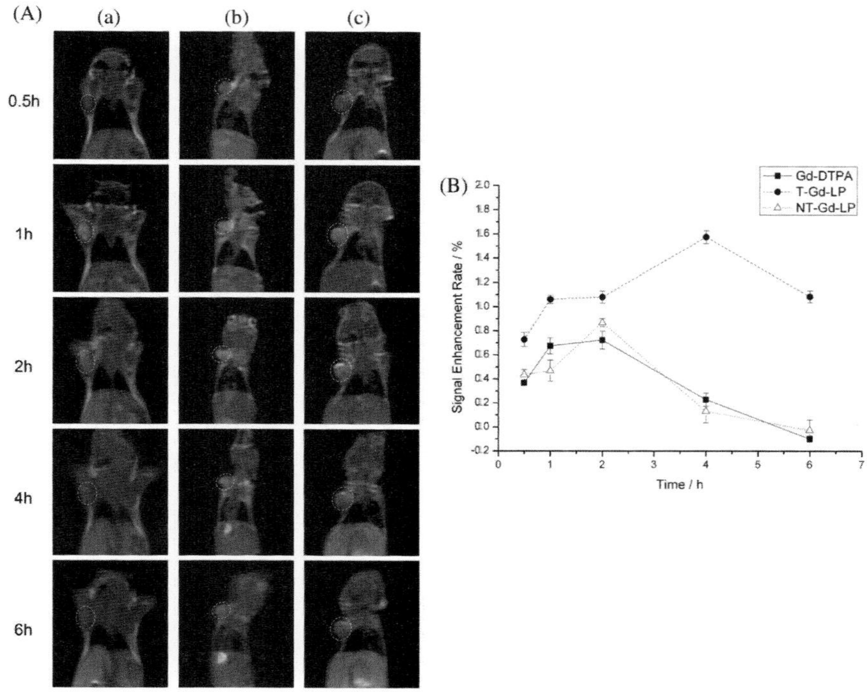

Figure 8.6 (A) MRI images of A549 tumor-bearing mice injected with (a) Pure Gd-DTPA, (b) RGD-targeted paramagnetic liposomes and (c) non-targeted paramagnetic liposomes via tail vein. (B) Analysis of MRI signal enhancement rate in A549 tumor-bearing mice shows that targeted liposomes were accumulated in higher amounts in the tumor. Reproduced with permission from Ref. 106.

An angiogenesis targeted contrast agent (T-Gd-LP) was synthesized in the liposomal form by first incorporation of RGD-coupled lipopeptide into lipid bilayers forming liposomes.[106] Gd-DTPA was then encapsulated in the aqueous interior phase of liposomes (99.1 ± 9.2 nm). A549 tumor bearing female BALB/C nude mice were injected with T-Gd-LP and imaged at 1.5 T (Figure 8.6). The agent showed tumor targeting by the agent as it depicted better binding than the non-targeted liposomes. T-Gd-LP also displayed more accumulation in the tumor than pure GD-DTPA and non-targeted liposomes.

In 2010, Tan *et al.* developed a fibrin-fibronectin targeted dendrimer based contrast agent, CLT1-G2-(Gd-DOTA),[107] by incorporating the previously developed CLT1-(Gd-DTPA)[42] using silsesquioxane-based dendrimers[108] with a high functionality for conjugation. This probe had r_1 of 7.92 $mM^{-1}s^{-1}$ and r_2 of 10.34 $mM^{-1}s^{-1}$ measured at 25 °C and 3 T. The *in vivo* investigation was performed in MDA-MB-231 xenograft bearing female athymic nude mice. T_1 weighted images were acquired before and 2, 5, 10, 15, 20 and 25 minutes after injection of 0.03 mmol Gd/kg of targeted and non-targeted contrast agents. Significant contrast enhancement was seen from both targeted and non-targeted agents 10 minutes post injection but only targeted agent resulted in maintaining the enhancement at 60 minutes as compared to the non-targeted agent. The enhancement seen from non-targeted agent was confined mostly to the periphery of the tumor, while targeted agent resulted in a more uniform enhancement over the entire region of the tumor. The agent was eliminated through the renal route and accumulated in the bladder. In 2012, Tan *et al.* investigated the efficacy of CLT1-G2-(Gd-DOTA) in orthotopic prostate cancer models.[109] This *in vivo* study demonstrated that the agent binds specifically to the prostate tumors and not to the normal tissues.

Paramagnetic CEST agents encapsulated in liposomes are called lipoCEST agents.[110] LipoCEST agents are highly sensitive, they can be surface modified with specific biomarkers and are biocompatible due to the presence of lipid bilayers. Flament *et al.* developed a RGD targeting Tm(III)-complex containing lipoCEST agent.[111] The RGD tripeptide was conjugated to the surface of the liposomes. U87MG malignant gliomas were grown orthotopically in nude mice and imaged at 7 T. The pre and post-contrast CEST images showed that RGD-lipoCEST was retained in the tumor and show contrast difference for a longer time as compared to the control lipoCEST agent.

Angiogenesis can also be studied by targeting the vascular endothelial growth factor (VEGF). VEGF plays an important role in tumor angiogenesis, progression and metastasis. Therefore it is a potential target for developing targeted imaging agents. Vascular endothelial growth factor receptor (VEGFR) is overexpressed in several tumor cells and on tumor neovasculature and presents itself as a good target for diagnosis and therapy. Several anti-VEGF therapies have been developed, for instance anti-VEGF monoclonal antibody bevacizumab and VEGF receptor tyrosine kinase inhibitors sorafenib, pazopanib and sunitinib.[112–115] Liu *et al.*

developed a multifunctional polymeric nanoparticle system based MRI contrast agent that targets VEGFR.[116] The developed agent (target pH-sensitive theranostic nanoparticle, TPTN) is pH sensitive, tumor targeting and has sorafenib-loading and delivery capacity as well. As compared to earlier theranostic agents this agent has better release profile for both the drug and Gd^{3+} which makes it an efficient therapeutic and imaging agent. The agent was conjugated with VEGFR antibodies to make it tumor selective.[117] The agent has a relaxivity r_1 of 17.3 $mM^{-1}s^{-1}$ measured at 3.0 T. H22 (overexpresses VEGFR) tumor bearing mice were injected with TPTN, non-targeted pH-sensitive theranostic nanoparticle (PTN), and Magnevist at a dose of 40 μmol Gd^{3+} per kg, respectively. *In vivo* study demonstrated that both PTN and TPTN accumulate in the tumor and result in contrast enhancement as compared to Magnevist ($p < 0.05$) 60 minutes post injection. TPTN showed higher contrast enhancement ($p < 0.05$ till 90 minutes post injection) than PTN and greater antitumor efficacy than intravenous delivery of sorafenib.

8.2.3.2 *Extravascular targeting nano-platform probes*

As described, dendrimers have a very high surface functionality and this property has been utilized in making folate targeted dendrimeric contrast agent. Rapidly dividing cells require folic acid for cell proliferation and maintenance.[118–120] In 2008, Swanson *et al.* developed an agent where generation 5 (G5) PAMAM dendrimers were conjugated with folate receptors and GD-DOTA.[121] The agent, G5-FA-(Gd-DOTA), had a relaxivity r_1 of 26.0 $mM^{-1}s^{-1}$ per Gd(III) (2 T, 20°C). G5-FA-(Gd-DOTA) was injected at a dose of 0.029 mmol-Gd/kg in SCID mice bearing KB human epithelial tumors (overexpressing folic acid receptor) and the images acquired showed a significant enhancement in the tumor and also in kidney and liver as they express folic acid in high amount. In 2010, Cheng *et al.* developed a dendrimer nanocluster based contrast agent which was composed of Gd-labeled PAMAM dendrimers cross linked to form larger nano carriers functionalized with folate for targeted imaging.[122] The r_1 relaxivity of this agent at was found to be 12.3 mM^{-1} s^{-1} per Gd^{3+} at 1.41 T (60 MHz) and at 40°C. *In Vivo* analysis confirmed the target specificity of this agent for folic acid expressing KB tumor xenografts (Figure 8.7).

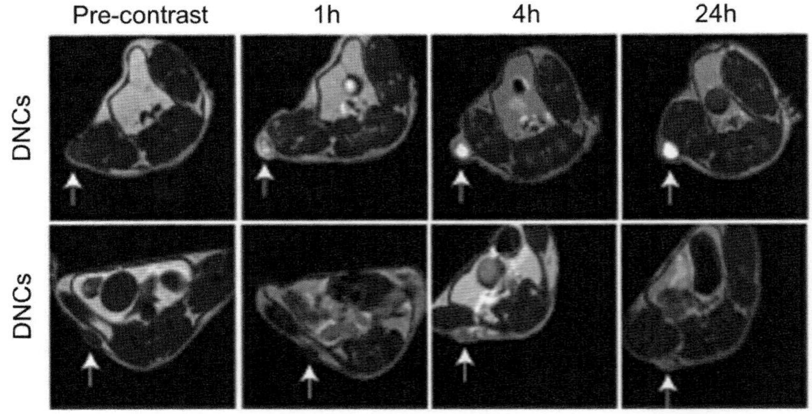

Figure 8.7 MR images of nude mice with subcutaneous tumor xenografts pre-injection and 1, 4 and 24 hours post injection with folate receptor-targeted dendrimer nano-clusters (DNCs). Top row: KB cell tumor xenografts before and after i.v. injection with folate receptor-targeted DNCs. Second row: T6–17 cell tumor xenografts (i.e. folate receptor-negative) before and after i.v. injection with folate receptor-targeted DNCs. Significant enhancement was only seen in folate receptor positive tumors. Reproduced with permission from Ref. 122.

Dendrimer based folate targeted MR contrast agents suffer from high clearance rate, thus in order to improve the retention time Nakamura *et al.* developed a novel MRI/fluorescence agent by the addition of cationic cholesterol derivatives to PEG-lipid micelles.[123] The nanoparticles were prepared by a modified version of protocol defined by Beilvert *et al.*[124] The contrast agent composed of Gd-DTPA-bis-stearylamide and PEG_{2000}-DSPE with the inclusion of folate-PEG_{5000}-DSPE copolymer to make it folate targeted and Cy7-labeled PEG2000-DSPE for fluorescence . The resultant F(5000)-NPs were 76.0 nm in size. KB tumors bearing female athymic BALB/c nu/nu mice maintained on folate-deficient rodent diet were imaged in a 9.4 T for *in vivo* analysis. T_1 weighted images were acquired before and 24 hours after intravenous injection of the agent and depicted increased signal intensity in tumor region by 24 hours. The accumulation of F(5000)-NPs was 1.2-fold higher than the NPs. The relaxivity of this agent was found to be similar to Gd-DTPA. Kamaly *et al.* synthesized folate targeted bimodal liposomes for tumor by incorporating

DSPE-PEG(2000)folate with Gd-DOTA-DSA to target α–FR isoform of the folate receptor overexpressed in ovarian cancer.[125] Targeted liposomes had r_1 and r_2 relaxivities of 1.30 $mM^{-1}s^{-1}$ and 5.38 $mM^{-1}s^{-1}$, respectively at 4.7 T. IGROV-1 tumor-bearing Balb/c nude mice were injected with 200 μl of the targeted and non-targeted agent and images were acquired 2, 6 and 24 hours post injection. A 4-fold increase in the T_1 signal intensity was observed in mice injected with targeted liposomes 2 hours post injection and maintained till 16 hours, while it took 24 hours for the non-targeted liposomes to show a similar effect.

Ali *et al.* conjugated the Eu-DOTA-Gly_4 PARACEST agent to generation 5 PAMAM dendrimers for detection of gliomas.[126] Dendrimers facilitated the accumulation of agents in tumors *via* the EPR effect. The agent, Eu-G5, was later conjugated with a fluorescent dye to make it dual-modality and validate the results using *ex vivo* fluorescence.[127] The r_1 relaxivity for Eu-G5 was found to be 0.107 $mM^{-1}s^{-1}$ at 9.4 T and does not affect the image contrast directly. *In vivo* studies were performed in a 3 T scanner on mice bearing orthotopic U87 gliomas. The analysis of pre and post injection CEST MRI images showed a CEST effect of $\geq$ 5.2% and it was persistent throughout the imaging protocol while no such change was observed in the non-tumor regions on the brain.

Kim *et al.* developed a nanosized pH responsive cancer recognizable MRI contrast agent (CR-CAs).[128] The agent is essentially a micelle consisting of a cancer recognizable block copolymer PEG-p(l-His) and PEG-p(l-LA)-DTPA-Gd (~40 nm). The agent stays as a stable micelle in pH 7.4 i.e. normal tissues and breaks down into positively charged water-soluble polymers in acidic pH of 6.5 i.e. tumor tissues. The breakdown of the micelle exposes the Gd^{3+} from the micelle core to the acidic tumor microenvironment, which results in an increased T_1-relaxivity. Once inside the tumor tissue the protonation of the pH responsive imidazole groups p(l-His) causes accumulation of the positively charged CR-CAs . The r_1 and r_2 of CR-CAs (PEG-p(L-LA)-DTPA-Gd:PEG-p(L-His)) at 4.7 T were estimated to be 8.56 $mM^{-1}s^{-1}$ and 26.61 $mM^{-1}s^{-1}$, respectively, at pH 7.5 and 12.01 $mM^{-1}s^{-1}$ and 19.47 $mM^{-1}s^{-1}$, respectively, at a pH of 6.5. CT26 tumors (~3 mm^3) in male BALB/c mice were injected intravenously with CR-CAs and pH insensitive contrast agent (Ins-CAs) and T_1 weighted images were acquired in a 1.5 T human clinical scanner (Figure 8.8).

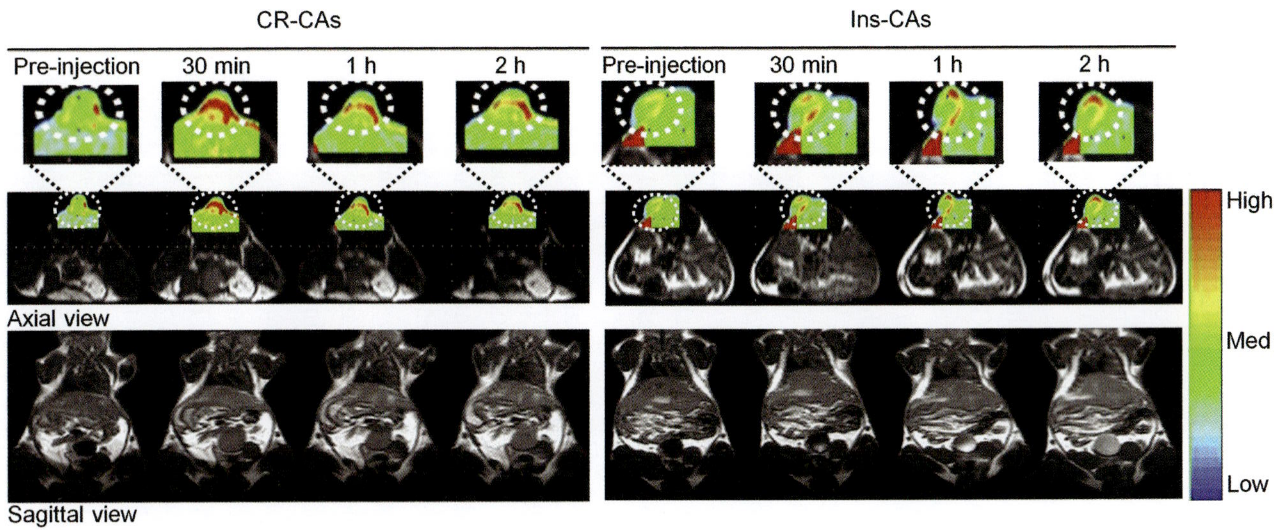

Figure 8.8 Temporal color-coded *in vivo* T1-weighted MR images of CT26 murine tumor bearing Balb/c mice after the intravenous injection of CR-CAs and Ins-CAs. Significant enhancement can be seen in tumors with CR-CAs after 30 minutes post injection as compared to tumors with Ins-CAs (control) Reproduced with permission from Ref. 128.

Tumors injected with CR-CAs showed an increase in contrast within few minutes and a 116% increase in the signal intensity was observed after 30 minutes post injection as opposed to Ins-CAs, which showed an enhancement < 50% throughout the experiment. Biodistribution analysis showed that the amount of CR-CAs was 1.5-fold higher than the Ins-CAs in tumors. The results show that this agent has good pH selectivity and can diagnose small tumors of ~3 mm^3 in few minutes.

Actively proliferating tumor cells have an increased uptake of amino acids for protein synthesis via amino acid transport systems as compared to the normal tissues.[129–131] Mehravi *et al.* used this property of tumor to develop an agent conjugated with methionine which is transported into the cells *via* methionine transport system.[132] Gd^{3+}-based mesoporous silica nanospheres (MSN) were conjugated with methionine (Met-MSNs-Gd^{3+}). The r_1 and r_2 values were measured after cell uptake by MRI (3 T), and were found to be 11.65 and 22.33 $mM^{-1}s^{-1}$ in cells, respectively. Murine mammary adenocarcinoma bearing BALB/c mice were injected with 5.0 μmol/kg body weight of the agent and T_1 weighted images were acquired 15 minutes later. A significant increase in the signal intensity was reported in the T_1 weighted images of tumors injected with the contrast agent but not quantified.

8.3 Iron based agents

Iron based contrast agents are predominantly nano-scale T_2 or T_{2*} shortening agents. They are typically iron oxide nanoparticles composed of magnetite or maghemite core and a dextran, citrate, or a polymer such as polyethylene glycol or polyvinyl alcohol coating. They can be classified[133–137] into: (1) superparamagnetic iron oxides (SPIO) (50–500 nm); (2) ultra-small superparamagnetic iron oxides (USPIO) (< 50 nm); (3) mono crystalline iron oxide (MION, < 10 nm) and (4) cross-linked iron oxide (CLIO < 50 nm) depending upon the size, coating and structure. MIONs are a sub-class of USPIOs with a monocrystalline iron oxide core and CLIOs are a sub-class of MIONs that have a shell crosslinked with a polysaccharide. The smaller size nanoparticle agents have prolonged blood half-life and have wider tissue distribution as compared to SPIOs. Accumulation of iron atoms results in a very high net magnetic moment that causes significant drop in the relaxation of the surrounding protons. A single particle contains thousands of Fe atoms which result in drastic drop in the T_2 and T_2^*, thus appearing dark/hypointense on a T_2 or T_2^* weighted image. Most of the iron oxide based agents are termed as blood pool agents as they tend to accumulate in the intravascular regions due to their high molecular weight. The high molecular weight also results in their uptake by macrophages and transfer to liver and spleen. Thus they find application in detection of macrophage infiltration, liver lesions and inflammation.

Iron oxide based agents are not as common in clinical settings as Gd^{3+} based agents due to their slower clearance rates. Few of the FDA approved SPIO contrast agents for clinical settings are sulfonated styrenedivinly-benzene copolymer coated Ferristene (OMP) (Abdoscan),[138] siloxane coated Ferumoxsil (AMI-121) (GastroMARK),[139] dextran coated Ferrixan (Ferucarbotran, SHU 555A) (Resovist)[140–142] and Ferumoxide (Feridex).[143,144] Resovist and Feridex are primarily used for liver MR imaging and are administered intravenously. The r_2 relaxivity of Resovist and Feridex is 151.0 and 98.3 $mM^{-1}sec^{-1}$, respectively. Abdoscan and GastroMARK are oral iron oxide based agents and are primarily used for imaging of diseases and tumors in gastrointestinal tract. The nanoparticles assembly allows for various surface modifications such as drug loading and surface conjugation with targeted molecules, making them dual functional agents. In this section, we describe various targeted iron-oxide nanoparticles for cancer imaging with *in vivo* applications. Table 8.4 summarizes the agents discussed below.

8.3.1 *Vascular targeting iron based probes*

We have previously discussed gadolinium based agents conjugated with cyclic pentapeptide c(RGDyK) for targeting angiogenesis. Melemenidis *et al.* developed an iron oxide "macroparticle" (2.8 μm MPIO) conjugated with multiple c(RGDyK) peptides for *in vivo* targeting of angiogenesis.[145] Blood T_2 relaxation rate for agent at 4.7 T and was found to be 74.6 s^{-1}. The pre and post contrast agent administration T_2* weighted images of female, C57Bl/6 mice with subcutaneous B16F10 melanoma and MC-38 colon carcinoma tumors showed significantly high hypointense signals in

Table 8.4 Applications of nanoscale iron based agents as actively targeted MRI contrast agents.

Contrast agent	Target	Targeting agent	Agent class	Application	Per Fe relaxivity ($mM^{-1}s^{-1}$)
c(RADyK)-MPIO (Ref. 145)	$\alpha_v\beta_3$ integrin	cyclic RGDfK (cRGD)	Peptide	Angiogenesis	Not reported
RGD-PASPIO (Ref. 146)	$\alpha_v\beta_3$ integrin	cyclic RGD	Peptide	Angiogenesis	r_2 = 105.5; r_2^* = 165.5 (3 T)
cRGD-SPPM (Ref. 148)	$\alpha_v\beta_3$ integrin	cyclic RGDfK (cRGD)	Peptide	Angiogenesis	r_2 = 406.6 (7 T)
folate-IO (Ref. 149)	Folate receptors	Folate	Ligand	Angiogenesis	not reported
SCIO-FA (Ref. 150)	Folate receptors	Folate	Ligand	Angiogenesis	r_2 = 46.3 (2 T)
uMUC-1 targeted CLIO-EPPT (Ref. 151)	Mucin-1 tumor-specific antigen	Glu-Pro-Pro-Thr (EPPT)	Peptide	Malignancy	r_1 = 26.43 r_2 = 53.44 (0.47 T)
NP-PEG-CTX (Ref. 157)	Matrix metalloproteinase-2 (MMP-2)	Chlorotoxin (CTX)	Peptide	Cell-migration	r_2 = 5.2 (4.7 T)
mAbMB-SPIONs (Ref. 159)	Mesothelin	Anti-mesothelin (mAbMB)	Antibody	Tumor detection	r_2 = 469.57 r_1 = 0.59 (7 T)
ADR-LEV (Ref. 161)	VEGF receptors	Anti-VEGF	Antibody	Angiogenesis	r_2= 190 (1.5 T)
VEGF165-aptamer–USPIO (Ref. 162)	VEGF receptors	VEGF165	Aptamer	Angiogenesis	Not reported
HFn-NPs (Ref. 164)	Transferrin receptor 1 (TfR1)	Heavy chain ferritin	Ligand	Tumor proliferation	r_2= 224 (4.7 T)

the post contrast images from both the tumor models as compared to control agent (c(RADyK)-MPIO). Analysis of the MR images between two tumor models showed that the agent showed higher retention in melanoma tumors, which had higher vessel mean vessel area as compared to colorectal cancer. This shows that the agent has ability to differentiate between vasculature of tumors that would be useful in studying vasculature heterogeneity. The micron size of the agent facilitates shorter blood half-life and rapid clearance resulting in enhanced contrast effects as compared to nanoparticle based agents. It also allows for increased payload of the iron oxide and more peptides. The agent is an improvement from previous $\alpha_v\beta_3$ integrin targeting iron oxide agents with respect to the faster clearance, better contrast enhancement and lesser non-specific binding.

Lee *et al.* developed an integrin-targeting agent by conjugating polyaspartic acid coated iron oxide nanoparticles with c(RGD) peptides to form DOTA-IO-RGD nanoparticles.[146] PASP is biodegradable, has low toxicity and compatibility with the peptide chains. It also has multifunctional modification capabilities for development of better agents. The agent was PEG coated to reduce non-specific uptake by the reticuloendothelial system (RES). The particles had a core diameter of 5 nm and a hydrodynamic diameter of 45 ± 10 nm and relaxivity of $r_2 = 105.5\ mM^{-1}s^{-1}$ and $r_2^* = 165.5\ mM^{-1}s^{-1}$ at 3 T and 20°C. *In vivo* studies were conducted on a U87MG human glioblastoma tumor model grown subcutaneously in nude mice. T_2 weighted images were acquired 4 hours pre and post injection of DOTA-IO-RGD, DOTA-IO (non-targeting) and DOTA-IO-RGD plus a blocking dose of c(RGDyK) (10 mg/kg) at a concentration of 300 μg of Fe. A strong negative contrast was seen only in tumors injected with the RGD conjugated agent, but not quantified. Kessinger *et al.* developed novel micelle based nanoparticles for integrin targeting.[147,148] The superparamagnetic polymeric micelles (SPPM) were surface functionalized with cyclic RGDfK (cRGD) peptide. The core diameter of the nanoparticle was 50–70 nm. The agent had r_2 relaxivity of $406.6\ mM^{-1}s^{-1}$ at 7 T. *In vivo* analysis in lung cancer subcutaneous tumors showed preferential accumulation in the tumors, which facilitated quantification of angiogenic vasculature in the tumor models. The agent was further analyzed *in vivo* by the use of T_2^*-weighted time-resolved MRI method with 1.3 seconds temporal resolution over 30 minutes to study the targeting kinetics of the nanoprobes in non-small cell A549 lung cancer, MDA-MB-231 breast cancer, or U87 glioblastoma cell lines. Analysis of changes in signal intensities between the pre and post injection images showed that the agent accumulated in the vascular hot spots of all the three tumor models 30 minutes after injection.

8.3.2 *Extravascular targeting iron based probes*

As mentioned earlier, folate receptors are overexpressed in various human tumors and Choi *et al.* developed a dextran coated (maghemite) iron oxide nanoparticles tethered with folate[149] to target the tumor. Dextran-coated IO (maghemite) particles were prepared by a precipitation method and then tethered with folate resulting in hydrophilic and 26 nm size particles (folate-IO). For *in vivo* studies, CD-1 male nude mice were implanted subcutaneously with KB tumor cells in the thigh and T_2^*-weighted gradient echo images were acquired. The folate-IOs are taken up by the folate receptors after they extravasate into the interstitium from the capillaries. The analysis of pre and 2.5 hours post injection T_2 weighted images showed a 38% decrease in the signal between pre and post contrast images of the tumor while for muscle tissue it only decreased by 13%.

In 2008, Shi *et al.* developed PAMAM dendrimer based, folic acid receptor-targeting, iron oxide nanoparticles.[150] The multilayered (PGA and PLL) generation 5 PAMAM dendrimers were surface functionalized with folic acid to form FA modified shell-crosslinked iron oxide (SCIO) NPs (r_2 relaxivity of 46.3 $mM^{-1}s^{-1}$ at 2T). SCID mice with FAR overexpressing KB tumors were imaged using two-dimensional (2D) and three-dimensional (3D) gradient echo sequences both before and after administration of imaging agent. The image contrast was analyzed as a function of time and it was observed that the contrast in tumors injected with the imaging agent decreased with time while no such change was observed in control agent injected tumors. Most significant decrease was seen 24 hours post injection. This agent was able to selectively accumulate in a tumor as small as 0.60 ± 0.15 cm^3. This is the first example of a shell-crosslinked multifunctional iron oxide NP with dendrimer chemistry for *in vivo* MR imaging studies.

Under-glycosylated mucin-1 tumor-specific antigen (uMUC-1) is a biomarker for early detection of malignant tumors. Moore *et al.* developed uMUC-1 targeting crosslinked superparamagnetic iron oxide nanoparticles (CLIO).[151] Glu–Pro–Pro–Thr (EPPT) peptides that recognize uMUC-1 were tethered to the dextran coating of the nanoparticle.[152,153] The r_1 and r_2 relaxivites of the agent at 0.47 T were 26.43 $mM^{-1}s^{-1}$ and 53.44 $mM^{-1}s^{-1}$ respectively at 37 °C.[151] Medarova *et al.* demonstrated the ability of the agent in detecting orthotopic model of pancreatic cancer.[154] NOD SCID mice surgically implanted with the uMUC-1-positive human pancreatic adenocarcinoma cell line were imaged and uMUC-1 targeted CLIO-EPPT agent administered tumors showed a significant decrease in the signal intensity while control had no change.

Chlorotoxin (CTX) is a peptide of 36 amino acids purified from the venom of the giant Israeli scorpion (*Leiurus quinquestriatus*).[155] CTX binds to the membrane-bound matrix metalloproteinase-2 (MMP-2), which is preferentially upregulated in tumors of the neuroectodermal origin such as gliomas and medullablastomas. Veiseh *et al.* were the first to report, in 2005, a novel glioma tumor targeting, dual modality (fluorescence/MR), iron oxide based contrast agent covalently bound with PEG and conjugated with CTX peptide *in vitro*.[156] In 2008, Sun *et al.* demonstrated the *in vivo* tumor targeting ability of NP-PEG-CTX.[157] The PEG coating results in increased blood circulation time, cell internalization and reduced non-specific macrophage uptake as PEG coating prevents agglomeration and protein adsorption. The small size of the CTX facilitates (36 amino acids) increased cell and tissue penetration, thus providing better contrast. For relaxivity analysis, 9L glioma cells were incubated with varying concentration of the nanoprobe (0–15 mM Fe) at 4.7 T and relaxivity maps were acquired. The r_2 relaxivity of the cells incubated with NP-PEG-CTX was found to be 5.2 $mM^{-1}s^{-1}$. Relaxivity maps of 9L rat gliosarcoma subcutaneous tumor bearing nude mice were acquired, pre and 0.3, 2, 12, and 24 hours post injection of NP-PEG-CTX at a dose of 6 mg Fe/kg. A significant R_2 change (14 s^{-1}) was observed at the 2 hours imaging time point and maximum R_2 change (~16 s^{-1}) was seen at the 12 hour time point. The R_2 change in tumors dropped to 10 s^{-1} at the 24-hour time point, while the surrounding muscle tissue only showed a R_2 change of less than ~5 s^{-1} at the 12-hour time point, which remained till 24 hours. The analysis of acquired images showed broad imaging window for the agent, 2–25 hours post injection, as a significant R_2 change (14–16 s^{-1}) was seen during that time point window. The study reported a long blood half-life, prolonged retention in tumors and low toxicity for NP-PEG-CTX that indicates suitability for imaging in clinical settings. The same group reported the drug loading and delivery capabilities of NP-MTX-CTX agent by conjugating the nanoparticles with a conventional chemotherapeutic agent, methotrexate (MTX).[158] The r_2 relaxivity of the agent NP-PEG-CTX at 4.7 T was found to be 637.8 $mM^{-1}s^{-1}$. The *in vivo* analysis of the tumor bearing mice injected with the agent NP-MTX-CTX showed prolonged retention of the agent for up to two weeks in the tumor. Thus, this is a promising agent for longer imaging time and sustained delivery of therapeutic agents. Further analysis of the therapy efficacy is still underway by the group.

Mesothelin is a cell-surface antigen over-expressed in mesotheliomas, pancreatic adenocarcinomas, and ovarian cancers. In 2012, Misri *et al.* developed a novel contrast agent with mesothelin targeting capabilities.[159] Anti-mesothelin antibodies (mAbMB) were conjugated to SPIONs to form mAbMB-SPIONs (76.6 nm). Relaxivity measurements were made at 7 T and were found to be $r_2 = 469.57$ $mM^{-1}s^{-1}$ and $r_1 = 0.59$ $mM^{-1}s^{-1}$. Male SCID mice with A431K5 (mesothelin+) tumors were injected with mAbMB-SPION and gradient echo imaging was performed at 24 and 72 hours post injection. Signal intensity decreased by 33% in tumors 24 hr post injection and by 9% 72 hours post injection. The results confirm the specificity of the agent in mesothelin expressing tumors. The higher relaxivity of the agent allowed for lower concentration yielding sufficient contrast.

Wang *et al.* developed a novel MRI reporter agent with anticancer properties by tethering dechlorinated cisplatin (CMDP) to the surface of OTPBA–SPION through carboxylate groups.[160] The stable superparamagnetic nanocomposite, CMDP–OTPBA–SPION (~60 nm), showed higher DNA binding capacity at an acidic pH (pH 5.2) and insignificant binding at pH 7.4. Thus has the potential to specifically bind to tumor regions due to their acidic environment. Another advantage of magnetic nanoparticles as drug-carriers is that they can be potentially guided to the target of interest, using external magnetic fields, for improved imaging and treatment. CMDP–OTPBA–SPION has a relaxivity r_2 of 141.9 mM^{-1} s^{-1} in phosphate-buffered saline in terms of Fe concentration. T_2-weighted images of

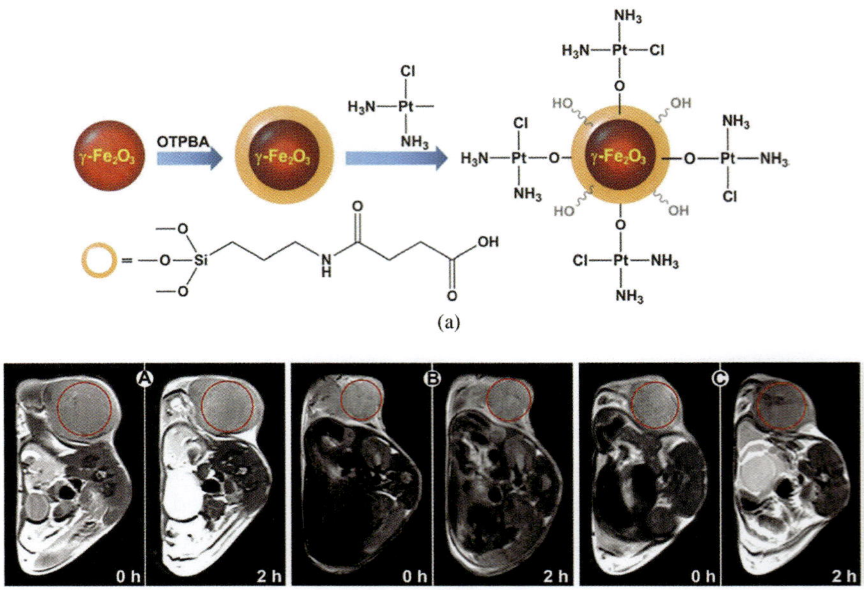

(b)

Figure 8.9 (a) Fabrication route to the CMDP–OTPBA–SPION nanocomposite. (b) T_2-weighted MRI of RM1 murine prostate cancer (red circles) implanted in C57BL/6J mice in (A) presence of only external magnetic field; (B) presence of only superparamagnetic nanocomposite CMDP–OTPBA–SPION; (C) in presence of the superparamagnetic nanocomposite CMDP–OTPBA–SPION and an external magnetic field. Noticeable darkening was seen in tumors injected with the superparamagnetic nanocomposite in conjunction with the application of external magnetic field. Reproduced with permission from Ref. 160.

RM1 murine prostate cancer (red circles) implanted in C57BL/6J mice injected with CMDP–OTPBA–SPION (5.0 mg kg^{-1}) in presence or absence of an external magnetic field were acquired (Figure 8.9). The tumor darkens significantly post 2 hours injection when the contrast agent is injected agent along with the presence of an external magnetic field, while only slightly when there is no external magnetic field, as compared to control tumors (no agent). Thus, CMDP–OTPBA–SPION was shown to be guided to the tumors with the aid of an external magnetic field.

Huang *et al.* reported another pH responsive anti-VEGF targeting superparamagnetic iron oxide contrast agent in 2013.[161] The amphiphilic polymer-anticancer drug conjugate, poly(aspartate)-graft-poly(ethylene glycol)-dodecylamine-hydrazone-(adriamycin-levulinic acid) or PASP-g-PEG-DDA-Hyd-(ADR-LEV) self assembles into micelles that are encapsulated with SPIONs and surface modified with an anti-VEGF antibody. The micelles demonstrated an r_2 relaxivity of 190 mM^{-1} s^{-1}. The micelle assembly showed stability at a pH of 7.4, while dissociation and drug release at a pH range of 4–6 (relevant tumor pH). *In vitro* cytotoxicity analysis of the targeted micelles showed a higher toxicity in Hep-G2 (human liver carcinoma cells) 24 hours post incubation as compared to the non-ADR conjugated micelles. For *in vivo* studies, T_2 maps were acquired before and 1, 12 and 24 hours after the injection of the polymeric micelles, anti-VEGF polymeric micelles and control (conjugate without IOs) in H22 (mouse hepatoma) tumor bearing mice. Significant drop in signal intensity was observed 1 hour post injection in the tumors injected with anti-VEGF polymeric micelles as compared to the other two agents. The results show that the anti-VEGF polymeric micelles had higher accumulation in the tumor regions as compared to the non-targeted micelles. The targeted and pH dependent release of the drug as well as MR effect shows that this agent is a potential drug delivery system for cancer theranostics.

You *et al.* developed a tumor targeted VEGF165-aptamer–USPIO probe by chemically crosslinking ultrasmall superparamagnetic iron oxide (USPIO) nanoparticles and the aptamer specific for VEGF165.[162] T_2-weighted images acquired of subaxillary liver cancer xenograft bearing mice administered with the agent showed significant decrease in the signal intensity post 3 hours. This agent shows faster accumulation and signal enhancement as compared to previously developed VEGF targeting USPIO agents such as anti-VEGF-NPs, which were able to show maximum signal decrease only 24 hours post injection.[163]

Transferrin receptor protein 1 (TfR1), a transmembrane glycoprotein that facilitates iron transport across cells, is overexpressed in certain cancer types. Cao *et al.* developed nanoprobes (M-Hfn) for imaging ferritin by conjugating H-chain (Hfn) subunit to transferrin receptor 1 (TfR1).[164] The core of M–HFn nanoparticles was composed of single crystal cores of magnetite (Fe_3O_4) with an average diameter of 5.2 ± 1.0 nm. The agent had r_2 relaxivity of 224 $mM^{-1}s^{-1}$ measured at 4.7 T. T_2*- and T_2-weighted images were continuously acquired before and up to 6 hours after intravenous injection of the M–HFn nanoparticles in TfR1-positive MDA-MB-231 tumor-bearing mice. The analysis between pre- and post-injection images showed that the agent was able to selectively accumulate in the tumor and allowed for detection of tumors that were not visible in pre injection MR images. In order to study the ability of the nanoprobes to cross the blood brain barrier and detect microscopic brain tumors, orthotopic xenograft models of TfR1-positive U87 tumors were imaged and it was observed that only M-Hfn particles were able to show focal hypointense areas (maximum diameter of ~1 mm) in the brain. Thus, this agent is able to detect pre-angiogenic tumors that are difficult to target due to the inability of other agents to cross the biological barriers.

8.4 Conclusion and future prospects

Contrast agents have single handedly improved the clinical application of MRI and contrast enhanced MRI is now considered the benchmark tool for detecting tumors. The next step in this evolution would be the translation of targeted MR agents to the clinical setting. Many of the agents reviewed here have tremendous potential in the clinic as the pre-clinical studies show that these agents would greatly improve the diagnosis and detection of various tumors. Early detection of cancer is critical for effective treatment and administration of targeted agents would make it possible. The agents discussed in this review have shown the ability to selectively target tumor and have the potential to detect earlier stages of cancer than was previously possible. We have discussed how altering the agent's size and targeting ligands improve the relaxivity, which in turn improves the imaging properties. Macromolecular and nano-platform based agents facilitate incorporation of a higher payload of contrast agent in the core as well as modification of the outer shell with multiple target ligands for high specificity with reduced dosage. Development of imaging agents with biodegradable moieties has improved their excretion as well as safety *in vivo*. Macromolecular and nano-platform agents can also serve the dual purpose of therapy as well as imaging by loading chemotherapeutic drugs in the core of the delivery agent. Over the past few years significant advancements have been made in the development of safer and effective MR imaging agents for cancer imaging. These advancements can facilitate extraction of information, non-invasively, about the tumor microenvironment, expression profile of proteins, genes, enzymatic activity and therapy response by MR. MRI generated information can be further used to develop novel and improved therapies tailored to individual patients and ensure maximum effect on the tumor. Decades of research on cancer have shown that it is very dynamic and differs starkly between individuals with the same tumor types and grades and thus personalized/precision medicine is the future of cancer therapeutics. Tumor-targeted ^{1}H MRI contrast agents can assist in the development and monitoring of this therapeutic strategy.

References

1. Jain RK. Transport of molecules, particles, and cells in solid tumors. *Ann Rev Biomed Engin,* 1999; 1: 241–263.
2. Hanahan D, Weinberg RA. Hallmarks of cancer: the next generation. *Cell,* 2011; 144: 646–674.
3. Fass L. Imaging and cancer: a review. *Mol Oncol,* 2008; 2: 115–152.
4. Wilson WR, Hay MP. Targeting hypoxia in cancer therapy. *Nat Revi Cancer,* 2011; 11: 393–410.
5. Blanco E, *et al.* Nanomedicine in cancer therapy: innovative trends and prospects. *Cancer Sci,* 2011; 102: 1247–1252.
6. Hellebust A, Richards-Kortum R. Advances in molecular imaging: targeted optical contrast agents for cancer diagnostics. *Nanomedicine,* 2012; 7: 429–445.
7. Torchilin, V. Tumor delivery of macromolecular drugs based on the EPR effect. *Adv Drug Deli Rev,* 2011; 63: 131–135.
8. Mouli SK, *et al.* Image-guided local delivery strategies enhance therapeutic nanoparticle uptake in solid tumors. *ACS Nano,* 2013; 7: 7724–7733.
9. Kiessling F, *et al.* Targeted ultrasound imaging of cancer: an emerging technology on its way to clinics. *Curr Pharm Design,* 2012; 18: 2184–2199.
10. Garcia C, Gebhart G, Flamen P. New PET imaging agents in the management of solid cancers. *Curr Opin Oncol,* 2012; 24: 748–755.
11. Van Dort ME, Rehemtulla A, Ross BD. PET and SPECT imaging of tumor biology: new approaches towards oncology drug discovery and development. *Curr Comput-Aid Drug Design,* 2008; 4: 46–53.

12. Morrow M, Waters J, Morris E. MRI for breast cancer screening, diagnosis, and treatment. *Lancet,* 2011; 378: 1804–1811.
13. Panebianco V, *et al.* Conventional imaging and multiparametric magnetic resonance (MRI, MRS, DWI, MRP) in the diagnosis of prostate cancer. *Q J Nucl Med Mol Imaging,* 2012; 56: 331–342.
14. Lu J, Feng F, Jin Z. Cancer diagnosis and treatment guidance: role of MRI and MRI probes in the era of molecular imaging. *Curr Pharm Biotechnol,* 2013; 14: 714–722.
15. Gore JC, Manning HC, Quarles CC, Waddell KW, Yankeelov TE. Magnetic resonance in the era of molecular imaging of cancer. *Magn Reson Imaging* 2011; 29: 587–600.
16. Haacke EM. *Magnetic Resonance Imaging: Physical Principles and Sequence Design.* Wiley, 1999.
17. Hingorani DV, Bernstein AS, Pagel MD. A review of responsive MRI contrast agents: 2005–2014. *Contr Media Mol Imag,* 2015; 10: 245–265.
18. Lee N, *et al.* Iron oxide based nanoparticles for multimodal imaging and magnetoresponsive therapy. *Chem Rev,* 2015; 115: 10637–10689.
19. Sinharay S, Pagel MD. Advances in magnetic resonance imaging contrast agents for biomarker detection. *Annu Rev Anal Chem,* 2016; 9: 95–115.
20. Atlas SW, Atlas SW. *Magnetic Resonance Imaging of the Brain and Spine.* 3rd Edn., Lippincott Williams & Wilkins, 2002.
21. Damadian R. Tumor detection by nuclear magnetic resonance. *Science,* 1971; 171: 1151–1153.
22. Lauffer RB. Paramagnetic metal-complexes as water proton relaxation agents for NMR imaging — theory and design. *Chem Rev,* 1987; 87: 901–927.
23. Caravan P, Ellison JJ, McMurry TJ, Lauffer RB. Gadolinium(III) Chelates as MRI contrast agents: structure, dynamics, and applications. *Chem Rev,* 1999; 99: 2293–2352.
24. Aime S, *et al.* High sensitivity lanthanide(III) based probes for MR-medical imaging. *Coordin Chem Rev,* 2006; 250: 1562–1579.
25. Caravan P. Strategies for increasing the sensitivity of gadolinium based MRI contrast agents. *Chem Soc Rev,* 2006; 35: 512–523.
26. Werner EJ, Datta A, Jocher CJ, Raymond KN. High-relaxivity MRI contrast agents: where coordination chemistry meets medical imaging. *Angew Chem,* 2008; 47: 8568–8580.

27. Glunde K, Artemov D, Penet MF, Jacobs MA, Bhujwalla ZM. Magnetic resonance spectroscopy in metabolic and molecular imaging and diagnosis of cancer. *Chem Rev,* 2010; 110: 3043–3059.
28. Villaraza AJ, Bumb A, Brechbiel MW. Macromolecules, dendrimers, and nanomaterials in magnetic resonance imaging: the interplay between size, function, and pharmacokinetics. *Chem Rev,* 2010; 110: 2921–2959.
29. Kamaly N, Miller AD, Bell JD. Chemistry of tumour targeted T1 based MRI contrast agents. *Curr Topics Medi Chem,* 2010; 10: 1158–1183.
30. Caravan P, Farrar CT, Frullano L, Uppal R. Influence of molecular parameters and increasing magnetic field strength on relaxivity of gadolinium- and manganese-based T1 contrast agents. *Contr Media Mol Imag,* 2009; 4: 89–100.
31. Rohrer M, Bauer H, Mintorovitch J, Requardt M, Weinmann HJ. Comparison of magnetic properties of MRI contrast media solutions at different magnetic field strengths. *Invest Radiol,* 2005; 40: 715–724.
32. Forsen S, Hoffman RA. Study of Moderately Rapid Chemical Exchange Reactions by Means of Nuclear Magnetic Double Resonance. *J Chem Phys,* 1963; 39: 2892.
33. Woods, M, Woessner DE, Sherry AD. Paramagnetic lanthanide complexes as PARACEST agents for medical imaging. *Chem Soc Rev,* 2006; 35: 500–511.
34. Weinmann HJ, Brasch RC, Press WR, Wesbey GE. Characteristics of gadolinium-DTPA complex: a potential NMR contrast agent. *Am J Roentgenol,* 1984; 142: 619–624.
35. Pintaske J, *et al.* Relaxivity of gadopentetate dimeglumine (Magnevist), gadobutrol (Giadovist), and gadobenate dimeglumine (MultiHance) in human blood plasma at 0.2, 1.5, and 3 Tesla. *Investig Radiol,* 2006; 41: 213–221.
36. Lauffer RB. *et al.* MS-325: a small-molecule vascular imaging agent for magnetic resonance imaging. *Acad Radiol,* 1996; 3(Suppl 2): S356–S358.
37. Turetschek K. *et al.* MRI assessment of microvascular characteristics in experimental breast tumors using a new blood pool contrast agent (MS-325) with correlations to histopathology. *J Magn Reson Imag,* 2001; 14: 237–242.
38. Huang CH, Tsourkas A. Gd-based macromolecules and nanoparticles as magnetic resonance contrast agents for molecular imaging. *Curr Topics Medi Chem,* 2013; 13: 411–421.

39. Desgrosellier JS, Cheresh DA. Integrins in cancer: biological implications and therapeutic opportunities. *Nat Rev Cancer,* 2010; 10: 9–22.
40. Park JA, *et al.* Gd-DOTA conjugate of RGD as a potential tumor-targeting MRI contrast agent. *Chembiochem,* 2008; 9: 2811–2813.
41. Park JA, *et al.* Improved tumor-targeting MRI contrast agents: Gd(DOTA) conjugates of a cycloalkane-based RGD peptide. *Biochem Biophys Res Commun,* 2014; 455: 246–250.
42. Ye F, *et al.* A peptide targeted contrast agent specific to fibrin-fibronectin complexes for cancer molecular imaging with MRI. *Bioconjug Chem,* 2008; 19: 2300–2303.
43. Pilch J, *et al.* Peptides selected for binding to clotted plasma accumulate in tumor stroma and wounds. *Proc Natl Acad Sci USA,* 2006; 103: 2800–2804.
44. Ye F, Jeong EK, Parker D, Lu ZR. Evaluation of CLT1-(Gd-DTPA) for Cancer MR Molecular Imaging in a Mouse Breast Cancer Model. *Bo Pu Xue Za Zhi,* 2011; 2: 325–330.
45. Wu X, Lindner D, Yu GP, Brady-Kalnay S, Lu ZR. MR molecular imaging of prostate cancer with a small molecular CLT1 peptide targeted contrast agent. *J Vis Exp*, 2013.
46. Zhou Z, Wu X, Kresak A, Griswold M, Lu ZR. Peptide targeted tripod macrocyclic Gd(III) chelates for cancer molecular MRI. *Biomaterials,* 2013; 34: 7683–7693.
47. Zhou Z, *et al.* MRI detection of breast cancer micrometastases with a fibronectin-targeting contrast agent. *Nat Commun,* 2015; 6: 7984.
48. Rotin D, Robinson B, Tannock IF. Influence of hypoxia and an acidic environment on the metabolism and viability of cultured cells: potential implications for cell death in tumors. *Cancer Res,* 1986; 46: 2821–2826.
49. Estrella V, *et al.* Acidity generated by the tumor microenvironment drives local invasion. *Cancer Res,* 2013; 73: 1524–1535.
50. Garcia-Martin ML, *et al.* High resolution pH(e) imaging of rat glioma using pH-dependent relaxivity. *Magn Reson Med,* 2006; 55: 309–315.
51. Raghunand N, Howison C, Sherry AD, Zhang S, Gillies RJ. Renal and systemic pH imaging by contrast-enhanced MRI. *Magn Reson Med,* 2003; 49: 249–257.
52. Liu G, LiY, Sheth VR, Pagel MD. Imaging *in vivo* extracellular pH with a single paramagnetic chemical exchange saturation transfer magnetic resonance imaging contrast agent. *Mol Imaging,* 2012; 11: 47–57.

53. Nwe K, Huang CH, Tsourkas A. Gd-labeled glycol chitosan as a pH-responsive magnetic resonance imaging agent for detecting acidic tumor microenvironments. *J Med Chem,* 2013; 56: 7862–7869.
54. Jonas CR, Ziegler TR, Gu LH, Jones DP. Extracellular thiol/disulfide redox state affects proliferation rate in a human colon carcinoma (Caco2) cell line. *Free Radical Bio Med,* 2002; 33: 1499–1506.
55. Policastro LL, Ibanez IL, Notcovich C, Duran HA, Podhajcer OL. The tumor microenvironment: characterization, redox considerations, and novel approaches for reactive oxygen species-targeted gene therapy. *Antioxid Redox Signal,* 2013; 19: 854–895.
56. Carrera C, *et al.* Synthesis and characterization of a Gd(III) based contrast agent responsive to thiol containing compounds. *Dalton Trans,* 2007; 4980–4987.
57. Digilio G, *et al.* Exofacial protein thiols as a route for the internalization of Gd(III)-based complexes for magnetic resonance imaging cell labeling. *J Med Chem,* 2010; 53: 4877–4890.
58. Menchise V, *et al. In vivo* labeling of B16 melanoma tumor xenograft with a thiol-reactive gadolinium based MRI contrast agent. *Mol Pharm,* 2011; 8: 1750–1756.
59. Raghunand N, *et al.* Redox-sensitive contrast agents for MRI based on reversible binding of thiols to serum albumin. *Magn Reson Med,* 2006; 55: 1272–1280.
60. Raghunand N, *et al.* Design, synthesis, and evaluation of 1,4,7,10-tetraazacyclododecane-1,4,7-triacetic acid derived, redox-sensitive contrast agents for magnetic resonance imaging. *J Med Chem,* 2010; 53: 6747–6757.
61. Guntle GP, *et al.* Tumor xenograft response to redox-active therapies assessed by magnetic resonance imaging using a thiol-bearing DOTA complex of gadolinium. *Transl Oncol,* 2012; 5: 190–199.
62. Brown JM. The hypoxic cell: a target for selective cancer therapy — eighteenth Bruce F. Cain Memorial Award lecture. *Cancer Res,* 1999; 59: 5863–5870.
63. Wardman P. Chemical radiosensitizers for use in radiotherapy. *Clin Oncol,* 2007; 19: 397–417.
64. Melillo G. Targeting hypoxia cell signaling for cancer therapy. *Cancer Metast Rev,* 2007; 26: 341–352.
65. Bischoff P, Altmeyer A, Dumont F. Radiosensitising agents for the radiotherapy of cancer: advances in traditional and hypoxia targeted radiosensitisers. *Exp Opin Ther Patents,* 2009; 19: 643–662.

66. Stewart GD, *et al.* The relevance of a hypoxic tumour microenvironment in prostate cancer. *BJU Intl* 2010; 105: 8–13.
67. Graves EE, Maity A, Le QT. The tumor microenvironment in non-small-cell lung cancer. *Semin Radiat Oncol,* 2010; 20: 156–163.
68. Ahmadi M, *et al.* Hypoxia modulates the activity of a series of clinically approved tyrosine kinase inhibitors. *Br J Pharm,* 2014; 171: 224–236.
69. Rojas-Quijano FA, *et al.* Synthesis and characterization of a hypoxia-sensitive MRI probe. *Chemistry,* 2012; 18: 9669–9676.
70. Sheldon PW, Foster JL, Fowler JF. Radiosensitization of C3H mouse mammary tumours by a 2-nitroimidazole drug. *Br J Cancer,* 1974; 30: 560–565.
71. Brown JM. Selective radiosensitization of the hypoxic cells of mouse tumors with the nitroimidazoles metronidazole and Ro 7-0582. *Radia Res,* 1975; 64: 633–647.
72. Gulaka PK, *et al.* GdDO3NI, a nitroimidazole-based T1 MRI contrast agent for imaging tumor hypoxia *in vivo. J Biol Inorg Chem,* 2014; 19: 271–279.
73. Iwaki S, *et al.* Development of hypoxia-sensitive Gd3+-based MRI contrast agents. *Bioorg Med Chem Lett,* 2012; 22: 2798–2802.
74. Ogan MD, *et al.* Albumin labeled with Gd-DTPA. An intravascular contrast-enhancing agent for magnetic resonance blood pool imaging: preparation and characterization. *Invest Radiol,* 1987; 22: 665–671.
75. Bhujwalla ZM, *et al.* Reduction of vascular and permeable regions in solid tumors detected by macromolecular contrast magnetic resonance imaging after treatment with antiangiogenic agent TNP-470. *Clin Cancer Res,* 2003; 9: 355–362.
76. Preda A, *et al.* MRI monitoring of Avastin antiangiogenesis therapy using B22956/1, a new blood pool contrast agent, in an experimental model of human cancer. *J Magn Reson Imag,* 2004; 20: 865–873.
77. Raman V, *et al.* Characterizing vascular parameters in hypoxic regions: a combined magnetic resonance and optical imaging study of a human prostate cancer model. *Cancer Res,* 2006; 66: 9929–9936.
78. Raatschen HJ, *et al.* Vascular permeability during antiangiogenesis treatment: MR imaging assay results as biomarker for subsequent tumor growth in rats. *Radiology,* 2008; 247: 391–399.
79. Dafni H, Kim SJ, Bankson JA, Sankaranarayanapillai M, Ronen SM. Macromolecular dynamic contrast-enhanced (DCE)-MRI detects reduced vascular permeability in a prostate cancer bone metastasis model following

anti-platelet-derived growth factor receptor (PDGFR) therapy, indicating a drop in vascular endothelial growth factor receptor (VEGFR) activation. *Magn Reson Med,* 2008; 60: 822–833.

80. Ye Z, *et al.* Synthesis and evaluation of a polydisulfide with Gd-DOTA monoamide side chains as a biodegradable macromolecular contrast agent for MR blood pool imaging. *Contr Media Mole Imag,* 2013; 8: 220–228.
81. Artemov D, Mori N, Ravi R, Bhujwalla ZM. Magnetic resonance molecular imaging of the HER-2/neu receptor. *Cancer Res,* 2003; 63: 2723–2727.
82. Chen K, Xie J, Chen X. RGD-human serum albumin conjugates as efficient tumor targeting probes. *Mol Imaging,* 2009; 8: 65–73.
83. Qiao J, *et al.* HER2 targeted molecular MR imaging using a *de novo* designed protein contrast agent. *PloS One,* 2011; 6: e18103.
84. Zhu X, Verma S. Targeted therapy in HER2-positive metastatic breast cancer: a review of the literature. *Curr Oncol,* 2015; 22, S19–S28.
85. Hynes NE, Stern DF. The Biology of Erbb-2 Neu Her-2 and Its Role in Cancer. *Bba-Rev Cancer,* 1994; 1198: 165–184.
86. Bhujwalla ZM, Artemov D, Natarajan K, Ackerstaff E, Solaiyappan M. Vascular differences detected by MRI for metastatic *versus* nonmetastatic breast and prostate cancer xenografts. *Neoplasia,* 2001; 3: 143–153.
87. Sun B, Halmos G, Schally AV, Wang X, Martinez M. Presence of receptors for bombesin/gastrin-releasing peptide and mRNA for three receptor subtypes in human prostate cancers. *The Prostate,* 2000; 42: 295–303.
88. Uchida K, *et al.* Expression of progastrin-releasing peptide and gastrin-releasing peptide receptor mRNA transcripts in tumor cells of patients with small cell lung cancer. *J Cancer res Clin Oncol,* 2002; 128: 633–640.
89. Wei L, *et al.* Protein-based MRI contrast agents for molecular imaging of prostate cancer. *Mol Imag Biol,* 2011; 13: 416–423.
90. Yang JJ, *et al.* Rational design of protein-based MRI contrast agents. *J Am Chem Soc,* 2008; 130: 9260–9267.
91. Kesharwani P, Jain K, Jain NK. Dendrimer as nanocarrier for drug delivery. *Prog Polym Sci,* 2014; 39: 268–307.
92. Noriega-Luna B, *et al.* Applications of Dendrimers in Drug Delivery Agents, Diagnosis, Therapy, and Detection. *J Nanomater*, 2014.
93. Wiener EC, *et al.* Dendrimer-based metal chelates: a new class of magnetic resonance imaging contrast agents. *Magn Reson Med,* 1994; 31: 1–8.

94. Longmire MR, Ogawa M, Choyke PL, Kobayashi H. Dendrimers as high relaxivity MR contrast agents. *Wiley Interdiscip Rev Nanomedi Nanobiotechnol* 2014; 6: 155–162.
95. Langereis S, de Lussanet QG, van Genderen MHP, Backes WH, Meijer EW. Multivalent contrast agents based on gadolinium-diethylenetriamine-pentaacetic acid-terminated poly(propylene imine) dendrimers for magnetic resonance imaging. *Macromolecules,* 2004; 37: 3084–3091.
96. Langereis S, *et al.* Evaluation of Gd(III)DTPA-terminated poly(propylene imine) dendrimers as contrast agents for MR imaging. *NMR Biomed,* 2006; 19: 133–141.
97. Pikkemaat JA, *et al.* Dendritic PARACEST contrast agents for magnetic resonance imaging. *Contr Media Mol Imag,* 2007; 2: 229–239.
98. Luo K, *et al.* Gadolinium-labeled peptide dendrimers with controlled structures as potential magnetic resonance imaging contrast agents. *Biomaterials,* 2011; 32: 7951–7960.
99. Torchilin VP. Structure and design of polymeric surfactant-based drug delivery systems. *J Control Release,* 2001; 73: 137–172.
100. Mi Y, Liu Y, Feng SS. Formulation of Docetaxel by folic acid-conjugated d-alpha-tocopheryl polyethylene glycol succinate 2000 (Vitamin E TPGS(2k)) micelles for targeted and synergistic chemotherapy. *Biomaterials,* 2011; 32: 4058–4066.
101. Unger E, Shen D, Wu GL, Fritz T. Liposomes as Mr contrast agents — pros and cons. *Magn Reson Med,* 1991; 22: 304–308.
102. Terreno, E, *et al.* Paramagnetic liposomes as innovative contrast agents for magnetic resonance (MR) molecular imaging applications. *Chem Biodivers,* 2008; 5: 1901–1912.
103. Na K, Lee SA, Jung SH, Shin BC. Gadolinium-based cancer therapeutic liposomes for chemotherapeutics and diagnostics. *Coll Surf B Biointerfaces,* 2011; 84: 82–87.
104. Sipkins DA, *et al.* Detection of tumor angiogenesis *in vivo* by alphaV-beta3-targeted magnetic resonance imaging. *Nat Med,* 1998; 4: 623–626.
105. Storrs RW, *et al.* Paramagnetic polymerized liposomes — synthesis, characterization, and applications for magnetic-resonance-imaging. *J Am Chem Soc,* 1995; 117: 7301–7306.

106. Li W, *et al.* RGD-targeted paramagnetic liposomes for early detection of tumor: *in vitro* and *in vivo* studies. *Euro J Radiol,* 2011; 80: 598–606.
107. Tan M, Wu, X, Jeong EK, Chen, Q, Lu ZR. Peptide-targeted Nanoglobular Gd-DOTA monoamide conjugates for magnetic resonance cancer molecular imaging. *Biomacromolecules* 2010; 11: 754–761.
108. Kaneshiro TL, Lu ZR. Targeted intracellular codelivery of chemotherapeutics and nucleic acid with a well-defined dendrimer-based nanoglobular carrier. *Biomaterials,* 2009; 30: 5660–5666.
109. Tan M. *et al.* MR molecular imaging of prostate cancer with a peptide-targeted contrast agent in a mouse orthotopic prostate cancer model. *Pharma Res,* 2012; 29: 953–960.
110. Ferrauto G, Delli Castelli D, Di Gregorio E, Terreno E, Aime S. LipoCEST and cellCEST imaging agents: opportunities and challenges. *Wiley Interdiscip Rev Nanomed Nanobiotechnol,* 2016; 8: 602–618.
111. Flament J, *et al. In Vivo* CEST MR imaging of U87 mice brain tumor angiogenesis using targeted LipoCEST contrast agent at 7 T. *Magn Reson Med,* 2013; 69: 179–187.
112. Ellis LM, Hicklin DJ. VEGF-targeted therapy: mechanisms of anti-tumour activity. *Nat Rev. Cancer,* 2008; 8: 579–591.
113. Lee YJ, *et al.* Differential effects of VEGFR-1 and VEGFR-2 inhibition on tumor metastases based on host organ environment. *Cancer Res,* 2010; 70: 8357–8367.
114. Liu DH, Liu FX, Liu ZH, Wang LL, Zhang, N. Tumor specific delivery and therapy by double-targeted nanostructured lipid carriers with anti-VEGFR-2 Antibody. *Mol Pharmaceut,* 2011; 8: 2291–2301.
115. Meadows KL, Hurwitz HI. Anti-VEGF therapies in the clinic. *Cold Spring Harb Perspect Med,* 2012; 2.
116. Liu Y, *et al.* Multifunctional pH-sensitive polymeric nanoparticles for theranostics evaluated experimentally in cancer. *Nanoscale,* 2014; 6: 3231–3242.
117. Sun Z, *et al.* VEGF-loaded graphene oxide as theranostics for multi-modality imaging-monitored targeting therapeutic angiogenesis of ischemic muscle. *Nanoscale,* 2013; 5: 6857–6866.
118. Antony AC. The biological chemistry of folate receptors. *Blood,* 1992; 79: 2807–2820.

119. Toffoli G, *et al.* Overexpression of folate binding protein in ovarian cancers. *Int J Cancer. (J Int Du Cancer)*, 1997; 74: 193–198.
120. Low PS, Henne WA, Doorneweerd DD. Discovery and development of folic-acid-based receptor targeting for imaging and therapy of cancer and inflammatory diseases. *Acc Chem Res,* 2008; 41: 120–129.
121. Swanson SD, *et al.* Targeted gadolinium-loaded dendrimer nanoparticles for tumor-specific magnetic resonance contrast enhancement. *Int J Nanomed,* 2008; 3: 201–210.
122. Cheng Z, Thorek DL, Tsourkas A. Gadolinium-conjugated dendrimer nanoclusters as a tumor-targeted T1 magnetic resonance imaging contrast agent. *Angew Chem,* 2010; 49: 346–350.
123. Nakamura T, Kawano K, Shiraishi K, Yokoyama M, Maitani Y. Folate-targeted gadolinium-lipid-based nanoparticles as a bimodal contrast agent for tumor fluorescent and magnetic resonance imaging. *Biol Pharm Bull,* 2014; 37: 521–527.
124. Beilvert A, *et al.* Tyrosine polyethylene glycol (PEG)-micelle magnetic resonance contrast agent for the detection of lipid rich areas in atherosclerotic plaque. *Magn Reson Med,* 2009; 62: 1195–1201.
125. Kamaly N, Kalber T, Thanou M, Bell JD, Miller AD. Folate receptor targeted bimodal liposomes for tumor magnetic resonance imaging. *Bioconjug Chem,* 2009; 20: 648–655.
126. Ali MM Yoo, B, Pagel MD. Tracking the relative *in vivo* pharmacokinetics of nanoparticles with PARACEST MRI. *Mol Pharm,* 2009; 6: 1409–1416.
127. Ali MM, *et al.* A nano-sized PARACEST-fluorescence imaging contrast agent facilitates and validates *in vivo* CEST MRI detection of glioma. *Nanomedicine (Lond),* 2012; 7: 1827–1837.
128. Kim KS, Park W, Hu J, Bae YH, Na KA cancer-recognizable MRI contrast agents using pH-responsive polymeric micelle. *Biomaterials,* 2014; 35: 337–343.
129. Hyde R, Taylor PM, Hundal HS. Amino acid transporters: roles in amino acid sensing and signalling in animal cells. *Biochemi J,* 2003; 373: 1–18.
130. Crich SG, *et al. In Vitro* and *in vivo* magnetic resonance detection of tumor cells by targeting glutamine transporters with Gd-based probes. *J Med Chem,* 2006; 49: 4926–4936.
131. del Amo EM, Urtti A, Yliperttula M. Pharmacokinetic role of L-type amino acid transporters LAT1 and LAT2. *Eur J Pharm Sci,* 2008; 35: 161–174.

132. Mehravi B, *et al.* Breast cancer cells imaging by targeting methionine transporters with gadolinium-based nanoprobe. *Mol Imag Biol,* 2014.
133. Weissleder R, *et al.* Superparamagnetic iron oxide: pharmacokinetics and toxicity. *Am J Roentgenol,* 1989; 152: 167–173.
134. Wang YX, Hussain SM, Krestin GP. Superparamagnetic iron oxide contrast agents: physicochemical characteristics and applications in MR imaging. *Eur Radiol,* 2001; 11: 2319–2331.
135. Di Marco M, *et al.* Physicochemical characterization of ultrasmall superparamagnetic iron oxide particles (USPIO) for biomedical application as MRI contrast agents. *Int J Nanomed,* 2007; 2: 609–622.
136. Wang YX. Superparamagnetic iron oxide based MRI contrast agents: Current status of clinical application. *Quant Imag Med Surg,* 2011; 1: 35–40.
137. Wunderbaldinger P, Josephson L, Weissleder R. Crosslinked iron oxides (CLIO): a new platform for the development of targeted MR contrast agents. *Acad Radiol,* 2002; 9(2): S304–S306.
138. Rieber A, *et al.* MRI in the diagnosis of small bowel disease: use of positive and negative oral contrast media in combination with enteroclysis. *Eur Radiol,* 2000; 10: 1377–1382.
139. Hahn PF, *et al.* First clinical trial of a new superparamagnetic iron oxide for use as an oral gastrointestinal contrast agent in MR imaging. *Radiology,* 1990; 175: 695–700.
140. Reimer P, *et al.* T1 effects of a bolus-injectable superparamagnetic iron oxide, SH U 555 A: dependence on field strength and plasma concentration — preliminary clinical experience with dynamic T1-weighted MR imaging. *Radiology,* 1998; 209: 831–836.
141. Ichikawa T, *et al.* Perfusion MR imaging with a superparamagnetic iron oxide using T2-weighted and susceptibility-sensitive echoplanar sequences: evaluation of tumor vascularity in hepatocellular carcinoma. *Am J Roentgenol,* 1999; 173: 207–213.
142. Reimer P, Balzer T. Ferucarbotran (Resovist): a new clinically approved RES-specific contrast agent for contrast-enhanced MRI of the liver: properties, clinical development, and applications. *Eur Radiol,* 2003; 13: 1266–1276.
143. Ros PR, *et al.* Hepatic MR imaging with ferumoxides: a multicenter clinical trial of the safety and efficacy in the detection of focal hepatic lesions. *Radiology,* 1995; 196: 481–488.

144. Seneterre E, *et al.* Detection of hepatic metastases: ferumoxides-enhanced MR imaging *versus* unenhanced MR imaging and CT during arterial portography. *Radiology* 1996; 200: 785–792.
145. Melemenidis S, *et al.* Molecular magnetic resonance imaging of angiogenesis *in vivo* using polyvalent cyclic RGD-iron oxide microparticle conjugates. *Theranostics,* 2015; 5: 515–529.
146. Lee HY, *et al.* PET/MRI dual-modality tumor imaging using arginine-glycine-aspartic (RGD)-conjugated radiolabeled iron oxide nanoparticles. *J Nucl Med,* 2008; 49: 1371–1379.
147. Kessinger CW, *et al. In vivo* angiogenesis imaging of solid tumors by alpha(v)beta(3)-targeted, dual-modality micellar nanoprobes. *Exp Biol Med (Maywood),* 2010; 235: 957–965.
148. Kessinger CW, *et al.* Investigation of *in vivo* targeting kinetics of alpha(v) beta(3)-specific superparamagnetic nanoprobes by time-resolved MRI. *Theranostics,* 2011; 1: 263–273.
149. Choi H, Choi SR, Zhou R, Kung HF, Chen IW. Iron oxide nanoparticles as magnetic resonance contrast agent for tumor imaging via folate receptor-targeted delivery. *Acad Radiol,* 2004; 11: 996–1004.
150. Shi XY, *et al.* Dendrimer-functionalized shell-crosslinked iron oxide nanoparticles for *in vivo* magnetic resonance imaging of tumors. *Adv Mater,* 2008; 20: 1671-+.
151. Moore A, Medarova, Z, Potthast A, Dai G. *In vivo* targeting of underglycosylated MUC-1 tumor antigen using a multimodal imaging probe. *Cancer Res,* 2004; 64: 1821–1827.
152. Perey L, *et al.* Tumor selective reactivity of a monoclonal antibody prepared against a recombinant peptide derived from the DF3 human breast carcinoma-associated antigen. *Cancer Res,* 1992; 52: 2563–2568.
153. Barratt-Boyes SM. Making the most of mucin: a novel target for tumor immunotherapy. *Cancer Immunol, Immunother,* 1996; 43: 142–151.
154. Medarova Z, Pham W, Kim Y, Dai G, Moore A. *In Vivo* imaging of tumor response to therapy using a dual-modality imaging strategy. *Int J Cancer,* 2006; 118: 2796–2802.
155. Lyons SA, O'Neal J, Sontheimer H. Chlorotoxin, a scorpion-derived peptide, specifically binds to gliomas and tumors of neuroectodermal origin. *Glia,* 2002; 39: 162–173.

156. Veiseh O, *et al.* Optical and MRI multifunctional nanoprobe for targeting gliomas. *Nano Lett,* 2005; 5: 1003–1008.
157. Sun C, *et al. In Vivo* MRI detection of gliomas by chlorotoxin-conjugated superparamagnetic nanoprobes. *Small,* 2008; 4: 372–379.
158. Sun C, *et al.* Tumor-targeted drug delivery and MRI contrast enhancement by chlorotoxin-conjugated iron oxide nanoparticles. *Nanomedicine,* 2008; 3: 495–505.
159. Misri R, Meier D, Yung AC, Kozlowski P, Hafeli UO. Development and evaluation of a dual-modality (MRI/SPECT) molecular imaging bioprobe. *Nanomedicine,* 2012; 8: 1007–1016.
160. Wang JZ, *et al.* A platinum anticancer theranostic agent with magnetic targeting potential derived from maghemite nanoparticles. *Chem Sci,* 2013; 4: 2605–2612.
161. Huang HL, *et al.* A novel anti-VEGF targeting and MRI-visible smart drug delivery system for specific diagnosis and therapy of liver cancer. *Macromol Biosci,* 2013; 13: 1358–1368.
162. You XG, *et al.* Molecular magnetic resonance probe targeting VEGF165: preparation and *in vitro* and *in vivo* evaluation. *Contr Media Mol Imag,* 2014; 9: 349–354.
163. Hsieh WJ, *et al. In Vivo* tumor targeting and imaging with anti-vascular endothelial growth factor antibody-conjugated dextran-coated iron oxide nanoparticles. *Int J Nanomed,* 2012; 7: 2833–2842.
164. Cao C, *et al.* Targeted *in vivo* imaging of microscopic tumors with ferritin-based nanoprobes across biological barriers. *Adv Mater,* 2014; 26: 2566–2571.

Chapter 9

SOLID LIPID NANOPARTICLES AND NANOSTRUCTURED LIPID CARRIERS AS ANTI-CANCER DELIVERY SYSTEMS FOR THERAPY AND DIAGNOSTICS

*Samuel V. Mussi and Vladimir P. Torchilin**

Center for Pharmaceutical Biotechnology and Nanomedicine,
360 Huntington Avenue, 140 The Fenway, Northeastern University,
Boston, MA 02115, USA
**v.torchilin@neu.edu*

Abstract

Solid lipid nanoparticles (SLN) and nanostructured lipid carrier (NLC) have been widely described as an alternative drug delivery system and they have some advantages such as simple industrial scale-up, low toxicity of the lipids and the controlled release of the drugs incorporated into the matrix. Many works have described the higher cytotoxicity of those lipid nanoparticles loaded with anticancer drugs compared with the free form of the respective drugs. They have shown better anticancer activity in mice in controlling the tumor volume growth and they have also shown a reduction of the toxicity of very toxic drugs, such as doxorubicin. Furthermore, recent works have described the addition of more functions to the SLN and NLC to increase the anticancer effect, such as long-circulation, co-loading of drugs, the combination with RNA/DNA, pH stimulus-sensitive drug release, incorporation of agents for imaging and the attachment of ligands for active targeting. Thus, in this chapter we will discuss about the potential of SLN and NLC for cancer therapy and diagnostics, and the recent trends in their development.

9.1 Introduction

Solid lipid nanoparticles (SLN) and nanostructured lipid carriers (NLC) are colloidal particles with size of 50–1000 nm. They have been described as drug delivery systems for delivery of pharmaceutical actives incorporating several compounds for different ways of administrations, such as parenteral. The literature has demonstrated the potential of SLN and NLC as promising candidates for delivery of anticancer and imaging agents. Several *in vitro* and *in vivo* studies have demonstrated the potential of these lipid nanoparticles in the increase of the efficacy of anticancer agents.[1] SLN loaded with doxorubicin were more cytotoxic than free drug against resistant cancer cell lines and these SLN are promising as an approach to overcome the chemoresistance of adriamycin-resistant breast cancer.[2–5] Koziara *et al.*, verified an increase in the uptake and cytotoxicity of SLN loaded with paclitaxel in the human glioblastoma cell line U-118.[6] They also evaluated the *in vitro* and *in vivo* activity against HCT-15. These SLN overcame drug resistance and increased the efficacy of the treatment. Other studies investigated the activity of NLC loaded with paclitaxel and doxorubicin against different cell lines. These NLC overcame the drug resistance of SKOV3-TR30 cells. NLC loaded with paclitaxel were 31 times more cytotoxic, and NLC loaded with doxorubicin were 2.5 times more effective than the respective free drugs.[7,8] SLN loaded with docetaxel were shown to be more effective than the commercial docetaxel, Taxotere®, *in vitro* against colorectal (C-26) and malignant melanoma (A-375) cell lines. In *in vivo* tests with C-26-implanted BALB/c mice, SLN loaded with docetaxel showed better tumor inhibitory efficacy and survival than Taxotere®.[9]

SLN carrying different antineoplastic agents (doxorubicin, paclitaxel, cholesteryl butyrate, and anti-VEGF antisense oligonucleotides) were tested in experimental animal models of cerebral gliomas and induced an effective anti-tumoral therapeutic response.[10] An explanation involves the accumulation of drugs encapsulated in SLN in the brain after intravenous administration. This was attributed to the transport of intact

SLN through the blood-brain barrier and also to the sustained release from the lipid matrix.[11] Pharmacokinetic studies showed a prolonged circulation up to 24 hours, a significant increase in $t_{1/2}$, C_{max} and low uptake of SLN by the reticuloendothelial system (RES). A lower concentration of doxorubicin was found in the heart when administered as SLN, indicating the possibility of lower cardiotoxicity compared to the commercial doxorubicin solution, which is one of the effects limiting the usefulness of doxorubicin.[12]

Thus, SLN and NLC can offer many benefits for cancer therapy. In the first topic of this chapter, we will briefly describe their characteristics, the composition and preparation methods. Next, our focus will be on the recent trends that have been proposed to further optimize chemotherapy using these drug delivery systems. They have been successfully multi-functionalized to capture a payload of drugs, to target specific cells and to release entrapped drugs in a controlled manner. We will describe these new trends, related to the addition of more functions to these nanoparticles, such as pegylation, co-encapsulation of anticancer agents, pH-sensitivity, magnetic nanoparticle loading and other agents for imaging and for attachment of ligands on the surface.

9.2 Characteristics, composition and methods of preparation

Müller *et al.* proposed SLN for the first time in 1993 that SLN are composed of a lipid matrix that is solid at body temperature and are derived from parenteral nutrition emulsions by replacement of the liquid lipid (oil) of the emulsion droplets with a lipid solid.[13] The use of solid lipids instead of liquid oils is a very attractive idea to achieve controlled drug release, because drug mobility in a solid lipid should be considerably lower compared with a liquid lipid. The main features of SLN with regard to parenteral application are the excellent physical stability, protection of incorporated labile drugs from degradation, ability to carry compounds with different physiochemical characteristics, higher drug stability, improved pharmacokinetics and drug activity, controlled drug release and good tolerability. Potential disadvantages such as insufficient loading capacity, drug expulsion after polymorphic transition during storage, relatively high water content of the dispersions (70–99.9%) and the presence of other colloids have been verified.[13–15] NLC has been described as an improved generation of SLN and they are composed of binary mixture of solid lipid and a liquid lipid. The NLC share the advantages of SLN and may overcome some of the disadvantages. The most important advantage of NLC over SLN is increased drug loading and it has been attributed to better accommodation of the drug in the liquid lipid that might be spatially distributed in the solid lipid matrix.[16,65]

It has been reported that SLN and NLC are probably spherical because a spherical subject has the smallest possible surface area-to-volume ratio. This may minimize the surface tension between the lipids and the aqueous phase, thereby leading to thermodynamic stability. However, it should be noted that the platelet shape of SLN and NLC was also verified by some groups. Thus far, there has been no concrete data revealing the detailed structure of SLN and NLC, but it is known that several factors such as the physicochemical characteristics of the ingredients and the production method may affect their structure and morphology.[17,18]

The main components of SLN and NLC are the drug, solid/liquid lipids, emulsifiers, and water. Other ingredients are generally added, such as osmotic agents, matrices for lyophilization, buffers, etc. One of the most interesting advantages of these nanoparticles is the low potential of acute and chronic toxicity resulting from the lipids. In general, physiological lipids with low cost are used and include triglycerides (e.g., tristearin, tripalmitin), hard fat types (e.g., glyceryl behenate, monostearate, stearic acid, palmitic acid, cetyl palmitate) and others (e.g., cholesterol). A large variety of ionic and nonionic emulsifiers of different molecular weight has been used to stabilize the lipid dispersion but the choice of emulsifier is more limited for parenteral administrations. Poloxamers, polysorbates, lecithin, and bile acids are the most used. In many cases, a combination of these might be more efficient.[14,15]

The physicochemical properties of the components and incorporated drugs have also been described as factors that can determine the release kinetics. It has been reported that, in general, it is seen a strong burst effect in the first minutes or hours and a subsequent prolonged release of the drug from SLN and NLC. Studies have described that the sustained release profiles might be correlated to a drug-enriched core model, and both the drug-enriched shell model and the solid solution model will have fast release kinetics. More details about these theoretical models can be found in References 19 and 20.

Another important advantage of SLN and NLC is the ease in manufacturing and industrial scale-up. Due to the similarity of the method of preparation to parenteral emulsions, SLN and NLC have gained attention as a promising alternative over other well-known drug nanocarriers, such as liposomes and polymeric nanoparticles. The most discussed methods to prepare SLN and NLC are high-shear homogenization followed by ultrasonication and high-pressure homogenization. The first described technique used to prepare SLN is the high-shear homogenization followed by ultrasonication. This method is carried out at temperatures above the melting point of the lipid. Since the lipids will be liquid at that temperature, an emulsion is prepared by adding the aqueous phase to the oil phase (bulk of lipid and drug) and homogenized using a high-shear mixing equipment (e.g., Ultra-Turrax). Then, the formed emulsion is sonicated using an ultrasound probe sonicator to further reduce the particle size. For this method, a bimodal size distribution may be obtained and metal contamination has to be considered. Using this method is also important to consider the degradation rate of the drug and the carrier due to high temperatures. In addition, the solubilities and HLB values of many surfactants can be reduced at higher temperatures, which might have a negative impact on homogenization efficacy.[13–16]

High-Pressure Homogenization (HPH) has been used for years for the production of nanoemulsions for parenteral nutrition. It has emerged as an important technique for the preparation of SLN and NLC specially because the scaling up represents no or minor problems in most cases. Another advantage is the narrow size distribution obtained. In this technique, a liquid is pushed with high pressure through a narrow gap, which is a few microns. Very high-shear forces disrupt the particles down to the submicron range.[21] Two approaches of the homogenization step have been discussed, the hot and the cold homogenization techniques.[10,22,23] For both, the drug is incorporated by dissolving or dispersing into the bulk of melted lipid. Hot homogenization is carried out at temperatures above the melting point of the lipid. First, an emulsion of the oil phase (melted lipid containing drug) and the aqueous phase at the same temperature is obtained using high-shear mixing equipment (Ultra-Turrax). Then, this emulsion is added to high-pressure homogenizer keeping the temperatures above the melting point of the lipid. The homogenization step can be repeated several times. The last step is the cooling of the sample to room temperature or below.

Cold homogenization is carried out with the lipid at solid state. After the incorporation of the drug into the bulk of melted lipid, the mixture is rapidly cooled and dispersed in a chilled emulsifier solution. The suspension is subjected to HPH at or below room temperature.[24] The cold procedure minimizes the thermal exposure of the sample, reduces the drug distribution into the aqueous phase during homogenization and avoids problems related to the complexity of the crystallization of the lipids. However, works have said that the hot homogenization has higher efficacy.

Other techniques that have also been reported include microemulsion, solvent diffusion and w/o/w double emulsion method. In the microemulsion method, a warm microemulsion is prepared under stirring before it is dispersed in a large volume of cold water, usually at a ratio varying in the range 1–20 v/v, respectively. The particle size, size distribution, and stability of the nanoparticles might be affected basically by the drug/lipid ratio and concentration of surfactant. The solvent evaporation method is based on the solubilization of lipid in water-immiscible organic solvent followed by the addition of the hydrophobic drug. Then, the organic solution is mixed into a solution of emulsifiers in water under stirring. Finally, the nanoparticles are obtained by removing the organic solvent with evaporation under low pressure. Despite this method avoiding heating, the organic solvent use is a disadvantage concerning the industrial scale and bio-safety. In the w/o/w double emulsion method, a highly concentrated drug solution is dispersed under vigorous stirring into a lipid phase and stabilized by a surfactant to form the primary w/o emulsion. Then, this emulsion is dispersed into a large volume of aqueous solution containing another kind of surfactant and the w/o/w emulsion is formed. The last step to obtain the nanoparticles can be an ultrafiltration or solvent evaporation. These methods are designated for the encapsulation of hydrophilic drugs. To find more details and information about the preparation methods of SLN and NLC see these references.[25–29]

9.3 Recent trends in use SLN and NLC for cancer therapy

9.3.1 *Pegylation-longevity in the blood*

The pegylation of nanoparticles to reduce the opsonization and to increase the time in blood circulation has been widely described. It has a relevant function in cancer therapy. Because the blood longevity of nanocarriers is important to promote their accumulation in solid tumors due to the defective vasculature (passive targeting *via* EPR effect).[30,31] Poly(ethylene)glycol (PEG) derivatives and other hydrophilic polymers have been employed to modify nanoparticle's surface.[32] What is important to highlight in this topic is the fact that some emulsifiers used to prepare SLN and NLC may have a hydrophilic chain, sometimes being pegylated, and may have a dual function: surfactant to produce the nanoparticles and pegylating agents. In Figure 9.1 there are some examples of these surfactants. As expected, the efficiency in increasing the longevity in the blood is correlated to the size of the hydrophilic chain of the emulsifier. Some studies have demonstrated the influence of surfactants in the modification of pharmacokinetics (Table 9.1).

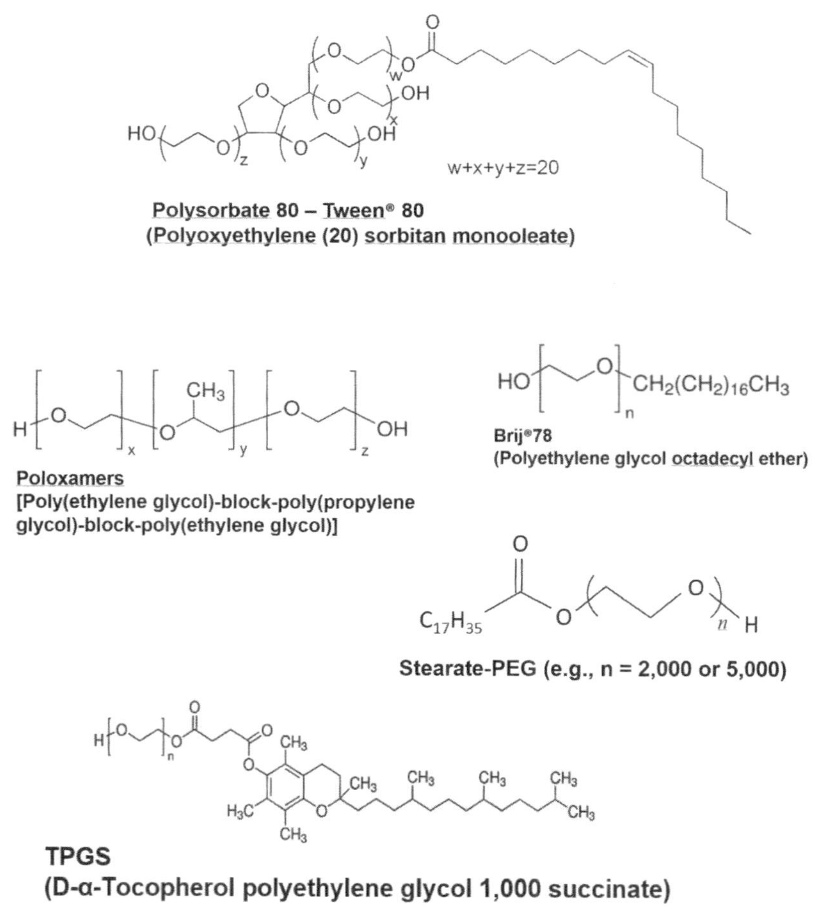

Figure 9.1 Examples of emulsifiers that can be used to prepare SLN and NLS and might offer a pegylating function.

Table 9.1 Works that describe the pegylation of SLN using emulsifiers.

Formulation	Drug/Pegylating agent	Reference(s)
SLN	Drug-free or doxorubicin/stearic acid-PEG 2000	33 12
NLC	Tamibarotene/NLC/stearic acid-PEG 2000	64
SLN	Paclitaxel/Pluronic® F68 and Brij78	22
SLN	Anthracyclines/Pluronic® F68	23

Long-circulating SLN made with two different surfactants, Pluronic® F68 (a kind of poloxamer) and Brij78, revealed marked differences in pharmacokinetic parameters, especially in the $t_{1/2}$ and AUC, in comparison with Cremophor® EL containing a solution of paclitaxel.[22] F68-SLN and Brij 78-SLN were long-circulating with a 7.4-fold and 3.6-fold higher $t_{1/2}$, respectively. This was attributed to the reduced uptake of SLN by the mononuclear phagocytic system and also to the lower clearance rate. SLN prepared with F68 had longer circulation than Brij 78 due to its longer PEG chain length. Another advantage of the pegylation using Pluronic® block copolymers may be better results in treatment of drug-resistant tumors. These block copolymers interact with multidrug-resistant cancer (MDR) tumors, resulting in drastic sensitization to various anticancer agents, particularly anthracycline antibiotics. Pluronic® affects several distinct drug resistance mechanisms including inhibition of drug efflux transporters.[23]

Drug-free pegylated SLN and non-pegylated SLN were prepared and injected intravenously in rats to study their tissue distribution and transport across the blood-brain barrier.[33] The average size of these lipid nanoparticles was below 100 nm. To pegylate the SLN, stearic acid-PEG 2000 at 0.15% w/v was added. Localization of SLN in the brain and in the cerebrospinal fluid (CSF) was higher for both formulations compared to the free drug. However, there was a lower uptake in the liver and lung for pegylated-SLN than for the non-pegylated-SLN, confirming the influence of pegylation.[33] Another study investigated the potential of the same pegylated and non-pegylated SLN for improved parenteral delivery of doxorubicin.[34] The pharmacokinetics and tissue distribution of doxorubicin in these SLN were compared with a commercial solution of doxorubicin. There was a significantly prolonged circulation, an increase of $t_{1/2}$ and of C_{max} with pegylated SLN.[34] The influence of stearic acid-PEG 2000 concentration for these SLN was also evaluated. The AUC (area under the curve) of doxorubicin increased as a function of the amount of pegylating agent present in the SLN. Doxorubicin was detected in blood six hours post-injection from both non-pegylated and pegylated SLN, while no doxorubicin was detectable after intravenous injection of doxorubicin solution. The highest concentration of doxorubicin in the blood was detected in pegylated SLN at the highest concentration of stearic acid-PEG 2000.[12]

9.3.2 *Co-loading of anticancer agents*

Some works have described the benefits of co-loading drugs in SLN and NLC (Table 9.2). SLN and NLC seem to be interesting as an alternative to co-load anticancer agents but so far not many studies have been published. A challenging issue could be the drug loading capacity, which can be harder when encapsulating more than one drug. The co-encapsulation has been proposed to achieve an uniform temporal and spatial co-delivery offering many advantages, such as more synchronized, controlled pharmacokinetics of the drugs, improved drug solubility and bioavailability and the potential to bypass mechanisms of multidrug resistance.[24,35–37]

A combination of anticancer drugs with some natural lipids, such as docosahexaenoic acid (DHA), tocotrienol, ceramide and others to increase the antitumor effect has been described. An interesting fact is that these

Table 9.2 SLN and NLC co-loading anticancer agents.

Formulation	Anticancer agents	Reference(s)
NLC	Simvastatin/tocotrienol	25
NLC	Doxorubicin/DHA (as triglyceride form)	38
SLN	Doxorubicin/ceramide/oligonucleotide (MBO-asGCS)	39
SLN	Paclitaxel/siRNA	40
SLN	Chemosensitizer GG918 (Elacridar)/doxorubicin	2
SLN	Doxorubicin/mitomycin	41 42

lipids (liquid or solid) constitute a part of the lipid matrix and may improve the encapsulation of the other combined drug. For example, NLC co-loaded with simvastatin and tocotrienol have been proposed as an anticancer alternative. The incorporation of tocotrienol, which is liquid, generated compartments into the solid matrix that promoted the entrapment of simvastatin. It was verified by differential scanning calorimetry and X-ray diffraction analysis that showed an absence of endothermic or crystalline peaks for the mixture of the solid lipid and tocotrienol. The researchers verified an enhancement in the antiproliferative effect for the co-encapsulation of drugs against malignant +SA mammary epithelial cells. The IC_{50} of the NLC loaded with simvastatin and tocotrienol was 34-fold lower than alpha-tocopherol nanoparticles, which was used as reference. Thus, the combination of tocotrienol and simvastatin entrapped in NLC supports the NLC formation and at the same time promotes an increase of the activity.[25]

We have proposed the co-encapsulation of doxorubicin and DHA (triglyceride form) in NLC. DHA is a liquid lipid and is part of the matrix lipid composition (Figure 9.2a). The nanoparticles had no aggregation in

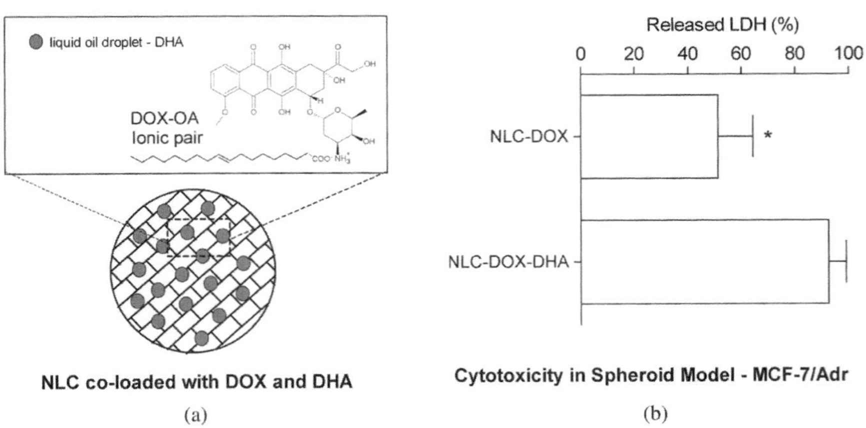

Figure 9.2 (a) Schematic of NLC co-loaded with doxorubicin (as ionic pair with oleic acid) and doxorubicin (triglyceride form — liquid lipid). The ionic pair was proposed to increase the doxorubicin loading in the lipid matrix, since it is more lipophilic than doxorubicin. (b) Cytotoxicity of NLC co-loaded with doxorubicin and docosahexaenoic acid (NLC-DOX-DHA) compared with NLC loaded only with doxorubicin (NLC-DOX) in MCF-7/Adr as spheroid model.[38]

the presence of serum and we verify a controlled drug release in DMEM, with no more than 40% release up to 48 hours. The NLC loaded with doxorubicin and DHA showed a stronger activity against MCF-7/Adr cells by the co-encapsulation of drugs. In monolayer model, the doxorubicin uptake as encapsulated drug was higher than free drug in MCF-7/Adr, suggesting the bypassing of P-glycoprotein bomb efflux. For spheroids, the NLC loaded with doxorubicin and DHA showed a prominent cytotoxicity and a greater penetration of doxorubicin in the 3D structure compared with NLC loaded only with doxorubicin (Figure 9.2b).[38]

Another alternative that has been described is the combination of drugs with oligonucleotides and siRNA, which have been attached to the surface of the nanoparticles. SLN co-loaded with doxorubicin, ceramide and the mixed backbone antisense glucosylceramide synthase oligonucleotide (MBO-asGCS) have been proposed.[39] The sphingolipid ceramide has antitumor activity by triggering apoptosis in cancer cells and it has been reported to increase the effect of doxorubicin. However, the cells can inactivate the ceramide by glycosylation. Thus, the combination with MBO-asGCS would block this glycosylation, could increase the intracellular concentration of ceramide and enhance apoptosis. The SLN were prepared with a cationic charge to attach the anionic molecule of MBO-asGCS by ionic interaction. The researchers verified a significant increase of the cytotoxicity in the drug resistant cancer cell line NCI/ADR-RES by co-loading of doxorubicin, ceramide and MBO-asGCS.[39] SLN for co-delivery of paclitaxel and siRNA was developed by Yu *et al.*[40] The SLN was PEGylated with methoxypolyethylene glycol 2000-distearoyl phosphatidylethanolamine (mPEG-DSPE). The positive charge needed to attach the siRNA was obtained with the incorporation of 3β(N-(N′,N′-dimethylaminoethane)-carbamoyl) cholesterol (DC-Chol). The SLN increased the cellular uptake and the attachment of the therapeutic siRNA, human MCL1-specific siRNA (siMCL1) and resulted in reduction of the MCL1 mRNA levels. The siMCL1 complexed to SLN loaded with paclitaxel showed a greater *in vitro* anticancer effect against KB cells than SLN loaded only with paclitaxel. In KB cell-xenografted mice, this SLN loaded with paclitaxel and complexed with siMCL1 significantly reduced the growth of tumors. Thus, it verified the potential of SLN for the co-delivery of lipophilic anticancer drug and therapeutic siRNA.[40]

SLN co-loaded with doxorubicin and the chemosensitizer GG918 (Elacridar) were also proposed. The *in vitro* results showed the greatest doxorubicin uptake and cytotoxicity against the human MDR breast cancer cell line MDA435/LCC6/MDR1 for SLN co-loaded with both drugs.[2,3] Other SLN co-loaded with doxorubicin and mitomycin were more effective in killing MDR cells with up to 30-fold lower doses than the free drugs and it was also attributed to the co-encapsulation.[41] These SLN co-loaded with doxorubicin and mitomycin were evaluated *in vivo* and demonstrated enhanced efficacy compared to liposomal doxorubicin with a 3-fold increase in animal life span, a 10–20% tumor cure rate, undetectable normal tissue toxicity and decreased tumor angiogenesis.[42]

9.3.3 *pH-sensitivity for drug release at lowered pH*

Many studies have described the use of nanoparticles pH-sensitive to delivery drugs in tumor regions and it has been one of the most exploited alternative of targeting. It is well-known that tumors stroma has an acidic pH about 6.8. In addition, upon cellular uptake, drug-loaded nanocarriers will be subjected to an intracellular pH of 5.9–6.2 in early endosomes and pH of 5.0–5.5 in late endosomes and lysosomes. With this in mind, researchers have proposed strategies to release entrapped drugs under those conditions, providing a target that may improve therapeutic efficacy by cancer site-specific delivery and controlled release.[43,44] However, not much has been investigated for SLN and NLC and so far two strategies have been proposed to increase the drug release in low pH (Table 9.3). One is the increase of the doxorubicin hydrophilicity and its release from the lipid core. The other is the preparation of nanoparticles with a lipid or polymer that can have the structure modified in low pH, destabilizing the particle and releasing the entrapped compounds.

Table 9.3 pH sensitive SLN.

Formulation	Drug/Agent	Reference(s)
SLN	Doxorubicin free base	5
SLN	Triamcinolone acetonide/n-glutaryl phosphatidylethanolamine coating	26

Subedi *et al.* (2009) proposed SLN loaded with doxorubicin as free base,[5] which is more lipophilic than the hydrochloride form and has an enhanced entrapment efficiency. The pKa of the doxorubicin molecule is 8.3, and under lower pH it can be protonated, increasing the hydrophilicity and reducing the interaction with the lipid matrix. Indeed, they verified higher doxorubicin release at pH of 5 as compared with pH of 7.4. We have investigated a similar strategy by obtaining an ionic pair of doxorubicin with docosahexaenoic acid (DHA). DHA is an omega 3 polyunsaturated fatty acid that shows synergic effect with anthracyclines.[38] We proposed the formation of an ionic pair between the cationic doxorubicin molecule and the anionic DHA *in situ*, during the SLN preparation, to improve drug encapsulation and anticancer activity. The ionic pair doxorubicin-DHA was more lipophilic and resulted in very high efficiency entrapment (~100%) and drug loading of 31 mg/g. We also verified a higher doxorubicin release (~80%) at pH 5 as compared to pH of 7.4. In this case, the doxorubicin is already cationic but the DHA is anionic and at low pH it may be protonated, undoing the ionic interaction and releasing the doxorubicin (Figure 9.3). *In vitro* evaluations showed higher uptake for drugs encapsulated in SLN compared to the free drugs as well as higher cytotoxicity against the human lung cancer cell line, A549.[38]

In another study, researchers used the phospholipid derivative n-glutaryl phosphatidylethanolamine (n-glut) to modify the outer shell of SLN. SLN were loaded with the anti-inflammatory drug triamcinolone that has been used to inflammatory disorders and cancer of the lymph nodes. The drug release results showed increased release under acidic conditions (pH 4 to 6) compared to pH 7.4. They attributed this to the fact that the n-glut is an acidic pH-sensitive lipid that hydrolyzes at low pH and enhances the release of the entrapped drug.[26]

9.3.4 *Imaging agents*

SLN are under investigation as a pharmaceutical tool to stably carry various types of diagnostic agents, such as superparamagnetic iron oxide,[45] technetium-99 (^{99m}Tc),[46] ^{64}Cu[47] and quantum dots[41,48] (Table 9.4).

Superparamagnetic SLN have been proposed for magnetic resonance imaging (MRI) contrast.[45] Iron oxide was encapsulated in SLN and

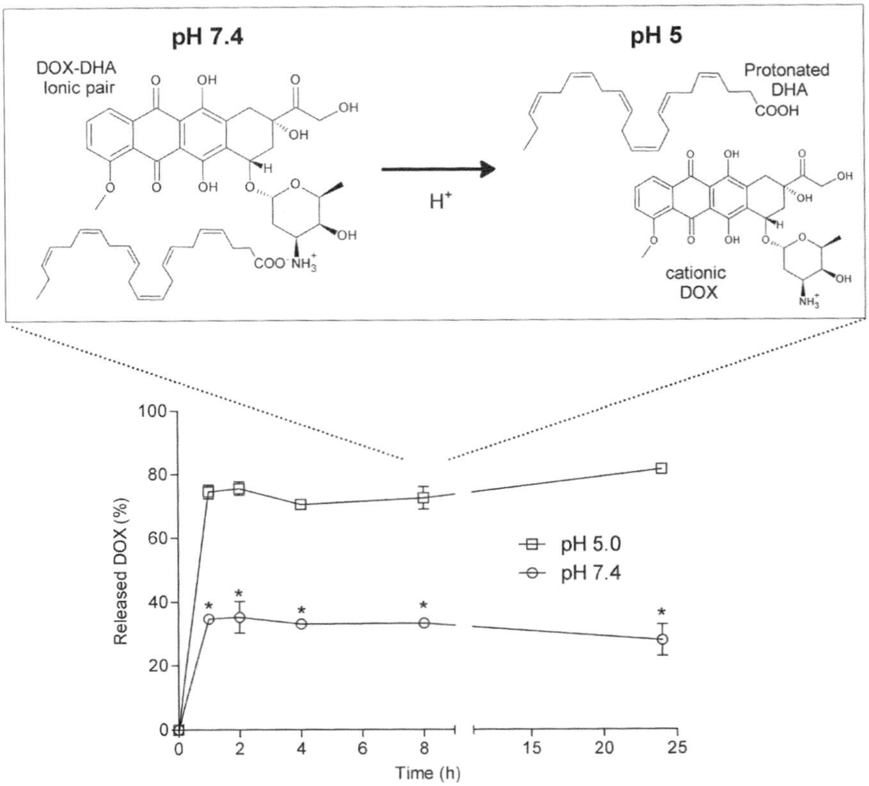

Figure 9.3 Doxorubicin release from SLN at pH 7.4 and pH 5.0. The high encapsulation of doxorubicin into the lipid matrix was achieved by an *in situ* ionic pair formation that resulted in a lipophilic salt ((1) DOX-DHA Ionic pair), as previously described in [35]. At low pH of 5, the DHA is protonated (2) and will undo the ionic interaction with doxorubicin (3). Due to the hydrophilic character of the cationic doxorubicin, its diffusion from the lipid matrix would be faster and result in enhanced drug release (4).

Table 9.4 SLN loaded with imaging agents.

Formulation	Drug/Agent	Reference(s)
SLN	Superparamagnetic iron oxide	45
SLN	^{99m}Tc	46
SLN	^{64}Cu	47
SLN	Paclitaxel/siRNA/quantum dot	48

in vitro analysis of the relaxometric properties showed similar results to the commercial superparamagnetic iron oxide nanoparticles (Endorem®). The authors also evaluated the *in vivo* MRI of the central nervous system (CNS) with SLN and verified a slower blood clearance than Endorem®. MRI and uptake by CNS data showed that both contrast agents lasted up to the end of the experiment (135 minutes). Thus, these results showed that the SLN are able to bypass the BBB and suggest the potential for use as a CNS MRI contrast agent.

Assessment of inhaled radiolabelled SLN biodistribution was described by Videira *et al.*[46] SLN were radiolabelled with ^{99m}Tc using the lipophilic chelator D, L-hexamehylpropyleneamine oxime (HMPAO). After inhalation of ^{99m}Tc-HMPAO-SLN, it was seen in the biodistribution studies a high uptake in lymphatics and high rate of distribution in peri-aortic, axillar and inguinal lymph nodes. Gamma scintigraphic images also showed a strong deposition of ^{99m}Tc-HMPAO and ^{99m}Tc-HMPAO-SLN on the lung but ^{99m}Tc-HMPAO was rapidly cleared *via* systemic circulation. They verified four hours after inhalation a higher accumulation of ^{99m}Tc in lung for ^{99m}Tc-HMPAO-SLN and it was attributed to the higher particle retention. Thus, they claim ^{99m}Tc-HMPAO-SLN as an alternative for targeting colloidal carriers to the lymphatics and for using as a lymphoscintigraphic agent.

Andreozzi *et al.*, proposed a SLN radiolabeled with ^{64}Cu for *in vivo* positron emission tomography (PET) studies.[47] A copper specific chelator, 6-(p-(bromoacetamido)benzyl)-1,4,8,11-tetraazacyclotetradecane-*N,N′,N″,N‴*-tetraacetic acid (BAT) conjugated with a synthetic lipid was incorporated into SLN and successfully radiolabeled with ^{64}Cu. Stability assays confirmed that copper remains associated with the SLN over a 48 hours time period. They verified a ^{64}Cu circulating in the blood after three hours of injection and an accumulation in the liver. This pattern of biodistribution was attributed to the ^{64}Cu associated with SLN. The results indicate the potential for a quantitative evaluation and *in vivo* PET imaging for ^{64}Cu-SLN.

SLN loaded with quantum dots for anticancer theranostics and delivery of a paclitaxel-siRNA combination were recently proposed.[48] Quantum dots and paclitaxel were incorporated into the SLN lipid matrix while the anionic siRNA molecules were electrostatically complexed with the outer cationic surface. The SLN loaded with paclitaxel-siRNA were efficiently taken up into human lung carcinoma cells and exhibited synergistic anticancer activity. The strong fluorescence from quantum dots within SLN enabled *in situ* visualization of the intracellular translocation of SLN into cancer cells and suggests that this formulation could be useful as a multifunctional and optically traceable nanocarrier for efficient anticancer theranostics. However, the study was just an *in vitro* evaluation, and further investigations, first of all, *in vivo* studies, should be done to verify the real potential of this formulation as a diagnostic tool. Indeed, the quantum dots are being successfully used for the sentinel lymph node mapping and tumor angiogenesis imaging. The emission color of quantum dots can be precisely tuned from the ultraviolet to the near-infrared, and they are extremely bright and photostable.[29]

An integrin-targeted near infrared light-emitting SLN has been developed by Shuhendler *et al.*[41] Tumor cells and tumor vascular endothelium overexpress α(v) β(3) integrin receptors and thus are useful targets for imaging, chemotherapy and anti-angiogenic therapy. SLN were conjugated with the α(v) β(3) integrin-specific ligand, cyclic Arg–Gly–Asp (cRGD), and loaded with near infrared quantum dots. Live animal imaging was performed using nude mice bearing xenografted orthotopic human breast tumors or dorsal window chamber breast tumors. The active tumor targeting with cRGD showed a "double-edged sword" effect: specific RGD-SLN accumulation in tumor blood vessels and increased tumor residence time. Thus, the RGD-SLN may perhaps be useful for tumor neovascular-specific therapies and imaging applications.

9.3.5 *Attachment of targeting ligands to the surface of lipid nanoparticles*

Currently, one of the biggest challenges in cancer therapy is to obtain a drug or a carrier that are able to reach specifically the tumor area, minimizing the accumulation and toxicity in healthy tissues and organs. Thus, the success of the treatment with higher efficacy and lower toxicity depends on the ability of the therapeutics to achieve their target sites. The attachment of ligands to the surface of nanoparticles to improve the specificity for cancer cells has emerged as an attractive tool and it has been widely described.[49] This addition makes them able to selectively bind in cancer cells and enhance the internalization. The endocytosis mediated by receptor has been extensively explored as a way to increase internalization of nanoparticles. These ligands are thought and designed to bind specifically in receptors over-expressed in cancer cells and minimally in healthy tissues. Among many molecules and structures described as ligands, there are antibodies, antibody fragments, proteins, small molecules (sugar, folate, vitamins and others), aptamers and peptides.[50] Furthermore, many works proposed a modification of the nanoparticles surface with PEG derivatives for enhance the circulation time in the blood and at the same time for incorporation of ligands. The attachment of ligands at the end of the PEG spacer arm keeps them out of the PEG brush, reducing the steric impediment and facilitating the binding to target receptors.[32] In Table 5 we can see some examples of SLN and NLC with ligands attached to the surface for active targeting.

A mannose coating of SLN, loaded with doxorubicin was proposed to target lectin-like receptors overexpressed in cancer cells.[51] The mannose was

Table 9.5 SLN and NLC modified with ligands for active targeting.

Formulation	Drug/Agent	Reference(s)
SLN	Integrin-specific ligand cyclic Arg-Gly-Asp (cRGD)/ quantum dots	41
SLN	Doxorubicin/mannose	51
SLN	Paclitaxel/octaarginine	32
SLN	Doxorubicin/ anti-EGFR	53
SLN	Dexamethasone/ transferrin/plasmid GFP	54
SLN	Transferrin/mannan/plasmid GFP	55
NLC	Docetaxel/folate-poly(PEG-cyanoacrylate-co-cholesteryl cyanoacrylate)	56
SLN	Paclitaxel/folate-poly(ethylene glycol) (PEG)-phosphatidylethanolamine	57
SLN	Docetaxel/ketoconazole/folate	58
NLC	Docetaxel/cell-penetrating peptide/folate	59
NLC	Paclitaxel/hyaluronic acid	28, 60

previously linked to stearylamine by a chemical reaction and then added to the SLN. The *ex vivo* evaluation showed that mannosylated-SLN were less hemolytic than the uncoated SLN and free doxorubicin. Mannosylated-SLN were more cytotoxic and were preferably taken up by A549 tumor cells. Pharmacokinetic studies revealed improved bioavailability, longer half-life and increased mean residence time of doxorubicin upon mannose conjugation. They also verified a higher concentration of doxorubicin in tumors for the mannosylated-SLN as well as lower damage to renal and hepatic tissues. Thus, the attachment of the small molecule mannose to the SLN surface seems to be promising to target cancer cells. However, studies concerning the higher efficacy *in vivo* of this formulation have to be investigated.

SLN loaded with paclitaxel and modified with octaarginine (R8-PTX-SLN) were proposed by Zhang *et al.*[52] The octaarginine was attached to the SLN through a linkage with stearic acid. The results showed a higher cytotoxicity against human lung cancer cells (A549) and an increase of the cell uptake. Kuo *et al.*[53] developed SLN loaded with doxorubicin and the anti-epithelial growth factor receptor, anti-EGFR, which was electrostatically attached to the catanionic (cationic and anionic) surface of the particle. Immunochemical staining revealed that the crosslinked anti-EGFR on the surface of SLN preserved a high specificity for EGFR on U87MG cells, and it induced a growth-inhibition effect. Another study proposed a linkage of transferrin (Tf) on polyethylene glycol-phosphatidylethanolamine (PEG-PE) to obtain Tf-PEG-PE ligands. This polymer was used for the surface modification of a cationic SLN for dual gene (plasmid pEGFP) and dexamethasone delivery. *In vitro* and *in vivo* studies showed increase in transfection due to the attachment of Tf.[54]

To target cancer cells and macrophages in the liver, SLN were modified with dual ligands Tf and mannan (M). Tf and M were linked to polyethylene glycol-phosphatidylethanolamine (PEG-PE) and PE separately to get transferrin-PEG-PE (T-PEG-PE) and mannan-PE (M-PE) ligands for the surface modification of SLN. The *in vivo* transfection efficiency was verified using fluorescence protein plasmid (GFP) that was promoted in the presence of the dual ligands. The evaluation in tumor-bearing animal models showed that Tf-PEG-PE and M-PE functioned as excellent ligands to improve the cell targeting ability of the carriers.[55]

A folate-poly(PEG-cyanoacrylate-co-cholesteryl cyanoacrylate) (FA-PEG-PCHL) was synthesized to modify docetaxel-loaded NLC and to obtain a long blood circulating effect and improved targeting.[56] This is an amphiphilic copolymer able to locate at the nanoparticle interface, exposing the PEG chain to the aqueous phase with the folate moiety at the end. *In vivo* studies indicated significant differences in the pharmacokinetic profile and tissue distribution for FA-PEG-PCHL-modified docetaxel-loaded NLC compared to the plain docetaxel-loaded NLC. The PEGylation contributed to an increase of the AUC, $t_{1/2}$ and MRT, and there was a significant accumulation of docetaxel in tumors due to the folic acid attachment. Wu *et al.* verified that the modification of paclitaxel-loaded SLN with folate-poly(ethylene glycol) (PEG)-phosphatidylethanolamine exhibited a higher tumor inhibition rate in mice with sarcoma180 ascites tumors.[57] It was proposed the co-loading of docetaxel and ketoconazole in SLN grafted with folic acid, which was incorporated by a stearic acid and folic acid conjugate (SA-FA).[58] Both drugs are metabolized hepatically by the cytochrome P-450 system, and ketoconazole can inhibit p-gp efflux of docetaxel at the blood brain barrier. Thus, a potential drug-drug interaction for these agents can be useful. The authors verified that the combination of drugs in SLN plus attached folic acid resulted in increased brain uptake of docetaxel and it was 44 times higher than the commercial Taxotere®, which suggests promise for brain tumor treatment.

NLC loaded with paclitaxel were modified with tumor microenvironment-sensitive polypeptides (TMSP) and folate.[59] The aim of the ligands combination was to create a more selective and efficient system to deliver the drug in cancer cells. The TMSP and folate were conjugated to the distal ends of DSPE-PEG2000-maleimide and DSPE-PEG5000-amine, respectively (see Figure 9.3). The folate was attached to increase the endocytosis by the higher interaction with folate-receptors, which are overexpressed in some tumors. TMSP are shielded in the blood circulation but are activated by matrix metalloprotease-2 and -9 (MMP-2/9) that are enzymes oversecreted in tumor tissues. This activation occurs after a cleavage by the MMP-2/9 that enables the TMSP to penetrate into the cell. The authors confirmed this cleavage of TMSP by MMP-2/9 through an *in vitro* proteolysis assay. The cellular uptake and cytotoxic of F/TMSP-DTX-NLC were improved in HT-1080, KB, MCF-7 and A549 cells, and they were attributed to the attachment of folate and TMSP. The penetration into the interior of tumor spheroids was deeper for F/TMSP-DTX-NLC.

In vivo evaluations using a nude mice model with KB tumor cells showed a stronger antitumor efficacy and enhanced tumor cell apoptosis.[59]

NLC coated with hyaluronic acid (HA) and paclitaxel-loaded (PXT-NLC) were prepared. The HA was attached *via* electrostatic attraction to the cationic PTX-NLC. *In vitro* investigation using B16, CT26 and HCT116 cell lines showed that the cytotoxicity of HA-PTX loaded NLC (HA-PTX-NLC) against these three cancer cell lines was superior to that of Taxol®. The *in vivo* evaluation in B16-bearing Kunming mice showed an increased antitumor effect of HA-PTX-NLC compared with Taxol®. Furthermore, a prolonged circulation time for PTX in the blood was obtained plus an increased accumulation in the tumor.[60]

As one can see, there are quite a few studies describing various ligands, such as antibodies, antibody fragments, proteins, small molecules like folate, and peptides, attached to SLN or NLC loaded with anticancer compounds (with aptamers remaining probably the only ligand candidate not tested with SLN and NLC yet).

9.4 Prospects for SLN and NLC in cancer therapy

Many studies have described about multifunctional nanocarriers for improved cancer treatment.[27,28,32,61–63] According to Torchilin, an ideal multifunctional pharmaceutical nanocarrier would provide highly efficient and specialized system for delivery of drugs, genes, and diagnostic agents.[32] This system would have individual functions acting in synchronized way, with more optimized temporal, spatial deposition and release pattern of the anticancer agents.

Such promising multifunctional nanocarriers could possess many useful functions in the same carrier system such as (1) long-circulation (PEGylation), (2) co-loading of two or more drugs, (3) magnetic particles loaded into the carrier together with the drug to allow for carrier sensitivity towards an external magnetic field and its use as a contrast agent for magnetic resonance imaging, (4) pH sensitivity to release drug in acidic areas of tumors, (5) ligands attached to the surface for active targeting (monoclonal antibody, peptides, proteins, small molecules, aptamers), (6) contrast agents for imaging purposes (heavy metals — ^{111}In, ^{99m}Tc, Gd, Mn for gamma- or MR imaging application), (7) cell-penetrating capacity (cell-penetrating peptide — CPP) for carrier-enhanced uptake, (8) a

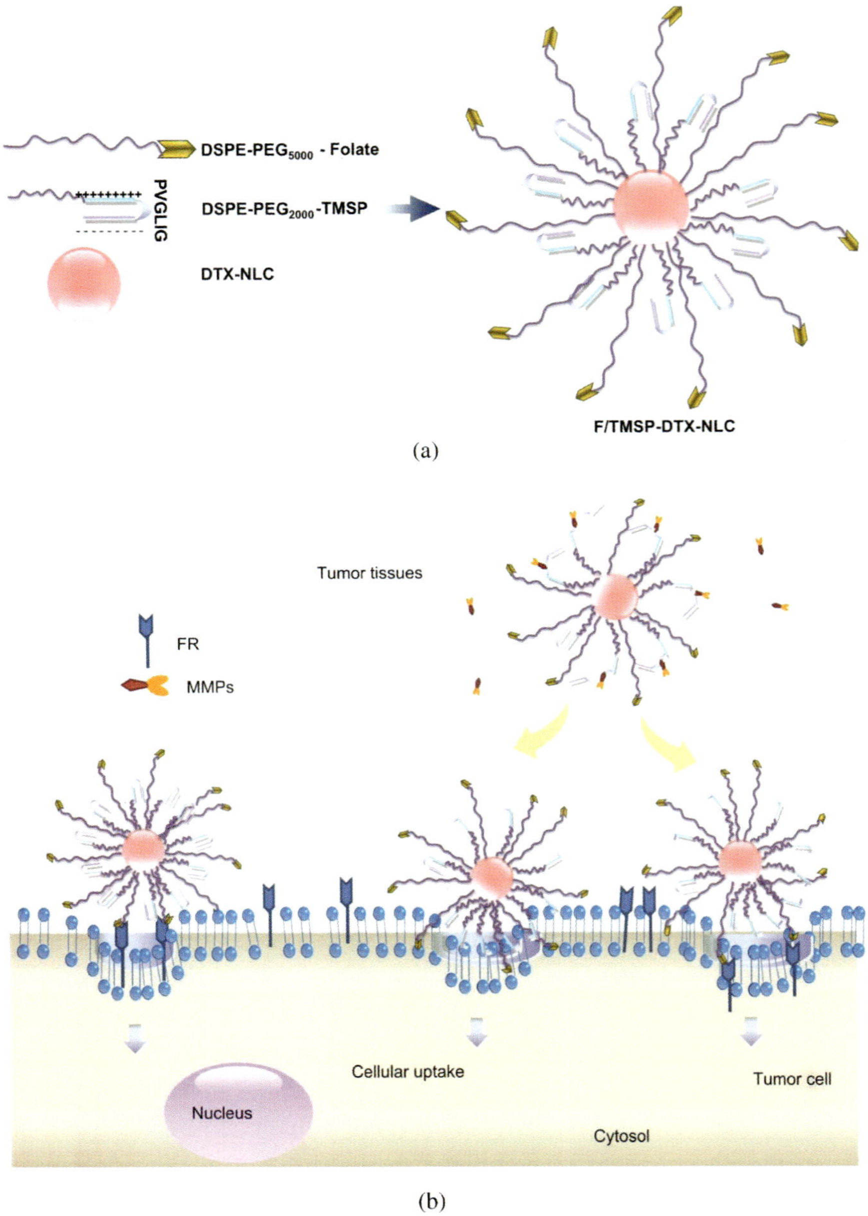

Figure 9.4 (a) A schematic of the dual-ligand DTX-NLC simultaneously modified with folate and TMSP (F/TMSP-DTX-NLC). (b) The schematic of the multivalent interactions of F/TMSP-DTX-NLC with cancer cells. Reprinted from Ref. 59, Copyright (2013), with permission from Elsevier (License number 3133680405511).

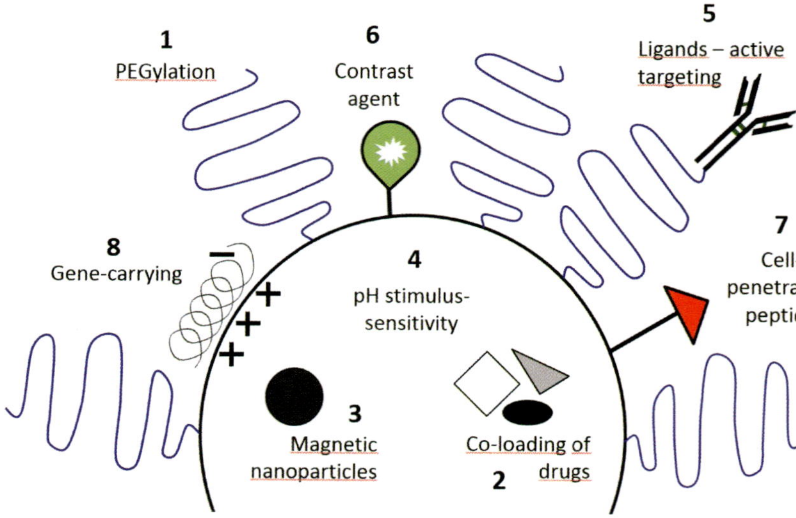

Figure 9.5 The schematic structure of an idealized multifunctional pharmaceutical nanocarrier. (1) Long-circulation — PEGylation; (2) co-loading of two or more drugs; (3) loading of magnetic particles; (4) pH stimulus-sensitivity; (5) ligands attached to the surface for active targeting; (6) contrast nanocarrier for imaging purposes; (7) cell-penetrating peptide; and (8) gene-carrying.

gene-carrying nanocarrier such as a lipoplex or polyplex (DNA or RNA complexed by the carrier *via* the carrier surface positive charge). (See Figure 9.4)[32,62]

All these functions have been addressed at least somewhat in this review and are summarized in Tables 9.1–9.5. The nanoparticles described have each been developed with a few of the outlined functions mentioned. However, the combination of all functions would be very challenging. Much effort has been done with liposomes and polymeric nanoparticles to obtain multifunctional nanocarriers. However, the advantages of SLN and NLC over the others nanosystems qualify them as promising candidates in the development of combined therapeutic and diagnostic multifunctional pharmaceutical nanocarrier systems.

References

1. Joshi MD, Müller RH. Lipid nanoparticles for parenteral delivery of actives. *Eur J Pharm Biopharm*, 2009; 71(2): 161–172.

2. Wong HL, Bendayan R, Rauth AM, Wu XY. Simultaneous delivery of doxorubicin and GG918 (Elacridar) by new polymer-lipid hybrid nanoparticles (PLN) for enhanced treatment of multidrug-resistant breast cancer. *J Control Release*, 2006; 116(3): 275–284.
3. Wong HL, Rauth AM, Bendayan R, Manias JL, Ramaswamy M, Liu Z, Erhan SZ, Wu XY. A new polymer-lipid hybrid nanoparticle system increases cytotoxicity of doxorubicin against multidrug-resistant human breast cancer cells. *Pharm Res*, 2006; (7): 1574–1585.
4. Kang KW, Chun MK, Kim O, Subedi RK, Ahn SG, Yoon JH, Choi HK. Doxorubicin-loaded solid lipid nanoparticles to overcome multidrug resistance in cancer therapy. *Nanomedicine*, 2010; 6(2): 210–213.
5. Subedi RK, Kang KW, Choi HK. Preparation and characterization of solid lipid nanoparticles loaded with doxorubicin. *Eur J Pharm Sci,* 2009; 37(3–4): 508–513.
6. Koziara JM, Lockman PR, Allen DD, Mumper RJ. Paclitaxel nanoparticles for the potential treatment of brain tumors. *J Control Release,* 2004; 99(2): 259–269.
7. Koziara JM, Whisman TR, Tseng MT, Mumper RJ. In-vivo efficacy of novel paclitaxel nanoparticles in paclitaxel-resistant human colorectal tumors. *J Control Release*, 2006; 112(3): 312–319.
8. Zhang XG, Miao J, Dai YQ, Du YZ,Yuan H, Hu FQ. Reversal activity of nanostructured lipid carriers loading cytotoxic drug in multi-drug resistant cancer cells. *Int J Pharm*, 2008; 361(1–2): 239–344.
9. Mosallaei N, Jaafari MR, Hanafi-Bojd MY, Golmohammadzadeh S, Malaekeh-Nikouei B. Docetaxel-loaded solid lipid nanoparticles: preparation, characterization, *in vitro*, and *in vivo* evaluations. *J Pharm Sci,* 2013; 102(6): 1994–2004.
10. Yang SC, Lu LF, Cai Y, Zhu JB, Liang BW, Yang CZ. Body distribution in mice of intravenously injected camptothecin solid lipid nanoparticles and targeting effect on brain. *J Control Release,* 1999; 59(3): 299–307.
11. Kaur IP, Bhandari R, Bhandari S, Kakkar V. Potential of solid lipid nanoparticles in brain targeting. *J Control Release*, 2008; 127(2): 97–109.
12. Zara GP, Cavalli R, Bargoni A, Fundarò A, Vighetto D, Gasco MR. Intravenous administration to rabbits of non-stealth and stealth doxorubicin-loaded solid lipid nanoparticles at increasing concentrations of stealth agent: pharmacokinetics and distribution of doxorubicin in brain and other tissues. *J Drug Target,* 2002; 10(4): 327–335.

13. R.H. Müller, C. Schwarz, W. Mehnert, J.S. Lucks, Production of solid lipid nanoparticles (SLN) for controlled drug delivery, *Proc Int. Symp. Control. Release Bioact. Mater*, 1993; 20: 480–481.
14. Müller RH, Mäder K, Gohla S. Solid lipid nanoparticles (SLN) for controlled drug delivery — a review of the state of the art. *Eur J Pharm Biopharm*, 2000; 50 (1): 161–177.
15. Mehnert W, Mäder K. Solid lipid nanoparticles: production, characterization and applications. *Adv Drug Deliv Rev,* 2001; 47(2–3): 165–196.
16. Müller, R.H., Radtke, M. and Wissing, S.A. Nanostructured lipid matrices for improved microencapsulation of drugs. *Int. J. Pharm,* 2002, 242(1–2): 121.
17. Jores, K., Mehnert, W., and Mader, K., Physicochemical investigations on solid lipid nanoparticles and on oil-loaded solid lipid nanoparticles: A nuclear magnetic resonance and electron spin resonance study. *Pharm. Res*, 2003; 20: 1274–1283.
18. Jores K, Mehnert W, Drechsler M, Bunjes H, Johann C, Mäder K. Investigation on the structure of solid lipid nanoparticles (SLN) and oil-loaded solid lipid nanoparticles by photon correlation spectroscopy, field-flow fractionation and transmission electron microscopy. *J. Control. Release*, 2004; 95: 217–227.
19. Rosenblatt KM, Douroumis D, Bunjes H. Drug release from differently structured monoolein/poloxamer nanodispersions studied with differential pulsepolarography and ultrafiltration at low pressure. *Pharm Sci,* 2007; 96(6): 1564–1575.
20. Bunjes H. Lipid nanoparticles for the delivery of poorly water-soluble drugs. *J Pharm Pharmacol*, 2010; 62(11): 1637–1645.
21. Yuan H, Miao J, Du YZ, You J, Hu FQ, Zeng S. Cellular uptake of solid lipid nanoparticles and cytotoxicity of encapsulated paclitaxel in A549 cancer cells. *Int J Pharm,* 2008; 348(1–2): 137–145.
22. Chen DB, Yang TZ, Lu WL, Zhang Q. *In vitro* and *in vivo* study of two types of long-circulating solid lipid nanoparticles containing paclitaxel. *Chem Pharm Bull* (Tokyo). 2001.
23. Kabanov AV, Batrakova EV, Alakhov VY. Pluronic block copolymers for overcoming drug resistance in cancer. *Adv Drug Deliv Rev*, 2002; 54(5): 759–779.

24. Livi L, Meattini I, Cardillo Cde L, Mangoni M, Greto D, Petrucci A, Rampini A, Bruni A, Galardi A, Cataliotti L, Biti G. Non-pegylated liposomal doxorubicin in combination with cyclophosphamide or docetaxel as first-line therapy in metastatic breast cancer: a retrospective analysis. *Tumori,* 2009; 95(4): 422–426.
25. Ali H, Shirode AB, Sylvester PW, Nazzal S. Preparation, characterization, and anticancer effects of simvastatin-tocotrienol lipid nanoparticles. *Int J Pharm*, 2010; 389(1–2): 223–231.
26. Kashanian S, Azandaryani AH, Derakhshandeh K. New surface-modified solid lipid nanoparticles using N-glutaryl phosphatidylethanolamine as the outer shell. Int J *Nanomedicine*, 2011; 6: 2393–2401.
27. Milane L, Ganesh S, Shah S, Duan ZF, Amiji M. Multi-modal strategies for overcoming tumor drug resistance: hypoxia, the Warburg effect, stem cells, and multifunctional nanotechnology. *J Control Release,* 2011; 155(2): 237–247.
28. Yao J, Zhang L, Zhou J, Liu H, Zhang Q. Efficient simultaneous tumor targeting delivery of all-trans retinoid acid and Paclitaxel based on hyaluronic acid-based multifunctional nanocarrier. *Mol Pharm.* 2013; 10(3): 1080–1091.
29. Bentolila LA, Ebenstein Y, Weiss S. Quantum dots for in vivo small-animal imaging. *J Nucl Med*, 2009; 50(4): 493–496.
30. Maeda H, Wu J, Sawa T, Matsumura Y, Hori K. Tumor vascular permeability and the EPR effect in macromolecular therapeutics: a review. *J Control Release*, 2000; 65(1–2): 271–284.
31. Torchilin V. Tumor delivery of macromolecular drugs based on the EPR effect. *Adv Drug Deliv Rev*, 2011; 63(3): 131–135.
32. Torchilin VP. Multifunctional nanocarriers. *Adv Drug Deliv Rev,* 2006; 58(14): 1532–1555.
33. Podio V, Zara GP, Carazzonet M, Cavalli R, Gasco MR. Biodistribution of stealth and non-stealth solid lipid nanospheres after intravenous administration to rats. *J Pharm Pharmacol,* 2000; 52(9): 1057–1063.
34. Fundarò A, Cavalli R, Bargoni A, Vighetto D, Zara GP, Gasco MR. Non-stealth and stealth solid lipid nanoparticles (SLN) carrying doxorubicin: pharmacokinetics and tissue distribution after i.v. administration to rats. *Pharmacol Res*, 2000; 42(4): 337–343.

35. Jassem J, Pieńkowski T, Płuzańska A, Jelic S, Gorbunova V, Mrsic-Krmpotic Z, Berzins J, Nagykalnai T, Wigler N, Renard J, Munier S, Weil C. Central & Eastern Europe and Israel Pacitaxel Breast Cancer Study Group. Doxorubicin and paclitaxel versus fluorouracil, doxorubicin, and cyclophosphamide as first-line therapy for women with metastatic breast cancer: final results of a randomized phase III multicenter trial. *J Clin Oncol*, 2001; 19(6): 1707–1715.
36. Parhi P, Mohanty C, Sahoo SK. Nanotechnology-based combinational drug delivery: an emerging approach for cancer therapy. *Drug Discov Today*, 2012; (17–18): 1044–1052.
37. Okuda T., Kidoaki S. Multidrug delivery systems with single formulation — current status and future perspective. *J. Biomater. Nanobiotechnol*, 2012; 3: 50–60.
38. Mussi SV, Silva RC, Oliveira MC, Lucci CM, Azevedo RB, Ferreira LA. New approach to improve encapsulation and antitumor activity of doxorubicin loaded in solid lipid nanoparticles. *Eur J Pharm Sci,* 2013; 48(1–2): 282–290.
39. Siddiqui A, Gupta V, Liu YY, Nazzal S. Doxorubicin and MBO-asGCS oligonucleotide loaded lipid nanoparticles overcome multidrug resistance in adriamycin resistant ovarian cancer cells (NCI/ADR-RES). *Int J Pharm,* 2012; 431(1–2): 222–229.
40. Yu YH, Kim E, Park DE, Shim G, Lee S, Kim YB, Kim CW, Oh YK. Cationic solid lipid nanoparticles for co-delivery of paclitaxel and siRNA. *Eur J Pharm Biopharm* 2012; 80(2): 268–273.
41. Shuhendler AJ, Prasad P, Leung M, Rauth AM, Dacosta RS, Wu XY. A novel solid lipid nanoparticle formulation for active targeting to tumor $\alpha(v)$ $\beta(3)$ integrin receptors reveals cyclic RGD as a double-edged sword. *Adv Healthc Mater,* 2012; 1(5): 600–608.
42. Prasad P, Shuhendler A, Cai P, Rauth AM, Wu XY. Doxorubicin and mitomycin C co-loaded polymer-lipid hybrid nanoparticles inhibit growth of sensitive and multidrug resistant human mammary tumor xenografts. *Cancer Lett,* 2013; 334(2): 263–273.
43. Reddy JA, Low PS. Enhanced folate receptor mediated gene therapy using a novel pH-sensitive lipid formulation. *J Control Release,* 2000; 64(1–3): 27–37.
44. Andresen TL, Jensen SS, Jørgensen K. Advanced strategies in liposomal cancer therapy: problems and prospects of active and tumor specific drug release. *Prog Lipid Res*, 2005; 44(1): 68–97.

45. Peira E, Marzola P, Podio V, Aime S, Sbarbati A, Gasco MR. In vitro and in vivo study of solid lipid nanoparticles loaded with superparamagnetic iron oxide. *J Drug Target,* 2003; 11(1): 19–24.
46. Videira MA, Botelho MF, Santos AC, Gouveia LF, de Lima JJ, Almeida AJ. Lymphatic uptake of pulmonary delivered radiolabelled solid lipid nanoparticles. *J Drug Target,* 2002; 10(8): 607–613.
47. Andreozzi E , Seo JW, Ferrara K, Louie A. Novel method to label solid lipid nanoparticles with 64cu for positron emission tomography imaging. *Bioconjug Chem*, 2011; 22(4): 808–818.
48. Bae KH, Lee JY, Lee SH, Park TG, Nam YS. Optically traceable solid lipid nanoparticles loaded with siRNA and paclitaxel for synergistic chemotherapy with in situ imaging. *Adv Healthc Mater*, 2013; 2(4): 576–584.
49. Van Vlerken LE, Vyas TK, Amiji MM. Poly(ethylene glycol)-modified nanocarriers for tumor-targeted and intracellular delivery. *Pharm Res,* 2007; 24(8): 1405–1414.
50. Wang M, Thanou M. Targeting nanoparticles to cancer. *Pharmacol Res,* 2010; 62(2): 90–99.
51. Jain A, Agarwal A, Majumder S, Lariya N, Khaya A, Agrawal H, Majumdar S, Agrawal GP. Mannosylated solid lipid nanoparticles as vectors for site-specific delivery of an anti-cancer drug. *J Control Release*, 2010; 148(3): 359–367.
52. Zhang YL, Zhang ZH, Jiang TY, Ayman-Waddad, Jing-Li, Lv HX, Zhou JP. Cell uptake of paclitaxel solid lipid nanoparticles modified by cell-penetrating peptides in A549 cells. *Pharmazie*, 2013; 68(1): 47–53.
53. Kuo YC, Liang CT. Catanionic solid lipid nanoparticles carrying doxorubicin for inhibiting the growth of U87MG cells. *Colloids Surf B Biointerfaces*, 2011; 1; 85(2): 131–137.
54. Wang W, Zhou F, Ge L, Liu X, Kong F. Transferrin-PEG-PE modified dexamethasone conjugated cationic lipid carrier mediated gene delivery system for tumor-targeted transfection. *Int J Nanomedicine*, 2012; 7: 2513–2522.
55. Jing F, Li J, Liu D, Wang C, Sui Z Dual ligands modified double targeted nano-system for liver targeted gene delivery. *Pharm Biol,* 2013; 51(5): 643–649.
56. Zhao X, Zhao Y, Geng L, Li X, Wang X, Liu Z, Wang D, Bi K, Chen X. Pharmacokinetics and tissue distribution of docetaxel by liquid chromatography-mass spectrometry: evaluation of folate receptor-targeting amphiphilic copolymer modified nanostructured lipid carrier. *J Chromatogr B Analyt Technol Biomed Life Sci* 2011; 879(31): 3721–3727.

57. Wu L,Tang C, Yin C. Folate-mediated solid-liquid lipid nanoparticles for paclitaxel-coated poly(ethylene glycol). *Drug Dev Ind Pharm,* 2010; 36(4): 439–448.
58. Venishetty VK, Komuravelli R, Kuncha M, Sistla R, Diwan PV. Increased brain uptake of docetaxel and ketoconazole loaded folate-grafted solid lipid nanoparticles. *Nanomedicine*, 2013; 9(1): 111–121.
59. Gao W, Xiang B, Meng TT, Liu F, Qi XR. Chemotherapeutic drug delivery to cancer cells using a combination of folate targeting and tumor microenvironment-sensitive polypeptides. Biomaterials. 2013; 34(16): 4137–4149.
60. Yang XY, Li YX, Li M, Zhang L, Feng LX, Zhang N. Hyaluronic acid-coated nanostructured lipid carriers for targeting paclitaxel to cancer. *Cancer Lett,* 2013; 334(2): 338–345.
61. Koning GA, Krijger GC. Targeted multifunctional lipid-based nanocarriers for image-guided drug delivery. *Anticancer Agents Med Chem,* 2007; 7(4): 425–440.
62. Torchilin V. Multifunctional and stimuli-sensitive pharmaceutical nanocarriers. *Eur J Pharm Biopharm.* 2009; 71(3): 431–444.
63. Xu Y, Karmakar A, Heberlein WE, Mustafa T, Biris AR, Biris AS. Multifunctional magnetic nanoparticles for synergistic enhancement of cancer treatment by combinatorial radio frequency thermolysis and drug delivery. *Adv Healthc Mater*, 2012; 1(4): 493–501.
64. Liu X , Zhang Z, Jiang Y, Hu Y, Wang Z, Liu J, Feng R, Zhang J, Huang G. Novel PEG-grafted nanostructured lipid carrier for systematic delivery of a poorly soluble anti-leukemia agent Tamibarotene: characterization and evaluation. *Drug Deliv*. 2015 Feb; 22(2): 223–229.
65. Müller RH, Petersen RD, Hommoss A, Pardeike J.Nanostructured lipid carriers (NLC) in cosmetic dermal products. *Adv Drug Deliv Rev*, 2007; 59 (6): 522–530.

INDEX

Frontiers in Nanobiomedical Research

(Continuation of series card page)

Vol. 6 *Handbook of Immunological Properties of Engineered Nanomaterials*
Second Edition
(In 3 Volumes)
Volume 1: Key Considerations for Nanoparticle Characterization Prior to Immunotoxicity Studies
Volume 2: Haematocompatibility of Engineered Nanomaterials
Volume 3: Engineered Nanomaterials and the Immune Cell Function
Second Edition
edited by Marina A Dobrovolskaia (Leidos Biomedical Research Inc., USA) and Scott E McNeil (Leidos Biomedical Research Inc., USA)

Vol. 5 *The World Scientific Encyclopedia of Nanomedicine and Bioengineering I: Nanotechnology for Translational Medicine: Tissue Engineering, Biological Sensing, Medical Imaging, and Therapeutics*
(A 4-Volume Set)
editor-in-chief: Donglu Shi (East Hospital, Tongji University School of Medicine, China & University of Cincinnati, USA)
Volume 1: Noble Metal Nanoparticles for Biomedical Applications
edited by Yu Cheng (East Hospital, Tongji University School of Medicine, China)
Volume 2: Application of Nano-structured Materials and Devices in Chemical and Biological Sensing
edited by Jia Huang (School of Materials Science and Engineering Tongji University, China)
Volume 3: Nanotechnology in Gene Delivery
Yarong Liu (University of Southern California, USA)
edited by Pin Wang (University of Southern California, USA)
Volume 4: Nano Imaging: From Fundamental Principles to Translational Medical Applications
edited by Bingbo Zhang (Tongji University School of Medicine, China)

Vol. 4 *Bio-Inspired Nanomaterials and Applications: Nano Detection, Drug/Gene Delivery, Medical Diagnosis and Therapy*
edited by Donglu Shi (University of Cincinnati, USA)

Vol. 3 *Handbook of Nanobiomedical Research: Fundamentals, Applications and Recent Developments*
(In 4 Volumes)
Volume 1: Materials for Nanomedicine
Volume 2: Applications in Therapy
Volume 3: Applications in Diagnostics
Volume 4: Biology, Safety and Novel Concepts in Nanomedicine
edited by Vladimir Torchilin (Northeastern University, USA)

Vol. 2 *Tissue Regeneration: Where Nano-Structure Meets Biology*
edited by Qing Liu (3D Biotek, USA & Tongji University, China) and Hongjun Wang (Stevens Institute of Technology, USA)

Vol. 1 *Handbook of Immunological Properties of Engineered Nanomaterials*
edited by Marina A. Dobrovolskaia (SAIC-Frederick, Inc., USA) and Scott E. McNeil (SAIC-Frederick, Inc., USA)